Peripheral Benzodiazepine Receptors

NEUROSCIENCE PERSPECTIVES

Editor: Peter Jenner
         Pharmacology Group
         Biomedical Sciences Division
         King's College London
         Manresa Road
         London SW3 6LX

Titles in this series:

Peripheral Benzodiazepine Receptors

edited by

E. Giesen-Crouse
Institut de Recherches Servier
Suresnes, France

ACADEMIC PRESS
Harcourt Brace Jovanovich, Publishers
London   San Diego   New York
Boston   Sydney   Tokyo   Toronto

ACADEMIC PRESS LIMITED
24/28 Oval Road,
London NW1 7DX

United States Edition published by
ACADEMIC PRESS INC.
San Diego, CA 92101

A catalogue record for this book
is available from the British Library

ISBN 0-12-282630-2

Typeset by P & R Typesetters Ltd, Salisbury, UK
Printed and Bound in Great Britain by
TJ Press Ltd, Padstow, Cornwall

Contents

**Colour plates 1–3 referring to chapters 3 and 13 respectively
appear between pages 242–243**

Contributors

Shalom Bar-Ami Rappaport Family Institute for Research in the Medical Sciences and Department of Obstetrics and Gynecology, Rambam Medical Center, Haifa, Israel

Gordon T. Bolger Department of Pharmacology, Bio-Méga Inc., 2100 rue Cunard, Laval, Quebec, H7S 2G5 Canada

J. J. Bourguignon Laboratoire de Pharmacochimie Moléculaire, Centre de Neurochimie CNRS, 5 rue Blaise Pascal, 67084 Strasbourg Cedex, France

Robert H. Dodd Institut de Chimie des Substances Naturelles, CNRS, 91198 Gif-sur-Yvette Cedex, France

Moshe Gavish Rappaport Family Institute for Research in the Medical Sciences and Department of Pharmacology, Bruce Rappaport Faculty of Medicine, Technion-Israel Institute of Technology, Haifa, Israel

Leif Hertz Departments of Pharmacology and Anesthesia, University of Saskatchewan, Saskatoon, Saskatchewan, S7N 0W0 Canada

Ivan Izquierdo Centro de Memoria, Departamento de Bioquimica, Instituto de Biociencias, Universidade Federal do Rio Grande do Sul, 90049 Porto Alegre RS, Brasil

Karl E. Krueger Fidia-Georgetown Institute of the Neurosciences and Department of Anatomy and Cell Biology, Georgetown University School of Medicine, 3900 Reservoir Road, Washington DC 20007, USA

Maryse Lenfant Institut de Chimie des Substances Naturelles, CNRS, 91198 Gif-sur-Yvette Cedex, France

Guiseppe Marano Laboratorio de Farmacologia, Istituto Superiore di Sanità, V. le Regina Elena 299, 00161 Roma, Italy

Jorge H. Medina Laboratorio de Neuroreceptores, Instituto de Biologia Celular, Facultad de Medicina, Universidad de Buenos Aires, Paraguay 2155 (1121) Buenos Aires, Argentina

Tiziana Mennini Istituto di Ricerche Farmacologiche 'Mario Negri', Via Eritrea 62, 20157 Milano, Italy

R. Myers MRC Cyclotron Unit, Hammersmith Hospital, Ducane Road, London W12 0HS, UK

Vassilios Papadopoulos Department of Anatomy and Cell Biology, Georgetown University School of Medicine, 3900 Reservoir Road, Washington, DC 20007, USA

Anthony L. Parola College of Medicine and Department of Pharmacology, The University of Arizona, Health Sciences Center, Tucson, Arizona 85724, USA

Marek Pawlikowski Institute of Endocrinology, Medical Academy of Lodz, Poland

Veijo Saano Department of Pharmacology and Toxicology, University of Kuopio, P.O. Box 1627, SF-70211 Kuopio, Finland

Ronit Weizman Tel Aviv Community Mental Health Center and Sackler Faculty of Medicine, Tel Aviv University, Israel

Henry I. Yamamura College of Medicine and Department of Pharmacology, The University of Arizona, Health Sciences Center, Tucson, Arizona 85724, USA

Series Preface

The driving force for the production of this series lies in my own inability to keep up with the advances occurring in those areas of neuroscience in which I am especially interested. So many times I have been frustrated by being unable to find a current review of an important research area. Even when I resort to bothering colleagues who are experts in a particular field, I am told, more often than not, that such an overview does not exist. In my own area of expertise I frequently send away students empty-handed who have asked me to direct them to a definitive article on a well researched topic.

Although regrettable, perhaps this situation is not surprising since the neurosciences are one of the most diverse and rapidly advancing areas in the biological sphere. By definition, research in the neurosciences encompasses anatomy, pathology, biochemistry, physiology, pharmacology, molecular biology, genetics and therapeutics. Indeed, there are few individuals capable of maintaining a grasp of the literature in all these aspects of their own research interests let alone in other fields.

My answer was to establish *Neuroscience Perspectives* and to develop gradually a series of individual edited monographs dealing in depth with issues of current interest to those working in the neuroscience area. Each volume is being designed to bring a multidisciplinary approach to the subject matter by pursuing the topic from the laboratory to the clinic. As a consequence I have asked the editors of the individual volumes to produce a balanced critique of their topic which will be read, understood and enjoyed by as wide an audience as possible within the realm of neuroscience.

The choice of the topics for the series is a difficult matter. In the first instance these were largely dictated by my own interests or by my awareness of important and fundamental work being undertaken by colleagues. More recently, I have been recruiting subject matter and editors through attending a variety of diverse symposia in the neuroscience area. However, the choice of topics should reflect the needs of the audience reached by the series. So I invite you to let me know of areas which you feel are of importance and to give me suggestions for individuals who would be keen to edit a book for *Neuroscience Perspectives*.

Finally, it only remains to thank those individuals at Academic Press who have already worked for several years to develop *Neuroscience Perspectives*. In particular, Dr. Carey Chapman who has the unenviable task of recruiting the editors that I suggest and then harassing them for the completed work. My hope is that the series will fill the gap that I perceive and provide for my colleagues in the neurosciences a collection of interesting books which will become reference volumes in their field. I hope you will enjoy *Neuroscience Perspectives*.

Peter Jenner

Preface

Mitochondrial benzodiazepine receptors (MBR), also known as peripheral benzodiazepine receptors, omega, or p-sites, were first discovered in the 1970s, more or less concomitantly with their central counterparts. Although ligands for both receptors may be structurally very similar (for example diazepam, 4-chlorodiazepam), the structure and function of the receptors are totally unrelated.

Over the years, numerous research groups have invested much time and effort to uncover the mystery of the biochemistry of the receptor, the chemistry of its ligands, and its pharmacological or (patho)-physiological role. Yet, the exact function of the MBR is still an enigma; however, the amount of information accumulated is such that a breakthrough in the understanding of this system appears imminent.

Our present knowledge is drawn from very diverse research areas, and is sometimes incomplete or conflicting, and therefore not easy to summarize. A comprehensive review of this information was a real challenge for all of us and the persevering interest and encouragement of Dr C. Chapman of Academic Press and of Prof. P. Jenner, the Series Editor, is gratefully acknowledged here.

The biochemical characteristics of the MBR, its molecular structure and coupling to ion transporters (L. Hertz, H. Yamamura and A. Parola), and its binding to endogenous and synthetic ligands (J.-J. Bouguignon) are summarized in the first three chapters.

The physiology–pharmacology is described in the following ten chapters. Two aspects were considered: the effect brought about by binding of ligands to the MBR on one hand, and the variation of receptor numbers in certain pathological states on the other.

A host of different effects of MBR ligands have been reported in the literature; many of them, such as effects on cell growth and differentiation (V. Saano), on memory (I. Izquierdo and J. H. Medina), and on smooth muscle reactivity and cardiovascular effects occur at high concentrations only. Recently, effects on stereoidogenesis and in the immune system, produced by nanomolar concentrations have been reported (M. Pawlikowski, K. E. Krueger and V. Papadopoulos). In many cases, MBR ligands have modulatory effects on other physiological systems.

The MBR density is sensitive to certain pathological situations such as stress (T. Mennini, M. Lenfant and R. H. Dodd), endocrinological disorders (M. Gavish, S. Bar-Ami and R. Weizman), and states of peripheral and central inflammation (R. Myers).

MBR numbers increase drastically both upon administration of IL_1 or LPS and in inflammatory states of experimental arthritis and reactive gliosis. MBR ligands administrated *in vivo* modify the immune response of the animal. MBR seem to work at the interface of cytokine and neurotransmitter signalling, with possibly oxygen radical and nitric oxide, and the heme containing enzymes necessary for their synthesis, regulated by the MBR.

These experimental findings, together with the recently reported altered expression of neutrophil MBR in chronic granulomatous disease and the activation of human neutrophil NADPH-oxidase by a monoclonal antibody against MBR, suggest that the MBR is an important trigger of the neuro-immune system. The therapeutic potential of MBR-ligands probably lies in this area.

Eva Giesen-Crouse

BIOCHEMICAL AND CHEMICAL CHARACTERISTICS OF THE MITOCHONDRIAL BENZODIAZEPINE RECEPTOR AND ITS LIGANDS

CHAPTER 1

MOLECULAR PROPERTIES OF MITOCHONDRIAL BENZODIAZEPINE RECEPTORS

Anthony L. Parola and Henry I. Yamamura

College of Medicine, Department of Pharmacology, The University of Arizona, Health Sciences Center, Tucson, Arizona 85724, USA

Table of Contents

1.1 Introduction

The mitochondrial benzodiazepine receptor (MBR), also commonly referred to as the peripheral benzodiazepine receptor, was first described as a diazepam binding site in rat peripheral tissues that was pharmacologically distinct from the central benzodiazepine receptor (CBR) associated with γ-aminobutyric acid (GABA)-regulated chloride channels of the neural crest (Braestrup and Squires, 1977; Regan et al., 1981; Schoemaker et al., 1981). Whereas the mechanism for CBR modulation of GABA-mediated chloride flux is well established (for review see Costa and Guidotti, 1979; Knapp et al., 1990; Olsen and Tobin, 1990), the mechanism(s) mediating the effects of MBR activation remains unknown. Although the MBR is found within the central nervous system, it is much more abundant in peripheral tissues (Schoemaker et al., 1981, 1983; Marangos et al., 1982; Gehlert et al., 1983), particularly in secretory tissues (Anholt et al., 1985, 1986a). Unlike the plasma membrane-associated CBR, the MBR is primarily localized in the mitochondrial outer membrane but in certain cases can also be associated with non-mitochondrial membranes (O'Beirne and Williams, 1984; Olson et al., 1988; Mukherjee and Das, 1989; O'Beirne et al., 1990).

Molecules smaller than 10 000 Da, including proteins, are freely permeable to the outer mitochondrial membrane due to the abundance of large channels formed by mitochondrial porins, also referred to as mitochondrial voltage-dependent anion channels (VDAC) (Benz, 1990; Dihanich, 1990). There is some evidence MBR are functionally associated with VDAC. Effects of MBR on mitochondrial functions include the mediation of intermembrane cholesterol transport (Besman et al., 1989; Mukhin et al., 1989; Krueger and Papadopoulos, 1990) and of mitochondrial respiration (Hirsch et al., 1989a; Larcher et al., 1989), but the underlying mechanism(s) responsible for these effects are unknown.

The rat MBR can be schematically represented with two ligand-binding domains: an isoquinoline (IQ) site, identified by high-affinity binding of the prototype isoquinoline, PK 11195 [dissociation constant (K_d) approximately 1–5 nM], and a benzodiazepine (BZ) site identified by binding of 1-alkyl-1,4-benzodiazepines such as Ro5-4864 (4'-chlorodiazepam) (Figure 1). High-affinity PK 11195 binding is diagnostic of the MBR, but the affinity of benzodiazepines for MBR is species dependent and varies over several orders of magnitude (Cymerman et al., 1986; Awad and Gavish 1987; Eshleman and Murray, 1989). In rats and mice, benzodiazepines potently inhibit PK 11195 binding and the dissociation constant of Ro5-4864 is similar to that of PK 11195, approximately 5 nM. The rank order of benzodiazepine potency to reversibly and competitively displace [^{3}H]PK 11195 is Ro5-4864 ($K_i \approx 5$ nM) > diazepam ($K_i \approx 30$ nM) $\geqslant$ flunitrazepam $\gg$ clonazepam ($K_i \approx$

$2\,\mu\text{M}$). This type of MBR, typified by the rat receptor, is herein designated MBR_r (for rat or rat-like). MBR have low-affinity benzodiazepine interactions in most non-rodent species. The decreased benzodiazepine affinities vary from several fold to several orders of magnitude. Bovine or bovine-like receptors, herein designated MBR_b, have a low affinity for 1,4-benzodiazepines including Ro5-4864, diazepam and clonazepam ($K_i \approx 2\,\mu\text{M}$). The heterogeneity in benzodiazepine binding characteristics found between species has not been observed within the same species. Human MBR is an MBR_b-like receptor (Broaddus and Bennett, 1990). Alpidem, an imidazopyridine, has the highest affinity for both types of MBR, but it is not selective (Langer *et al.*, 1990). Alpidem binds to CBR with similarly high affinity and binding is allosterically linked to $GABA_A$ receptors. Conserved functional characteristics often reflect common structural elements of receptors. Species differences in benzodiazepine affinities may be due to structural differences in common binding site elements or to differences in the components comprising the MBR. These species differences and similarities can be exploited to determine MBR structure–function relationships.

Determining the molecular organization of the MBR has used several different but complementary experimental approaches including: (1) differential perturbation of the ligand binding sites with chemical and enzyme reagents; (2) irreversible ligands which covalently attach to the MBR or associated proteins; (3) MBR purification; and (4) cloning and expression of MBR proteins. All of these approaches have provided information regarding the molecular properties of the MBR. Species differences in ligand affinities have also identified components that determine MBR ligand specificity. Some insight has been gained concerning associated proteins which may be important for MBR effector mechanisms (see Section 1.2).

1.2   Chemical and enzymatic inhibition of MBR

Chemical reagents that modify amino acid functional groups and enzymes such as proteases, glycosidases, and lipases can alter protein function. These changes can result in loss or reduction of function. To identify critical features of the MBR IQ and BZ binding domains, chemical and enzymatic reagents have been used to inhibit both sites or selectively inhibit one binding site, while leaving the other virtually unaffected. These have been useful reagents for differentiating the requirements for ligand interaction at the BZ and IQ sites. A list of reagents that modify MBR ligand interactions is shown in Table 1.

5

Table 1 Chemical and enzymatic inhibition of MBR.

Inhibitory reagent	Ligand site affected	Type of inhibition	Comments
Anions	BZ	Competitive	Rank order of potency: $I^- > Br^- > Cl^- > F^-$
Arachidonic acid	BZ	Competitive	Phospholipids with similar effects
Cyclosporin A	BZ	Non-competitive	
Detergents	BZ $\gg$ IQ	Competitive	Includes Triton X-100, digitonin, CHAPS, and cholate
Diethylpyrocarbonate (DEPC)	IQ	Non-competitive	Species-specific effect, only if high affinity BZ site also present
4,4'-Diisothiocyanostilbene-2,2'-disulphonic acid (DIDS)	BZ	Competitive	Effect enhanced by Na_2SO_4
Dithiothreitol (DTT)	IQ, BZ	Unknown, decreases binding to both sites	Pre-incubation with PK 11195 enhances effect
Konig's polyanion	IQ, BZ	Competitive	Polyanion also a selective inhibitor of VDAC
Phospholipase A_2 (PLA2)[a]	BZ	Competitive	Similar effect with mellitin
Pyruvate dehydrogenase (PDH)	IQ, BZ	Non-competitive	Used soluble form of enzyme

[a] Mellitin, an activator of endogenous PLA2, has a similar effect to exogenously added PLA2.

1.2.1　Isoquinoline site inhibition

Several chemicals and one enzyme inhibit [^{3}H]PK 11195 binding, but only diethylpyrocarbonate (DEPC) is selective (Benavides *et al.*, 1984; Skowronski *et al.*, 1987). DEPC, a histidine-specific reagent, non-competitively inhibits up to 70% of MBR$_r$ IQ sites while minimally affecting the affinity and number of BZ sites. Although the effect of DEPC is variable, the maximal 70% inhibition indicates that a subpopulation of MBR$_r$ are DEPC-insensitive. Pre-incubation with PK 11195 or Ro5-4864 protects the IQ site from DEPC modification. Protection by PK 11195 or Ro5-4864 is evidence that the IQ and BZ sites overlap, or that these ligands induce a DEPC-insensitive conformation. Selective PK 11195 inhibition by DEPC suggests that different residues are required for IQ and BZ interactions. DEPC has little or no effect on MBR$_b$ receptors (Eshleman and Murray, 1989; Parola and Laird, 1991). [^{3}H]PK 11195 binding to trout brain (MBR$_b$-like receptors) was reduced by only 15%, while bovine adrenal MBRs were unaffected by 2 mM DEPC, a concentration that inactivated 70% of rat liver MBR$_r$ IQ sites.

DEPC sensitivity indicates there is a histidine within the MBR$_r$ that specifically interacts with isoquinolines. However, MBR$_b$ and a subpopulation of MBR$_r$ are refractory to DEPC. This insensitivity suggests either that histidine is not critical for IQ interactions in all MBRs, or that there is a conformation that does not permit modification of a critical binding site histidine(s), although this residue remains accessible to PK 11195. DEPC does not affect the bovine receptor and there is no evidence for a histidine within the MBR$_b$-IQ site.

1.2.2　Benzodiazepine site inhibition

Most chemicals that affect benzodiazepine interactions are competitive inhibitors, with the notable exceptions of Konig's polyanion and cyclosporin A. The enzymes phospholipase A$_2$ (PLA2) and pyruvate dehydrogenase (PDH) are competitive and non-competitive inhibitors, respectively, of the BZ site. While PLA2 is selective for the BZ site, PDH also non-competitively inhibits the IQ site. Although the mechanism of benzodiazepine inhibition has not been discerned for either enzyme, both are endogenous mitochondrial enzymes (PLA2 in the outer membrane and PDH in the inner membrane), and may be functionally associated with the MBR *in vivo*.

Arachidonic acid decreases the potency of Ro5-4864 and diazepam to displace [^{3}H]PK 11195 by approximately 20-fold while increasing the number of IQ sites (Skowronski *et al.*, 1987; Beaumont *et al.*, 1988). PLA2 selectively decreases Ro5-4864 affinity by approximately five-fold (Havoundjian *et al.*, 1986; Doble *et al.*, 1987a; Mantione *et al.*, 1988). The effects of PLA2 may be due to direct interaction of the enzyme with MBR_r or to indirect effects mediated via the production of unsaturated fatty acids, including arachidonic acid. Endogenous PLA2 can be stimulated by mellitin, resulting in a similar effect on Ro5-4864 binding to that observed using exogenous PLA2 (Doble et al., 1987a).

4,4'-Diisothiocyanostilbene-2,2'-disulphonic acid (DIDS) is an anion transport inhibitor which reduces Ro5-4864 affinity by 2- to 3-fold and does not affect the IQ site (Lueddens and Skolnick, 1987). Disodium sulphate enhances the potency of DIDS by five-fold. Several anions also selctively inhibit the BZ site with the rank order of potency: $I^- > Br^- > Cl^- > F^-$; the same rank order anions permeate known chloride channels. The rank order of anion inhibitory potency and disruption by anion transport inhibitors indicates that the MBR_r BZ site may be coupled to an anion channel.

Detergents reduce the affinity of both the BZ and IQ sites, but the effect on the BZ site is much greater (Awad and Gavish, 1988, 1989a,b). Cyclosporin A, also known to disrupt hydrophobic interactions, has a 36-fold greater potency to non-competitively inhibit Ro5-4864 than PK 11195 (Hirsch et al., 1989b). The effects of detergents and cyclosporin A on BZ binding suggest that hydrophobic interactions are involved, possibly including protein–protein or protein–lipid interactions.

1.2.3 Molecular properties determined from chemical and enzymatic inhibition

There are three other agents which non-selectively inhibit the MBR_r; Konig's polyanion, dithiothreitol (DTT), and PDH. DTT inhibits both sites and the mechanism of inhibition has not been discerned (Skowronski et al., 1987). However, decreased binding was not observed with other reducing agents. Exogenous PDH non-competitively inhibits MBR_r through an unknown mechanism (Daval et al., 1989). Konig's polyanion, an inhibitor of VDAC, was equipotent at inhibiting both Ro5-4864 and PK 11195 interactions (Hirsch et al., 1989b).

Inactivation studies indicate that the IQ and BZ sites are not identical since they can be differentially perturbed. PK 11195 and Ro5-4864 are reversible competitive inhibitors, indicating the BZ and IQ sites share common binding domains. This arrangement of the sites was proposed by Skowronski et al. (1987). A model of the MBR_r consistent with inactivation studies is shown in Figure 1. First, a histidine residue(s) may be important for IQ binding, at least for a subpopulation of MBR_r. Arachidonic acid competitively and selectively inhibits benzodiazepine bonding. Exogenous PLA2 and mellitin stimulation of endogenous PLA2 lowers the MBR_r affinity for benzodiazepines,

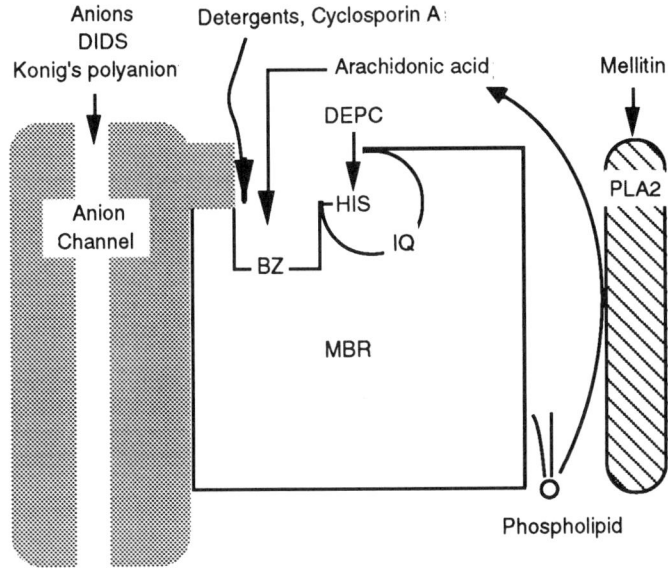

Figure 1 Schematic representation of MBR_r organization. The isoquinoline (IQ) and benzodiazepine (BZ) sites are depicted as overlapping but not identical since they can be differentially perturbed. Diethylpyrocarbonate (DEPC) selectively inactivates IQ binding, indicating that a histidine is localized within the IQ site. Ro5-4864 and PK 11195 can protect the IQ site from DEPC-modification. Arachidonic acid inhibits the BZ site and PLA2, an endogenous mitochondrial outer membrane enzyme, may inhibit the BZ site by the metabolism of phospholipids to yield unsaturated fatty acids (such as arachidonic acid) or by directly interacting with the MBR. Detergents and cyclosporin A disrupt BZ interactions by perturbing hydrophobic interactions. Inhibition by anions, DIDS, and Konig's polyanion suggests that MBR may be coupled to an anion channel such as a mitochondrial VDAC.

possibly by increasing the local arachidonic acid concentration. PLA2 may be functionally or directly coupled to MBRs. Anions inhibit benzodiazepine binding either directly by interfering with ligand–receptor interactions or indirectly through a mechanism involving functional coupling to an anion channel. The effects of DIDS and Konig's polyanion are consistent with the coupling of the BZ site to such a channel. VDAC is an abundant channel in the outer mitochondrial membrane and association with the MBR_r is further supported by irreversibly labelling the BZ site (see Section 1.3.2). There is no evidence to associate MBR_b with either PLA2 or an anion channel since these specifically affect benzodiazepine interactions, nor is there evidence implicating a histidine within the MBR_b IQ site.

1.3 MBR irreversible ligands

Irreversible ligands are useful in identifying proteins that either constitute a ligand-binding site or are closely associated. Photoreactive ligands are particularly useful since production of the reactive species can be easily controlled. Specificity is further enhanced by forming the ligand–receptor complex prior to photolysis. For other irreversible ligands such as alkylating agents, specificity is conferred by the affinity of the ligand and by reactivity of residues within the binding site. Several irreversible IQ and BZ radiolabelled ligands have been used to identify MBR-associated proteins. For a summary of MBR irreversible ligands, refer to Table 2.

1.3.1 Irreversible isoquinolines

Photoactivable and alkylating PK 11195 analogues have been synthesized including PK 14105, a p-fluoronitrophenyl photoreactive derivative, and 11a, a racemic N-alkylisothiocyanate alkylating analog. The $(+)$- and $(-)$-isomers comprising the racemic mixture of 11a are identified as 11b and 11c, respectively. Near ultraviolet light renders PK 14105 susceptible to nucleophilic substitution of the activated fluorine (Doble *et al.*, 1987b). Nanomolar concentrations of PK 14105 reversibly binds to the MBR with a five-fold lower affinity (K_d approximately 10 nM) than PK 11195, but irreversibly inactivates nearly all occupied receptors upon photolysis. The pH optimum of 9.0 suggests the presence of several possible nucleophilic amino acids that may react with the photoreactive ligand including cysteine and lysine, but none has yet been identified. Both isoquinoline and benzodiazepine binding

Table 2 MBR irreversible ligands.

Ligand	Chemical class	Mechanism of reactivity	Labelled proteins (kDa)	
			MBR$_r$	MBR$_b$
PK 14105	IQ	Photolysis	18 (specific) 31–35 (non-specific)	18 (specific) 45 (non-specific)
11a	IQ	Alkylation	ND	ND
Flunitrazepam	BZ	Photolysis	30–35 (non-specific)	ND
AHN 086	BZ	Alkylation	27–30 (non-specific)	None

ND, not determined.

are non-competitively inhibited by PK 14105 photolabelling. An 18 kDa membrane-bound protein, herein referred to as the isoquinoline binding protein (IBP), is specifically and covalently labelled by [^{3}H] PK 14105. MBR_r photolabelling is blocked with either PK 11195 or Ro5-4864, while only PK 11195 effectively inhibits MBR_b labelling (Doble *et al.*, 1987b; Skowronski *et al.*, 1988; Broaddus and Bennett, 1990; Parola and Laird, 1991). Several proteins are non-specifically labelled by [^{3}H] PK 14105 including a 45 kDa protein in bovine adrenal cortex mitochondria and 31–35 kDa mitochondrial proteins from a Chinese hamster ovary (CHO) cell line (Riond *et al.*, 1989; Parola and Laird, 1991). The IBP is coincident with high-affinity PK 11195 sites and is a characteristic MBR component in all species regardless of the affinity for benzodiazepines. There are two major differences between receptor inactivation with DEPC and PK 14105. First, whereas there are DEPC-insensitive IQ sites, all IQ sites are modifiable by PK 14105. Second, whereas DEPC discriminates between the IQ and BZ sites, PK 14105 does not and so simultaneously inactivates both.

The racemic alkylating IQ ligand, 11a, has an affinity similar to PK 11195 (Newman *et al.*, 1987). The (−)-isomer, 11c, is approximately 3-fold more potent than the (+)-isomer, 11b. Preliminary studies show nanomolar concentrations of 11a competitively inhibit both [^{3}H] PK 11195 and [^{3}H] Ro5-4864 and the potency of the alkylating ligand was decreased at pH > 8.0. Protein(s) alkylated by 11a have not been identified.

1.3.2 Irreversible benzodiazepines

Two irreversible benzodiazepines that interact with MBR_r are flunitrazepam, a photoligand that also labels CBR, and AHN 086, an alkylating analogue of Ro5-4864. Covalent modification with irreversible benzodiazepines results in competitive inhibition of the MBR BZ domain with only minimal changes to the IQ site. Flunitrazepam labels a 30–35 kDa protein in peripheral tissues lacking CBR, such as rat kidney (Snyder *et al.*, 1987, 1990). Mitochondrial VDAC are an abundant outer mitochondrial membrane protein of this size. VDAC purified from [^{3}H] flunitrazepam-photolabelled rat kidney mitochondria was found to be radiolabelled and therefore may be a component of or reside near the MBR_r BZ domain. [^{3}H] AHN 086 labels a 30 kDa protein from rat pineal mitochondria, but not bovine pineal mitochondria (that lacks high affinity Ro5-4864 sites) demonstrating [^{3}H] AHN 086 specificity for MBR_r (McCabe *et al.*, 1989). Covalent labelling with either [^{3}H] flunitrazepam or [^{3}H] AHN 086 cannot be inhibited with MBR-specific drugs including PK 11195 or Ro5-4864. However, pre-incubation with AHN 086 blocks [^{3}H] AHN 086 labelling of the 30 kDa protein. Both irreversible benzodiazepines, utilizing different mechanisms of reactivity, covalently modify protein(s) of similar sizes. The relationship between the 30 kDa protein(s) labelled by AHN 086 and the 30–35 kDa proteins labelled by flunitrazepam is unknown.

1.3.3 Receptor components identified using irreversible ligands

Irreversible ligands that interact with the MBR identify at least two different components: the IBP, an 18 kDa protein identified by [^{3}H]PK 14105 photolabelling, and a 30–35 kDa protein(s) labelled by [^{3}H]flunitrazepam and [^{3}H]AHN 086, which has tentatively been identified as a mitochondrial VDAC (see Figure 2). Whereas purification and subsequent cloning of the IBP demonstrates that this is a novel protein (see Section 1.5), the identity of the 30–35 kDa proteins is unknown, but may be a previously characterized outer mitochondrial membrane protein(s). The specificity of PK 14105 photolabelling of both MBR_r and MBR_b IBP, the non-competitive inhibitory nature of PK 14105 at both the MBR_r IQ and BZ sites, and the coincident distribution of high-affinity PK 11195 sites and the IBP indicates the IBP is a required MBR component. In contrast, the relationship of the 30–35 kDa protein(s) labelled with irreversible benzodiazepines to MBR_r is unknown.

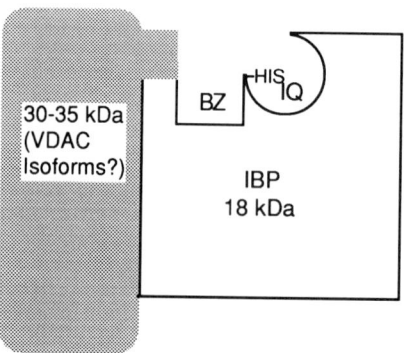

Figure 2 MBR receptor components identified using irreversible ligands. Photo-activated [^{3}H]PK 14105 labels the IBP, a protein coincident with MBR in all species regardless of the BZ affinity. [^{3}H]PK 14105 photoincorporation into the IBP can be inhibited with either Ro5-4864 or PK 11195 for MBR_r or PK 11195 for MBR_b. Other proteins are non-specifically labelled including a 31–35 kDa species in CHO cells and a 45 kDa protein in bovine adrenal mitochondria. Irreversible BZ label proteins ranging from 30–35 kDa. Pharmacological specificity of MBR_r covalent modification cannot be demonstrated for either flunitrazepam, a photoaffinity ligand, or AHN 086, an alkylating agent. A rat mitochondrial VDAC has been identified as a protein photolabelled by flunitrazepam. Irreversible BZ do not modify MBR_b, which character-istically lack a high-affinity BZ site.

At least one of these proteins modified by benzodiazepines has been identified as a mitochondrial VDAC. The 30–35 kDa proteins may have a role in modulating the affinity state of the BZ site but may not be required for BZ interactions for several reasons including: (1) irreversible BZ labelling cannot be blocked with MBR_r-specific ligands; and (2) BZ and IQ can interact with MBR_r even after irreversible BZ labelling. However, inactivation of the IBP with PK 14105 also inactivates the BZ site, suggesting that the IBP is required for a functional BZ site. The different inhibitory properties of the alkylating and photoreactive isoquinolines suggests 11a may not label the IBP. Identification of the protein labelled by 11a will be useful in further delineating the relationship between the IQ and BZ binding sites and the constituent MBR proteins. The relationship between the 31–35 kDa proteins non-specifically labelled by [^{3}H] PK 14105 and the 30–35 kDa proteins labelled by irreversible benzodiazepines is unknown.

1.4   MBR solubilization and purification

Several methods have been employed to estimate the molecular weight of the minimal components required for functional MBR including: (1) radiation inactivation of the BZ and IQ sites; (2) size exclusion chromatography of the functional detergent-solubilized MBR; (3) purification and reconstitution; and (4) purification of irreversibly labelled MBR components and associated proteins. While most of these methods require receptor solubilization, radiation inactivation can estimate the molecular weight *in situ*.

1.4.1   Solubilization

The MBR is composed of an integral membrane protein requiring detergents for solubilization. Digitonin is the most commonly used detergent, but other detergents successfully employed include 3-[(3-cholamidopropyl)-dimethylammonio]-1-propane sulphonate (CHAPS), sodium cholate and Triton X-100 (Martini *et al.*, 1983; Benavides *et al.*, 1985; Gavish and Fares, 1985; Anholt *et al.*, 1986b; Awad and Gavish, 1989a, b; Parola *et al.*, 1989). Solubilized MBR_r retains both high-affinity IQ and BZ sites while solubilized MBR_b retains high-affinity IQ sites and the characteristic inability to bind benzodiazepines (Awad and Gavish, 1989b). The species differences in ligand binding are maintained on the solubilized form of the receptor, suggesting that this distinction is not due to the membrane environment. While the soluble MBR_r maintains its characteristic rank order of benzodiazepine potency to displace [^{3}H]PK 11195, the affinity for all benzodiazepines is decreased by

13

detergents (see Section 1.2.2). The detergent-solubilized MBR_r can be functionally reconstituted into phospholipid vesicles (Anholt *et al.*, 1986b).

1.4.2 Molecular weight estimations of functional MBR

The native molecular weight has been estimated by several methods including radiation inactivation of both IQ and BZ sites and size-exclusion chromatography of the detergent-solubilized MBR. Radiation inactivation yields estimations of 21, 23 and 34 kDa for Ro5-4864, PK 11195, and diazepam binding sites, respectively (Paul *et al.*, 1981; Doble *et al.*, 1985). The estimations of the BZ site size of 21 kDa and 34 kDa (for the Ro5-4864 and diazepam, respectively) are somewhat disparate. A 23 kDa estimation for the IQ site is in agreement with the 18 kDa size of the IBP obtained from [³H]PK 14105 photolabelling (see Section 1.3.2). Size estimations for the Ro5-4864 and PK 11195 binding site are consistent with the IBP monomer being the only required receptor component for determining ligand specificity. While both size determinations for the BZ site do not concur, the receptor is rather small—less than 35 kDa for both diazepam and Ro5-4864.

Elution of detergent-solubilized MBRs has been monitored during chromato-graphic procedures using two different detection methods: (1) monitoring elution of the functional solubilized binding sites, or (2) photolabelling with [³H]PK 14105, followed by solubilization under non-denaturing conditions, and monitoring elution of specifically radiolabelled proteins. Both methods can be used simultaneously. The digitonin-solublized rat and bovine MBRs have an approximate M_r 200 000 using size-exclusion chromatography (Martini *et al.*, 1983; Benavides *et al.*, 1985; Parola *et al.*, 1989; Parola and Laird, 1991). The MBR_r IQ and BZ sites are intimately associated and co-migrate during anion-exchange and gel-filtration chromatography (Mukhin *et al.*, 1990). Partial purification of the [³H]PK 14105-labelled, digitonin-solubilized MBR_r by anion-exchange chromatography revealed two species—an M_r 200 000 species that exists as an M_r 30 000 species at low salt concentrations. Molecular weight estimation of detergent-solubilized MBR may be inflated due to the adsorption of detergent to the hydrophobic protein(s).

1.4.3 Purification

Purification of the functional MBR_r and the photolabelled MBR IBP has provided additional structural information and facilitated molecular cloning of the IBP from several species. The IBP has been purified from [³H]PK 14105-labelled rat adrenal mitochondria and from similarly labelled CHO cell mitochondria by reverse-phase high pressure liquid chromatography (Antkiewicz-Michaluk *et al.*, 1988; Riond *et al.*, 1989). Both IBP were purified to apparent homogeneity, had similar amino acid compositions, and were at least partially N-terminal blocked. Purification of peptides generated by site-specific cleavage with either cyanogen bromide or endoprotease V8 and subsequent sequencing yielded amino acid sequence information used to construct oligonucleotide probes to obtain IBP genes. Separation of peptides generated by digestion of the photolabelled CHO mitochondrial receptor with endoprotease V8 indicates that much of the covalently attached [^{3}H]PK 14105 coelutes with P4, a peptide corresponding to approximately the first 29 N-terminal amino acids (Riond *et al.*, 1991). This result suggests that IQ interactions involve at least some of these N-terminal residues, and that an amino acid sufficiently nucleophilic to react with the photoactivated PK 14105 is found within P4. Two nucleophilic residues probably not modified by PK 14105 include cysteine and histidine, both of which are notably absent from P4.

Several preliminary reports indicate that the functional MBR$_r$ has been purified using several conventional low pressure chromatography methods and monitoring elution of functional receptor by reconstitution into phospholipid vesicles to recover ligand-binding activity (Snyder *et al.*, 1990). The receptor was a heterodimer of 18 kDa and 30 kDa subunits that co-migrate through a sizing column at 52 kDa. [^{3}H]PK 14105 photolabelling identifies the 18 kDa component as the IBP while the 30 kDa component has not been identified. Although the IBP was the only protein required for IQ and BZ binding, the role of the 30 kDa protein with respect to MBR$_r$ receptor–ligand interactions is unknown.

1.4.4 Receptor properties determined from solubilization and purification

The MBR$_r$ IQ and BZ sites consists of protein(s) with a total mass less than 55 kDa of which the IBP (18 kDa) has been identified. Purification and functional reconstitution indicates the MBR$_r$ is at most a heterodimer consisting of the 18 kDa IBP and a 30–35 kDa component. The 30–35 kDa protein may be a mitochondrial VDAC. Although both the BZ and IQ sites are associated with the IBP, the role of the 30 kDa protein with respect to the ligand-binding properties has not been resolved. This heterodimer can be dissociated from a larger complex of approximately 200 kDa solubilized by digitonin. The 200 kDa complex may represent an aggregation of the MBR heterodimer or an association of the heterodimer with additional as yet unidentified proteins. The acidic nature of the [^{3}H]PK 14105-labelled digitonin-solubilized complex determined from isoelectric focusing ($pI = 4.5$) (Doble *et al.*, 1987a) is in direct contrast to the basis nature of the IBP calculated from the deduced amino acid composition of cDNA clones ($pI > 9.0$) (see Section 1.5). The differences in estimated and calculated pI

cannot be accounted for by an association of the IBP with VDAC, which is also a basic protein.

1.5 Molecular biology of the MBR

Purification and partial sequencing of IBP has facilitated the cloning of its gene. Cloning and subsequent expression have been useful in delineating its function with respect to MBR ligand-binding properties. Analysis of the primary amino acid sequence determined from gene cloning can identify structural motifs, such as potential transmembrane helices, membrane insertion signal sequences, and consensus glycosylation or phosphorylation sites. Comparison of amino acid sequences of MBRs with differing ligand-binding characteristics may be useful in identifying dissimilar regions or residues that may be responsible for functional differences. The cDNA of IBPs from three species differing in their affinity for benzodiazepines by several orders of magnitude have been cloned and two, MBR_r and MBR_b IBPs, have been expressed.

1.5.1 Molecular cloning of the isoquinoline binding protein

Partial amino acid sequences obtained for several peptides of purified rat IBP were used to construct oligonucleotide probes to screen a rat adrenal cDNA library (Sprengel *et al.*, 1989). A 781 bp cDNA encoding a 169 amino acid polypeptide was obtained. The protein encoded by the cDNA was a novel basic (calculated $pI = 9.9$) 19 kDa protein that did not contain any significant homology with previously sequenced proteins associated with CBR. Using a similar approach, the amino acid sequence from peptides generated by endoprotease digestion of IBP purified from CHO cells was used to construct oligonucleotide probes for screening a U937 cell (human lymphoma cell line) cDNA library (Riond *et al.*, 1991). A 831 bp cDNA encoding a 169 amino acid polypeptide was cloned and was 79% positionally identical to the cloned rat IBP.

Several lines of evidence indicate that although the MBR_r and MBR_b are distinct, differing by three orders of magnitude in their affinity for Ro5-4864, the IBP subunits are homologous. [^{3}H]PK 11195 binds with similarly high affinity to both MBRs and [^{3}H]PK 14105 photolabelling identifies an IBP of similar size estimated under both denaturing and non-denaturing conditions in rat and bovine tissues. Because of these similarities, we used rat IBP cDNA to clone the bovine homologue (Parola *et al.*, 1991). Using the rat IBP cDNA as a probe, Southern blot analysis of *Eco*RI-digested bovine genomic DNA identified at a single 5.2 kbp DNA fragment. Since a bovine gene hybridized the rat IBP cDNA, the rat probe was then used to screen a fetal bovine adrenal cDNA library. An 821 bp cDNA clone was obtained that encoded a basic polypeptide of 169 residues which was homologous to the rat IBP and later proved to be the bovine IBP.

The IBP cDNA clones from all three species have a similar structure with approximately 60 bp of untranslated 5′ nucleotide sequence, approximately 240 bp of 3′ untranslated sequence, and an open reading frame of 510 bp. The rat and bovine IBP cDNAs used as probes in Northern blot analysis identify a 900 bp mRNA whose tissue abundance is qualitatively similar to the tissue density distribution of [^{3}H]PK 11195 sites, indicating that this mRNA species is associated with the MBR (Sprengel et al., 1989; Parola et al., 1991). A human gene which hybridized the human IBP cDNA has been localized to band q13.3 on the long arm of chromosome 22 (Riond et al., 1991). Other genes localized on the long arm of chromosome 22 include mitochondrial aconitase 2, cytochrome b$_5$ reductase, arylsulphatase A, and the polypeptide platelet-derived growth factor b.

1.5.2 Analysis of isoquinoline binding protein amino acid sequences

The cloned IBPs are hydrophobic and basic with a calculated pI > 9.4 for all three proteins. Alignment of the sequences shows 69% of 169 total residues are positionally conserved in all three clones, indicating their importance to the structure and function of the IBP (Figure 3). The tryptophan content, usually 1% for most proteins, is 7% for the IBP. Cysteine is absent from rat IBP while a single cysteine is found in bovine IBP and two are present in the human clone. Histidine was identified as a critical residue for MBR$_r$ IQ interactions and His-43 is conserved in all three clones. The most dissimilar region is the carboxyl terminus, which has several potential protein kinase phosphorylation sites in the rat and bovine proteins, but not in the human (Bairoch, 1989). All three clones terminate in glutamic acid. There is no canonical amino terminal signal sequence for membrane insertion or targeting to the outer mitochondrial membrane (Schatz, 1987), which is consistent with previously reported non-mitochondrial localizations.

Membrane topology can be predicted by hydropathic analysis of the linear amino acid sequence (Kyte and Doolittle, 1982). At least 17 consecutive hydrophobic residues are required to form a helical transmembrane region. While hydropathy plots are useful for identifying potential transmembrane helices, other membrane-spanning structures are not identified, such as the transmembrane β barrel of mitochondrial VDAC. Hydropathic analysis of bovine IBP reveals five potential transmembrane helices, which are also

```
                    I
Bovine   MAPPWVPAVGFTLLPSL......GGFLGAQYTRGEGFRWYASLQKPPWHPP
Human    --------M----A---        -C-V-SRFVH---L----G----S----
Rat      -SQS------L--V---        ---M--YFV----L---------S----
CrtK     .......MSL--FAVYFVACACA-AT--IFSP-A...--D--K--S-V--

              II                              III
Bovine   RWILAPIWGTLYSAMGYGSYMIWKELGGFSKEAVVPLGLYAGQLALNWAWP
Human    H-V-G-V-------------LV-------TEK-V------T----------
Rat      --T----------------I--------TE--M------T----------
CrtK     N-LFPVA-S---IL-SISAARVS.G-AMENEL--LG-AFW-Y-I-V-TL-T

                IV                           V
Bovine   PLFFGTRQMGWALVDLLLTGGMAAATAMAWHQVSPPAACLLYPYLAWLAFA
Human    -I---A-----------VS-A----TV--Y----L--R----------
Rat      -I---A---------M-VS-V-T--TL---R------R----------
CrtK     -I---LHRLAGMLV-V-LWLSVF--CVLFWS-DWLSGLMFV--VI-VTV--


Bovine   GMLNYRMWQDNQVRRSGRRLSE
Human    TT---CV-RDNHGWHG----P-
Rat      T----YV-R--SG--G-S--T-
CrtK     -A--FSV-RL-PGEKPITL...
```

Figure 3 Alignment of the cloned bovine, rat and human IBP and *R. capsulatus* CrtK protein-deduced amino acid sequences. Amino acids are represented using the standard single letter code. Sequences were visually aligned except for CrtK, which was aligned with the rat IBP in a previous publication (Baker and Fanestil, 1991). Residues positionally conserved, compared to the bovine sequence, are indicated by a hyphen. Residues predicted to reside in membrane spanning regions by hydropathic analysis (Kyte and Doolittle, 1982) of the bovine clone are underlined and are similar in the other clones.

predicted for rat and human IBP and the homologous *Rhodobacter capsulatus* CrtK protein using the same algorithm (not shown). The conserved hydrophobic regions of the IBP, the length of each putative transmembrane segment, and the prediction of only a single charged residue, Asp-111, in the fourth membrane-spanning helix of all three IBPs, suggest that IBP topology is well conserved. A model of bovine IBP topology is shown (Figure 4), but more evidence of the membrane organization of the proteins is required. The purpose of the model is merely to serve as a starting point for further structural determinations.

Comparisons of the IBP primary amino acid sequence with other previously sequenced proteins indicate that IBP is related to the bacterial CrtK protein (Baker and Fanestil, 1991). The CrtK protein is found in *R. capsulatus*, a photosynthetic bacteria believed to be a mitochondrial ancestor. Like the IBP,

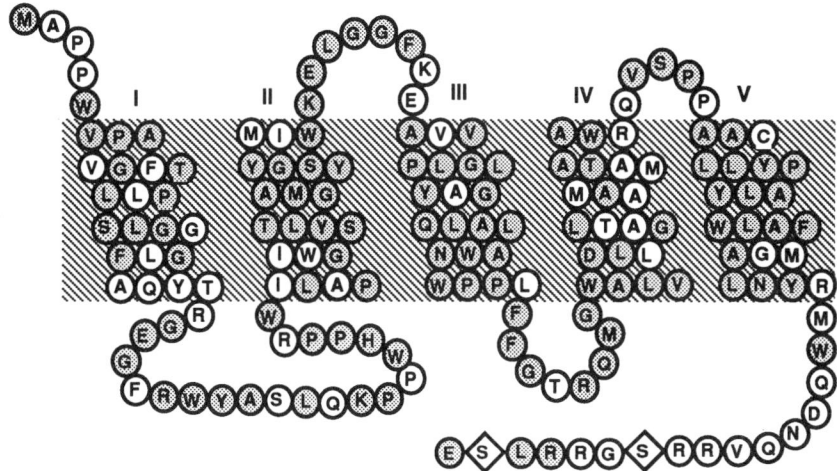

Figure 4 Proposed topology of the bovine IBP. Hydropathic analysis predicts 5 transmembrane helices (I-V) for the IBP. Shaded residues are conserved for all three cloned IBP (bovine, human, and rat). Possible sites for post-translational modifications predicted from the deduced primary sequence include a cAMP- or cGMP-dependent kinase site (ser-168) and a consensus PKC phosphorylation site (ser-163) (residues enclosed in diamonds). The location of phosphorylation sites along with the distribution of positively charged residues was used to orient the C-terminus in the cytosol. His-43, localized in the first cytosolic loop, is conserved in all three IBP, and may be important for IQ interactions.

the function of the CrtK protein is not well defined but may be involved in carotenoid synthesis. Since MBR_r and CBR both bind benzodiazepines, structural similarities in receptor components were suspected. Comparison of the IBP with CBR components indicates there is some similarity between transmembranes I, IV and V of the rat IBP and M1, M2, and M4 from subunits of the $GABA_A$/benzodiazepine receptors. However, the degree of similarity (41–59% conserved substitutions) was also observed for seemingly unrelated proteins, leading to the conclusion that any similarity was inconsequential (Sprengel et al., 1989).

1.5.3 Isoquinoline binding protein expression

Expression of cloned IBP is complicated by a constituent population of MBR in many mamalian cell lines. Two primate kidney lines have been used for IBP expression; 293-cells and COS-7 cells. Transfection of the MBR_r IBP cDNA into 293-cells resulted in a two-fold increase of both PK 11195 and Ro5-4864 sites which was accompanied by the appearance of high-affinity Ro5-4864 sites ($K_d = 8$ nM) not previously observed in the host cells ($K_d = 54$ nM for Ro5-4864) (Sprengel *et al.*, 1989). The MBR$_b$ has low affinity for Ro5-4864 ($K_i > 2.5\,\mu$M) and the cDNA encoding the MBR$_b$ IBP was transfected into COS-7 cells resulting in an 11-fold increase in the density of [^{3}H] PK 11195 binding sites (Parola *et al.*, 1991). While the affinity for [^{3}H] PK 11195 was essentially unchanged (K_d approximately 1 nM), the affinity for Ro5-4864 decreased 40-fold after transfection, from $K_i = 66$ nM to $K_i = 2500$ nM, and the host MBR was no longer discernible since it represented less than 10% of the total MBR population. A single affinity state for both PK 11195 and Ro5-4864 was observed in the transfected cells, presumably due to the predominance of the expressed bovine IBP. The increased expression of IBP was detected by photolabelling transfected COS-7 cells with [^{3}H] PK 14105. These results indicate that the affinity determinants of the IQ and BZ sites reside within the IBP. While COS-7 MBR are DEPC-sensitive, cells transfected with bovine IBP were DEPC-insensitive, indicating that DEPC sensitivity is a property of the IBP (A.L. Parola, unpubl. obs.). Although the IBP is a determinant of MBR ligand-binding properties, it may not itself be sufficient since other MBR components are present in the host cells.

1.5.4 Structure–function relationships determined by cloning

The IBP was cloned from three different species. All three clones were from cDNA libraries and are encoded by nuclear genes. Comparison of the primary amino acid sequences shows that the IBP is a highly conserved hydrophobic polypeptide of 169 residues that has a net positive charge. Hydropathic analysis predicts five transmembrane helices within the short span of 150 residues. Expression of the rat and bovine clones demonstrates that differences in benzodiazepine affinity are observed by expressing only the IBP. These studies suggest there is a fundamental difference in the ligand-binding properties conferred by IBPs of different species. The benzodiazepine and isoquinoline affinity determinants reside, at least in part, within the IBP. However, the participation of other proteins in determining MBR binding characteristics was not excluded in these expression studies. The importance of histidine for PK 11195 binding has been clearly demonstrated for MBR$_r$. His-43 is the only conserved histidine in all three clones and may participate in isoquinoline interactions. However, our results show DEPC sensitivity is conferred by the IBP and the role of histidine in isoquinoline binding to non-rodent MBR is untenable. There are several sites for possible post-translational modification of the IBP. The number of consensus C-terminal phosphorylation sites varies from two in bovine IBP and one in rat IBP to none in human IBP.

1.6 Conclusions

MBR in all species share functional characteristics that are reflected in common structural elements that comprise the receptor. High-affinity interactions ($K_d < 20$ nM) with isoquinolines related to PK 11195 characterize the MBR while the affinity for benzodiazepines is species dependent. Rat MBR has an approximately 1000-fold greater affinity for Ro5-4864 (the MBR-specific prototype benzodiazepine) than bovine MBR. The IBP is an MBR component identified by photolabelling with PK 14105, a photoactivable analogue of PK 11195. The IBP has been purified from several species and rat, bovine and human IBP have been cloned. Expression of cloned rat and bovine IBP reveals similar differences in the affinities for Ro5-4864 and other benzodiazepines to those differences found in tissues from the respective species. Although the IBP is necessary, it may not be sufficient for determining MBR ligand-binding characteristics. Since both expression systems used to date constitutively express MBR, an expression system devoid of MBR will be invaluable for further delineating the relationship between the IBP and the receptor. Proteins other than the IBP may also comprise, in part, the MBR. Mitochondrial VDAC is irreversibly labelled by flunitrazepam. AHN 086, an alkylating Ro5-4864 analogue, also labels mitochondrial proteins similar in size to VDAC. Agents that perturb anion channel function disrupt MBR ligand interaction, which further supports coupling of an anion channel, such as VDAC, to the MBR. Functional reconstitution of purified MBR would provide definitive evidence of the components required for MBR ligand binding. This approach may also be useful for delineating any relationship between mitochondrial VDAC or other anion channels and the MBR (see also Chapter 2).

Note added in proof

There have been several recent advances in understanding the molecular organization of the MBR. The cloned cDNA encoding a human MBR IBP has been expressed in yeast devoid of endogenous MBR (Riond et al., 1991). The expressed receptor has a low affinity for benzodiazepines, including Ro5-4864 and diazepam, and a high affinity for PK 11195 and related isoquinolines and isoquinoline binding is stereo-specific. The human IBP appears to confer the ligand binding properties characteristic of the human MBR when the IBP is expressed in yeast. Also purification of functional MBR from rat kidney mitochondria shows the native receptor is a heteromeric complex of at least 3 different proteins including the 18 kDa IBP, a 30 kDa voltage dependent anion channel, and a 32 kDa adenine nuleotide carrier

(McEnery *et al.*, 1992). The size of native receptor complex that binds both [^{3}H]PK 11195 and [^{3}H]Ro5-4864 with high affinity was estimated to be 70 kDa by gel filtration. The IBP is irreversibly labelled with [^{3}H]PK 14105 while both the voltage dependent anion channel and the adenine nucleotide carrier were labelled with the irreversible benzodiazepines [^{3}H]flunitrazepam and [^{3}H]AHN 086. These results suggest proteins other than the IBP participate in the ligand binding domains of the rat MBR. Apparently, while the IBP confers some ligand binding properties, other proteins are also required to form the ligand binding sites.

References

Anholt, R.R., DeSouza, E.B., Oster-Granite, M.L. and Snyder, S.H. (1985). Peripheral-type benzodiazepine receptors: autoradiographic localization in whole-body sections of neonatal rats. *J. Pharmacol. Exp. Ther.* **233**, 517–526.

Anholt, RR., Pedersen, P.L., DeSouza, E.B. and Snyder, S.H. (1986a). The peripheral-type benzodiazepine receptor. Localization to the mitochondrial outer membrane. *J. Biol. Chem.* **261**, 576–583.

Anholt, R.R., Aebi, U., Pedersen, P.L. and Snyder, S.H. (1986b). Solubilization and reassembly of the mitochondrial benzodiazepine receptor. *Biochemistry* **25**, 2120–2125.

Antkiewicz-Michaluk, L., Mukhin, A.G., Guidotti, A. and Krueger, K.E. (1988). Purification and characterization of a protein associated with peripheral-type benzodiazepine binding sites. *J. Biol. Chem.* **263** 17317–17321.

Awad, M. and Gavish, M. (1987). Binding of [^{3}H]Ro5-4864 and [^{3}H]PK 11195 to cerebral cortex and peripheral tissues of various species: species differences and heterogeneity in peripheral benzodiazepine binding sites. *J. Neurochem.* **49**, 1407–1414.

Awad, M. and Gavish, M. (1988). Differential effect of detergents on [^{3}H]Ro5-4864 and [^{3}H]PK 11195 binding to peripheral-type benzodiazepine-binding sites. *Life Sci.* **43**, 167–175.

Awad, M. and Gavish, M. (1989a). Heterogeneity between rat and calf peripheral-type benzodiazepine binding sites: differential sensitivity to Triton X-100. *J. Recept. Res.* **9**, 369–384.

Awad, M. and Gavish, M. (1989b). Sepcies differences and heterogeneity of solubilized peripheral-type benzodiazepine binding sites. *Biochem. Pharmacol.* **38**, 3843–3849.

Bairoch, A. (1989). *PROSITE: A Dictionary of Protein Sites and Patterns*, 2nd edn. University of Geneva, Switzerland.

Baker, M.E. and Fanestil, D.D. (1991). Mammalian peripheral-type benzodiazepine eceptor is homologous to CrtK of *Rhodobacter capsulatus*, a photosynthetic bacterium. *Cell* **65**, 721–722.

Beaumont, K., Skowronski, R., Vaughn, D.A. and Fanestil, D.D. (1988). Interactions of lipids with peripheral-type benzodiazepine receptors. *Biochem. Pharmacol.* **37**, 1009–1014.

Benavides, J., Begassat, F., Phan, T., Tur, C., Uzan, A., Renault, C., Dubroeucq, M.C., Gueremy, C. and LeFur, G. (1984). Histidine modification with diethyl-pyrocarbonate induces a decrease in the binding of an antagonist, PK 11195, but not of an agonist, Ro5-4864, of the peripheral benzodiazepine receptors. *Life Sci.* **35**, 1249–1256.

Benavides, J., Menager, J., Burgevin, M.C., Ferris, O., Uzan, A., Gueremy, C., Renault, C. and LeFur, G. (1985). Characterization of solubilized 'peripheral type' benzodiazepine binding sites from rat adrenals by using [³H]PK 11195, an isoquinoline carboxamide derivative. *Biochem. Pharmacol.* **34**, 167–170.

Besman, M.J., Yanagibashi, K., Lee, T.D., Kawamura, M., Hall, P.F. and Shively, J.E. (1989). Identification of des-(Gly-Ile)-endozepine as an effector of corticotropin-dependent adrenal steroidogenesis: stimulation of cholesterol delivery is mediated by the peripheral benzodiazepine receptor. *Proc. Natl. Acad. Sci. USA* **86**, 4897–4901.

Benz, R. (1990). Biophysical properties of porin pores from mitochondrial outer membranes of eukaryotic cells. *Experientia* **46**, 131–137.

Braestrup, C. and Squires, R.F. (1977). Specific benzodiazepine receptors in rat brain characterized by high-affinity [³H]diazepam binding. *Proc. Natl. Acad. Sci. USA* **74**, 3805–3809.

Broaddus, W.C. and Bennett, J.P. (1990). Peripheral-type benzodiazepine receptors in human glioblastomas: pharmalogical characterization and photoaffinity labeling of ligan recognition site. *Brain Res.* **518**, 199–208.

Costa, E. and Guidotti, A. (1979). Molecular mechanisms in the receptor actions of benzodiazepines. *Ann. Rev. Pharmacol. Toxicol.* **19**, 531–545.

Cymerman, U., Pazos, A. and Palacios, J.M. (1986). Evidence for species differences in 'peripheral' benzodiazepine receptors: an autoradioraphic study. *Neurosci. Lett.* **66**, 153–158.

Daval, J.L., Post, R.M. and Marangos, P.J. (1989). Pyruvate dehydrogenase interactions with peripheral-type benzodiazepine receptors. *J. Neurochem.* **52**, 110–116.

Dihanich, M. (1990). Biogenesis and function of eukaryotic porins. *Experientia* **46**, 153–161.

Doble, A., Benaides, J., Ferris, O., Bertrand, P., Menager, J., Vaucher, N., Burgevin, M.C., Uzan, A., Gueremy, C. and LeFur, G. (1985). Dihydropyridine and peripheral type benzodiazepine binding sites: subcellular distribution and molecular size determination. *Eur. J. Pharmacol.* **119**, 153–167.

Doble, A., Burgevin, M.C., Menager, J., Ferris, O., Begassat, F., Renault, C., Dubroeucq, M.C., Gueremy, C., Uzan, A. and Lefur, G. (1987a). Partial purification and pharmacology of peripheral-type benzodiazepine receptors. *J. Recept. Res.* **7**, 55–70.

Doble, A., Ferris, O., Burgevin, M.C., Menager, J., Uzan, A., Dubroeucq, M.C., Renault, C., Gueremy, C. and LeFur, G. (1987b). Photoaffinity labeling of peripheral-type benzodiazepine-binding sites. *Mol. Pharmacol.* **31**, 42–49.

Eshleman, A.J. and Murray, T.F. (1989). Differential binding properties of the peripheral-type benzodiazepine ligands [³H]PK 11195 and [³H]Ro5-4864 in trout and mouse brain membranes. *J. Neurochem.* **53**, 494–502.

Gavish, M. and Fares, F. (1985). Solubilization of peripheral benzodiazepine-binding sites from rat kidney. *J. Neurosci.* **5**, 2889–2893.

Gehlert, D.R., Yamamura, H.I. and Wamsley, J.K. (1983). Autoradiographic localization of 'peripheral' benzodiazepine binding sites in the rat brain and kidney using [³H]Ro5-4864. *Eur. J. Pharmacol.* **95**, 329–330.

Havoundjian, H., Cohen, R.M., Paul, S.M. and Skolnick, P. (1986). Differential sensitivity of 'central' and 'peripheral' type benzodiazepine receptors to phospholipase A2. *J. Neurochem.* **46**, 804–811.

Hirsch, J.D., Beyer, C.F., Malkowitz, L., Beer, B. and Blume, A.J. (1989a).

Mitochondrial benzodiazepine receptors mediate inhibition of mitochondrial respiratory control. *Mol. Pharmacol.* **35**, 157–163.

Hirsch, J.D., Beyer, C.F., Malkowitz, L., Loullis, C.C. and Blume, A.J. (1989b). Characterization of ligand binding to mitochondrial benzodiazepine receptors. *Mol. Pharmacol.* **35**, 164–172.

Knapp, R.J., Malatynska, E. and Yamamura, H.I. (1990). From binding studies to the molcular biology of GABA receptors. *Neurochem. Res.* **15**, 105–112.

Kreuger, K.E. and Papadopoulos, V. (1990). Peripheral-type benzodiazepine receptors mediate translation of cholesterol from outer to inner mitochondrial membranes in adrenocortical cells. *J. Biol. Chem.* **265**, 15015–15022.

Kyte, J. and Doolittle, R.F. (1982). A simple method for displaying the hydropathic character of a protein. *J. Mol. Biol.* **157**, 105–132.

Langer, S.Z., Arbilla, S., Tan, K., Lloyd, K.G., George, P., Allen, J. and Wick, A.E. (1990). Selectivity for omega-receptor subtypes as a strategy for the development of anxiolytic drugs. *Pharmacopsychiatry* **23**, 103–107.

Larcher, J.C., Vayssiere, J.L., Le Marquer, F.J., Cordeau, L.R., Keane, P.E., Bachy, A., Gros, F. and Croizat, B.P. (1989). Effects of peripheral benzodiazepines upon the O_2 consumption of neuroblastoma cells. *Eur. J. Pharmacol.* **161**, 197–202.

Lueddens, H.W. and Skolnick, P. (1987). 'Peripheral-type' benzodiazepine receptors in the kidney: regulation of radioligand binding by anions and DIDS. *Eur. J. Pharmacol.* **133**, 205–214.

Mantione, C.R., Goldman, M.E., Martin, B., Bolger, G.T., Lueddens, H.W., Paul, S.M. and Skolnick, P. (1988). Purification and characterization of an endogenous protein modulator of radioligand binding to 'peripheral-type' benzodiazepine receptors and dihydropyridine Ca^{2+}-channel antagonist binding sites. *Biochem. Pharmacol.* **37**, 339–347.

Marangos, P.J., Patel, J., Boulenger, J.P. and Clark-Rosenberg, R. (1982). Characterization of peripheral-type benzodiazepine binding sites in brain using [³H]Ro5-4864. *Mol. Pharmacol.* **22**, 26–32.

Martini, C., Giannaccini, G. and Lucacchini, A. (1983). Solubilization of rat kidney benzodiazepine binding sites. *Biochem. Biophys. Acta.* **728**, 289–292.

McCabe, R.T., Schoenheimer, J.A., Skolnick, P., Newman, A.H., Rice, K.C., Reig, J.A. and Klein, D.C. (1989). [³H]AHN 086 acylates peripheral benzodiazepine receptors in the rat pineal gland. *FEBS Lett.* **244**, 263–267.

McEnery, M.W., Snowman, A.M., Trifiletti, R.R. and Snyder, S.H. (1992). Isolation of the mitochondrial benzodiazepine receptor: association with the voltage-dependent anion channel and the adenine nucleotide carrier. *Proc. Natl. Acad. Sci. USA* **89**, 3170–3174.

Mukherjee, S. and Das, S.K. (1989). Subcellular distribution of 'peripheral type' binding sites for [³H]Ro5-4864 in guinea pig lung. Localization to the mitochondrial inner membrane. *J. Biol. Chem.* **264**, 16713–16718.

Mukhin, A.G., Papadopoulos, V., Costa, E. and Krueger, K.E. (1989). Mitochondrial benzodiazepine receptors regulate steriod biosynthesis. *Proc. Natl. Acad. Sci. USA* **86**, 9813–9816.

Mukhin, A.G., Zhong, P.Y. and Krueger, K.E. (1990). Cofractionation of the 17-kD PK 14105 binding site protein with solubilized peripheral-type benzodiazepine binding sites. *Biochem. Pharmacol.* **40**, 983–989.

Newman, A.H., Lueddens, H.W., Skolnick, P. and Rice, K.C. (1987). Novel irreversible ligands specific for 'peripheral' type benzodiazepine receptors: (±)-, (+)-, and (−)-1-(2-chlorophenyl)-*N*-(1-methylpropyl)-*N*-(2-isothiocyanatoethyl)-

3-isoquinolinecarboxamide and 1-(2-isothiocyanatoethyl)-7-chloro-1,3-dihydro-5-(4-chlorophenyl)-2H-1,4-benzodiazepin-2-one. *J. Med. Chem.* **30**, 1901–1905.

O'Beirne, G.B. and Williams, D.C. (1984). Binding of benzodiazepines to platelets from various species. *Biochem. Pharmacol.* **33**, 1568–1571.

O'Beirne, G.B., Woods, M.J. and Williams, D.C. (1990). Two subcellular localizations for peripheral-type benzodiazepine acceptors in rat liver. *Eur. J. Biochem.* **188**, 131–138.

Olsen, R.W. and Tobin, A.J. (1990). Molecular biology of $GABA_A$ receptors. *FASEB J.* **14**, 1469–1481.

Olson, J.M., Ciliax, B.J., Mancini, W.R. and Young, A.B. (1988). Presence of peripheral-type benzodiazepine binding sites on human erythrocyte membranes. *Eur. J. Pharmacol.* **152**, 47–53.

Parola, A.L. and Laird, H.E. II (1991). The bovine peripheral-type benzodiazepine receptor: a receptor with low affinity for benzodiazepines. *Life Sci.* **48**, 757–764.

Parola, A.L., Putnam, C.W., Russell, D.H. and Laird, H.E. II (1989). Solubilization and characterization of the liver peripheral-type benzodiazepine receptor. *J. Pharmacol. Exp. Ther.* **250**, 1149–1155.

Parola, A.L., Stump, D.G., Peperel, D.J., Krueger, K.E., Regan, J.W. and Laird, H.E. II (1991). Cloning and expression of a pharmacologically unique bovine peripheral-type benzodiazepine receptor isoquinoline binding protein. *J. Biol. Chem.* **266**, 14082–14087.

Paul, S.M., Kempner, E.S. and Skolnick, P. (1981). In situ molecular weight determination of brain and peripheral benzodiazepine binding sites. *Eur. J. Pharmacol.* **76**, 465–466.

Regan, J.W., Yamamura, H.I., Yamada, S. and Roeske, W.R. (1981). High affinity renal [³H]flunitrazepam binding: characterization, localization, and alteration in hypertension. *Life Sci.* **28**, 991–997.

Riond, J., Vita, N., LeFur, G. and Ferrara, P. (1989). Characterization of a peripheral-type benzodiazepine-binding site in the mitochondria of chinese hamster ovary cells. *FEBS Lett.* **245**, 238–244.

Riond, J., Leplation, P., Laurent, P., LeFur, G., Caput, D., Loison, G., Ferrara, P. (1991). Expression and pharmacological characterization of the human peripheral-type benzodiazepine receptor in yeast. *Eur. J. Pharmacol.* **208**, 307–312.

Riond, J., Mattei, M.G., Kaghad, M., Dumont, X., Guillemot, J.C., LeFur, G., Caput, D. and Ferrara, P. (1991). Molecular cloning and chromosomal localization of a human peripheral-type benzodiazepine receptor. *Eur. J. Biochem.* **195**, 305–311.

Schatz, G. (1987). Signals guiding proteins to their correct locations in mitochondria. *Eur. J. Biochem.* **165**, 1–6.

Schoemaker, H., Bliss, M. and Yamamura, H.I. (1981). Specific high affinity saturable binding of [³H]Ro5-4864 to benzodiazepine binding sites in rat cerebral cortex. *Eur. J. Pharmacol.* **71**, 473–475.

Shoemaker, H., Boles, R.G., Horst, W.D. and Yamamura, H.I. (1983). Specific high-affinity binding sites for [³H]Ro5-4864 in rat brain and kidney. *J. Pharmacol. Exp. Ther.* **225**, 61–69.

Skowronski, R., Beaumont, K. and Fanestil, D.D. (1987). Modification of the peripheral-type benzodiazepine receptor by arachidonate, diethylpyrocarbonate and thiol reagents. *Eur. J. Pharmacol.* **143**, 305–314.

Skowronski, R., Fanestil, D.D. and Beaumont, K. (1988). Photoaffinity labeling of peripheral-type benzodiazepine receptors in rat kidney mitochondria with [³H]PK 14105. *Eur. J. Pharmacol.* **148**, 187–193.

Snyder, S.H., Verma, A. and Trifiletti, R.R. (1987). The peripheral-type benzodiazepine receptor: a protein of mitochondrial outer membranes utilizing prophyrins as endogenous ligands. *FASEB J.* **1**, 282–288.

Snyder, S.H., McEnergy, M.W. and Verma, A. (1990). Molecular mechanisms of peripheral benzodiazepine receptors. *Neurochem. Res.* **15**, 119–123.

Sprengel, R., Werner, P., Seeburg, P.H., Mukhin, A.G., Santi, M.R., Grayson, D.R., Guidotti, A. and Krueger, K.E. (1989). Molcular cloning and expression of cDNA encoding a peripheral-type benzodiazepine receptor. *J. Biol. Chem.* **264**, 20415–20421.

CHAPTER 2

BINDING CHARACTERISTICS OF THE RECEPTOR AND COUPLING TO TRANSPORT PROTEINS

Leif Hertz

Departments of Pharmacology and Anaesthesia, University of Saskatchewan, Saskatoon, Saskatchewan S7N 0W0, Canada

Table of Contents

2.1 Introduction

Benzodiazepines are used therapeutically for their anxiolytic, sedative, anticonvulsant, muscle relaxant and amnestic effects. Several, but maybe not all, of these effects must be exerted on the central nervous system (CNS). In the CNS, they bind to two pharmacologically distinct types of benzodiazepine receptors. The first type is the 'central-type' of benzodiazepine receptor, which is present exclusively in the CNS and is localized on neurones. In this chapter, it will be referred to as the 'neuronal' or 'central' benzodiazepine receptor (CBR). The binding to this site is potently displaced by many benzodiazepines, including clonazepam, which is used clinically as an anticonvulsant. The second type of benzodiazepine receptor is the 'peripheral-type' of benzodiazepine binding site. This site is found in peripheral organs, including kidney, heart, several endocrine glands and erythrocytes, as well as in the CNS, and it is characterized by high-affinity binding of the benzodiazepine Ro5-4864, which has much less affinity for the CBR. In the CNS it is located on glial cells, mainly astrocytes. During recent years it has more and more commonly been referred to as the 'mitochondrial' benzodiazepine receptor (MBR). Regardless of whether or not this receptor is always located exclusively on mitochondria, this terminology will be used throughout this chapter.

In contrast to the CBR, the MBR is not connected to any GABA-regulated chloride channel, and the binding of benzodiazepines to this site is not enhanced in the presence of GABA (Hertz et al., 1980; Marangos et al., 1982a; Schoemaker et al., 1983). However, evidence has been found for the involvement of this receptor in the regulation of several physiologically important functions, for example calcium homeostasis, mitochondrial oxidation and lipid metabolism, including the secretion of steroid hormones. These different actions might be interrelated, or alternatively the MBR could be involved in regulatory mechanisms, *which vary among different cells and tissues*. Thus, as stated by Olson et al. (1988) 'it seems critical that until more is known about the peripheral-type site, questions of function should be examined independently in each tissue of interest.'

2.2   Cell and organ distribution

The first reports on binding sites for benzodiazepines, made in 1977 by several different groups (Bosmann et al., 1977; Braestrup and Squires, 1977; Mohler and Okada, 1977), created considerable excitement and enthusiasm. It was already at that time established that the prototype benzodiazepine, diazepam, bound with high affinity not only to receptors found in the CNS, but also to receptors found in various peripheral tissues such as kidney, liver and lung. A distinction was made between a 'central' and 'peripheral-type' (identical to the present terminology of CBR, and MBR, respectively).

Benzodiazepine receptor density is high in such endocrine tissues as the adrenal gland (LeFur et al., 1983a,b; De Souza et al., 1985) and testis (DeSouza et al., 1985). The ovary, uterus and placenta also have a high density of MBR (Fares et al., 1987). In the adrenal gland, the receptors are localized in the cortex with little, if any, labelling of the medulla (Benavides et al., 1983a) and the highest density in zona glomerulosa. In testis the receptors are mainly localized on the hormone-producing Leydig cells (Anholt *et al.*, 1985).

In blood, platelets have a very high density of binding sites for Ro5-4864 and PK 11195 but the binding is reduced or even lost after cell lysis (Benavides *et al.*, 1984a). Mammalian, including human, erythrocytes (Kendall and Nahorski, 1985; Olson *et al.*, 1988) also express MBR. Binding to heart has repeatedly been described (Davies and Huston, 1981; LeFur *et al.*, 1983a, b; Saano, 1986). Smooth muscle (Hullihan *et al.*, 1983), and especially vascular smooth muscle, shows a high density of MBR (French and Matlib, 1987).

In the CNS, MBR can easily be shown in membranes from both the brain (Schoemaker *et al.*, 1981; Marangos *et al.*, 1982a; Benavides *et al.*, 1983b) and spinal cord (Del Zompo *et al.*, 1983; Watanabe *et al.*, 1986) after labelling with the benzodiazepine ligand Ro5-4864, an agonist (Le Fur *et al.*, 1983c), or PK 11195 an isoquinoline benzodiazepine antagonist (at least in brain), often assumed to bind to the same site as Ro5-4864. However, peculiarly enough, Ro5-4864 is not able to displace diazepam or flunitrazepam binding from these preparations (Braestrup and Nielsen, 1983), suggesting that in the brain no binding sites for the MBR are labelled with these ligands. This paradox is probably due to a very rapid dissociation of both diazepam and flunitrazepam from the mitochondrial binding site in brain membranes, as will be discussed in more detail later.

In the rat brain, ventricular walls, choroid plexus and the olfactory bulb have the highest density of mitochondrial benzodiazepine binding sites (Anholt *et al.*, 1984; Benavides *et al.*, 1990), followed by brain stem, hypothalamus, cerebellum, cortex, striatum and hippocampus (Marangos *et al.*, 1982a; Schoemaker *et al.*, 1983). The astrocyte is probably not the only glial binding site for benzodiazepines; Myers (Chapter 13) discusses the possibility of MBR on microglia. There seems, however, to be a pronounced species variation between localization of the MBR in rat and human brain, and the cat brain might provide a much better model for mitochondrial benzodiazepine binding in man. In both, cat brain (Benavides *et al.*, 1984b) and human brain (Doble *et al.*, 1987), these binding sites are absent in white matter and so closely correlated with gray matter that one may get the incorrect impression that they are found on neurones rather than on adjacent glial cells.

Saturation kinetics for binding of Ro5-4864 or PK 11195 to CNS membranes have repeatedly been determined. Table 1 lists K_d and B_{max} values from such studies. This table consistently shows a high affinity (K_d 1–4 nM) binding with B_{max} values of 0.1–0.9 pmol/mg protein.

Another approach to obtain information about the MBR in brain has been the use of relatively intact astrocytes prepared by gradient centrifugation, or of astrocytes in primary cultures. This shows a very different picture (Table

Table 1 Saturation parameters for benzodiazepine binding to the mito-chondrial benzodiazepine receptor in CNS membranes or homogenates.

Ligand	Preparation	K_D (nM)	B_{max} (fmol/mg protein)	Reference
Ro5-4864	Brain membranes	1.6	245	Marangos et al. (1982a)
Ro5-4864	Rat brain membranes	1.9	245	Skerrit et al. (1982)
Ro5-4864	Brain membranes	1.1	205	Schoemaker et al. (1983)
Ro5-4864	Cortical membranes	2.1	184	LeFur et al. (1983a)
Ro5-4864	Cortical membranes	1.4	246	Weissman et al. (1984)
Ro5-4864	Cortical membranes	2.1	932	Weissman et al. (1984)
Ro5-4864	Cortical membranes	2.7	272	Villinger (1985)
Ro5-4864	Spinal cord membranes	3.5	267	Villinger (1985)
Ro5-4864	Spinal cord membranes	2.1	150	Villinger (1984)
Ro5-4864	Spinal cord membranes	2.9	691	Watanabe et al. (1986)
PK 11195	Cortical membranes	0.87	128	Benavides et al. (1983b)
PK 11195	Cortical membranes	4.3	400	Doble et al. (1987)

When several data are presented from the same authors or same publications, different species or regions were studied.

2) from that observed with homogenates or membranes. The binding of Ro5-4864 is at least 10 times higher than in membrane fractions (Table 1) and similar binding can be demonstrated of both diazepam and flunitrazepam (4–20 pmol/mg protein). The K_d values are only slightly higher than those for Ro5-4864 or PK 11195 to brain membrane preparations.

Some binding of flunitrazepam or diazepam can be observed even to astrocyte membranes or homogenates (Table 2; see also Braestrup et al., 1978; Gallager et al., 1981; McCarthy and Harden, 1981), but the B_{max} values are lower than in the intact cells. This is illustrated in Figure 1, which shows total and non-specific binding of flunitrazepam (a benzodiazepine with no

Table 2 Saturation parameters for benzodiazepine binding to the MBR in bulk prepared astrocytes, intact primary cultures of astrocytes or homogenates of cultured astrocytes.

Ligand	Preparation	K_D (nM)	B_{max} (fmol/mg protein)	Reference
Diazepam	Bulk prepared astrocytes	5.0	4390	Henn and Henke (1978)
Diazepam	Intact cultured astrocytes	<30	19 500	Hertz and Mukerji (1980)
Diazepam	Intact cultured astrocytes	13.8	13 000	Sher and Machen (1984)
Diazepam	Intact cultured astrocytes	13.4	8922	Bender and Hertz (1987a)
Diazepam	Intact cultured astrocytes	25.3	7515	Talwar and Sher (1987)
Flunitrazepam	Intact cultured astrocytes	6.6	6033	Bender and Hertz (1984)
Ro5-4864	Intact cultured astrocytes	6.7	12 228	Bender and Hertz (1985)
Diazepam	Cultured astrocyte membranes	5.0	100	Tardy et al. (1981)
Flunitrazepam	Cultured astrocyte homogenate	13.4	1027	Bender and Hertz (1984)
Flunitrazepam	Cultured astrocyte homogenate	5.0	1500	Tardy et al. (1985)

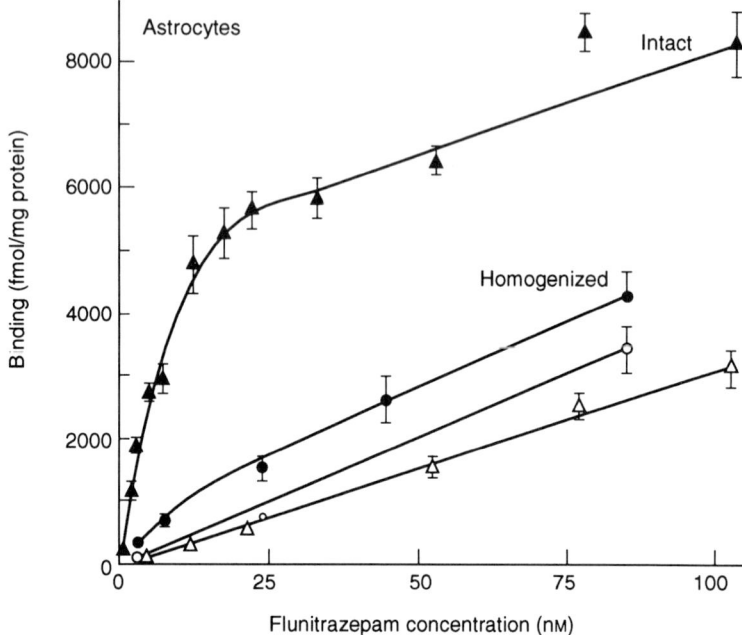

Figure 1 Saturation of [^{3}H]flunitrazepam binding to intact and homogenized astrocytes. Intact and homogenized cultures of astrocytes were incubated at 0–4 °C for 10 min with increasing concentrations of [^{3}H]flunitrazepam in the absence or presence of unlabelled flunitrazepam (50 μM). (▲) represents total binding and (△) non-specific binding to intact astrocytes; (●) and (○) represent corresponding binding to homogenized astrocytes. The K_m values were computed to be 6.6 ± 0.8 nM and 13.4 ± 2.2 nM in intact and homogenized cells, respectively, and the corresponding B_{max} values were 6.0 ± 0.3 and 1.0 ± 0.1 pmol/mg protein. From Bender and Hertz (1984).

demonstrable binding to astrocytes in membrane fractions from brain) to both intact and homogenized mouse astrocytes in primary cultures. The density of specific binding (total binding minus non-specific binding) is at all concentrations approximately six times higher in the intact than in the homogenized cells, whereas there is no major difference in affinity. The Hill coefficient (Bennett and Yamamura, 1985) was not significantly different from one in either intact or homogenized astrocytes, suggesting that only one binding site is involved and/or that there is no co-operativity (Bender and Hertz, 1984). Similar results have been reported by Sher and Machen (1984). The reason that a measurable binding was observed to the homogenates is that a centrifugation assay was used. In a filtration assay, both diazepam and

Table 3 Binding of benzodiazepines (1.8 nM) to various preparations of astrocytes in primary cultures.

Preparation	Specific binding (fmol/mg protein)		
	Flunitrazepam[a]	Diazepam[b]	Ro5-4864[c]
Intact astrocytes	1160 ± 102	1147 ± 39	3519 ± 225
Homogenized astrocytes			
Centrifugation assay	85 ± 22.2	51 ± 14	1226 ± 137
Filtration assay	29 ± 1.9	4.1 ± 1.0	328 ± 78

[a] From Bender and Hertz (1984).
[b] From Bender and Hertz (1986b).
[c] From Bender and Hertz (1985).

flunitrazepam showed negligible specific binding to astrocyte homogenates (Table 3). In contrast, some Ro5-4864 binding (0.3 pmol/mg protein at 1.8 nM Ro5-4864) could be demonstrated in a filtration assay (the method generally used in studies of binding to CNS membranes), and the binding observed using a centrifugation assay was only approximately three-fold lower than that observed in intact cells. This difference between the three benzo-diazepines might explain why binding of Ro5-4864 to astrocytes routinely can be demonstrated in brain membranes (Table 1), whereas there is no binding of flunitrazepam or diazepam, which can be displaced by Ro5-4864 or PK 11195. In contrast, binding of these ligands to neuronal membranes in homogenates is quite well recovered after filtration (McCarthy and Harden, 1981; Bender and Hertz, 1984).

2.3 Subcellular distribution

The most extensive studies of the subcellular distribution of the MBR have been made by Snyder and co-workers, using tissue from adrenal gland or kidney. They found that the subcellular distribution of the MBR closely correlated with mitochondrial markers, but not with markers for cytoplasm, plasma membrane, endoplasmic reticulum, lysosomes or nuclei (Anholt et al., 1986; Verma and Snyder, 1989). The mitochondrial localization was confirmed by Antkiewicz-Michaluk et al. (1988), who found a close correlation between PK 11195 binding sites and activity of succinate dehydrogenase, a mitochondrial enzyme, in adrenal gland, testes, lung, kidney, heart, skeletal muscle, liver and brain. However, especially in brain, there was a disproportionately large binding of PK 11195 in non-mitochondrial fractions, including both a non-mitochondrial pellet fraction and a synaptosomal plasma membrane and myelin fraction. Moreover, the density of binding sites for the MBR in brain is more than 100 times lower (per mg mitochondrial protein) than in adrenal gland, with a binding to kidney mitochondria which is in between (Table 4). Other groups have also reported relatively high levels of Ro5-4864 binding in 'nuclear' pellets of brain, compared with mitochondrial pellets (Marangos et al., 1982a; Schoemaker et al., 1983; Basile and Skolnick, 1986). Antkiewics-Michaluk et al. (1988) therefore concluded that there is strong evidence to suggest that mitochondria are primary sites of mitochondrial benzodiazepine binding, but that other membrane structures may also display these sites. A similar conclusion was reached by Doble et al. (1987), measuring binding of PK 11195 to human brain preparations. Furthermore, the B_{max} for mitochondrial binding of PK 11195 (Table 4) is only about 10% of that of diazepam, flunitrazepam or Ro5-4864 to intact cultured astrocytes. Although a substantial amount of brain mitochondria will originate from neurones, astrocytes constitute such a large fraction of brain cortical cell value

Table 4 Saturation parameters for benzodiazepine binding to MBR in adrenal gland, kidney and brain compared to the binding in intact cultured astrocytes, astrocytic membranes, brain cortical membranes and erythrocyte membranes.

Preparation	Benzodiazepine	K_D (nM)	B_{max} (pmol/mg protein)
Adrenal gland mitochondria (rat)[a]	PK 11195	1.3 ± 0.2	140 ± 4
Kidney mitochondria (rat)[a]	PK 11195	0.8 ± 0.4	13 ± 1.4
Brain mitochondria (rat)[a]	PK 11195	1.3 ± 0.8	1.0 ± 0.4
Intact cultured astrocytes (mouse)[b]	Diazepam	13.4	8.9
	Ro5-4864	6.7	12.3
	Flunitrazepam	6.6	6.0
Erythrocyte membranes[c]	PK 11195	3.9 ± 0.4	1.1 ± 0.1

[a] From Antkiewics-Michaluk et al. (1988).
[b] From Table 2.
[c] From Olson et al. (1988).

$(25-30\%)$ and respire so intensely (Hertz *et al.*, 1988; Hertz and Peng, 1992) that at least one quarter of brain mitochondria can be expected to be astrocytic. A maximum binding density in astrocytic mitochondria would, therefore, be ≈ 4 pmol/mg protein, which is still three times lower than the binding of PK 11195 to intact astrocytes, although mitochondria probably at most constitute $15-20\%$ of the cell volume. Thus, binding density in intact astrocytes appears to be at least 15 times larger than the estimated binding density in astrocytic mitochondria. This large difference could be caused by: (1) an underestimate of binding density in mitochondria for methodological reasons; (2) at least 15 times higher content of mitochondria in astrocytes in primary cultures than *in situ*; (3) 15 times higher content of mitochondria in mouse astrocytes than in rat astrocytes; and/or (4) substantial binding of mitochondrial benzodiazepine ligands to non-mitochondrial binding sites in astrocytes *in situ*. Another study of special interest is that of binding of PK 11195 to erythrocyte membranes (Olson *et al.*, 1988). Since these cells contain no mitochondria or remnants of mitochondrial membranes, and since the binding showed the typical pharmacological characteristics for binding of a mitochondrial benzodiazepine ligand, the distinct binding which was observed (B_{max} 1.1 pmol/mg protein—see Table 4) is a clear indication of mitochondrial benzodiazepine binding to non-mitochondrial sites. Studies using heart muscle (Doble *et al.*, 1985) have further supported the concept that not all mitochondrial benzodiazepine binding sites are found in mitochondrial membranes.

Further studies by the Snyder group have suggested that the MBR site must be located in the outer mitochondrial membrane, because treatment with progressively increasing concentrations of digitonin simultaneously released both monoamine oxidase, a marker of outer mitochondrial membranes, and peripheral benzodiazepine receptors, and at a digitonin concentration when cytochrome oxidase, which is localized at the inner mitochondrial membranes, was not released (Anholt *et al.*, 1986). They proposed that the protein representing the mitochondrial benzodiazepine site is a relatively non-specific, voltage-dependent anion channel (VDAC) which also is referred to as 'mitochondrial porin' (Verma and Snyder, 1989).

Mitochondrial porin is different from the mitochondrial membrane dihydropyridine binding site, described by Zernig *et al.* (1990), a dihydropyridine binding site which does not represent the L-type calcium channel, and to which dihydropyridines bind with micromolar affinity. In contrast to the mitochondrial benzodiazepine binding site (or mitochondrial porin), the mitochondrial dihydropyridine binding site is located on the inner mitochondrial membrane, as indicated by the fact that amiodarone, an inhibitor of an ion channel located at the inner mitochondrial membrane, interacts with dihydropyridines, with respect to both function (opening of the channel by amiodarone and competitive inhibition by dihydropyridines) and binding (potent inhibition of nitrendipine binding to mitochondrial membranes by amiodarone; Zernig *et al.*, 1990). Nevertheless, dihydropyridines do interact with binding of Ro5-4864 to the MBR in brain membranes and vice versa (see Section 6). Before completely abandoning the idea that the mitochondrial benzodiazepine binding site could be related to the amiodarone-sensitive ion channel across the inner mitochondrial membrane, it should be mentioned that Mukherjee and Das (1989) found strong evidence that a contamination of an outer mitochondrial membrane preparation with an inner mitochondrial fraction may occur. Moreover, Verma and Snyder (1989) have pointed out that the outer and the inner mitochondrial membranes fuse at specific contact points close to the localization of the mitochondrial porin. Finally, both Papadopoulos *et al.* (1991) and Hall (1991) have recently obtained evidence that the MBR may constitute part of a mitochondrial intermembrane apparatus, required for biosynthesis of steroids (see section 7).

Photolabelling with either flunitrazepam (a typical benzodiazepine, binding to both the MBR and the CBR in intact astrocytes and to the MBR in peripheral tissues), or the isoquinoline PK 11195 has shown striking differences. Although both ligands labelled proteins in purified outer mitochondrial membranes, flunitrazepam labelled two bands, at 35 and 30 kDa, whereas PK 11195 labelled a single band at 15–18 kDa. Thus these two chemically distinct ligands of the MBR might interact with different sites, or perhaps with different populations of receptors (Verma and Snyder, 1989), as discussed in Chapter 1.

2.4 Association and dissociation kinetics

The time-course for specific binding of PK 11195 (1.0 nM) to erythrocyte membranes (Olson *et al.*, 1988) is shown in Figure 2. The specific binding is 50%, completed in a few minutes, and reaches equilibrium at about 40 min. In some experiments, the rate of dissociation of specific binding was determined by adding 10 μM unlabelled PK 11195 after 120 min. This was followed by a rapid dissociation of the bound ligand. From the inserts in Figure 2 it can be seen that the binding both saturates and desaturates exponentially with an initial rate constant, K_{obs}, of 0.3/min, and a dissociation rate constant, K^{-1} of 0.1/min, i.e. a half-life ($t_{1/2}$) for dissociation of 7 min ($K \times t_{1/2} = 0.693$). Dissociation rate constants for Ro5-4864 or PK 11195 from brain membranes have been determined by Schoemaker *et al.* (1983), Weissman *et al.* (1984), Villinger (1985) and Benavides *et al.* (1983b) and are of the same order of magnitude (0.01–0.11/min). A similar rate constant for dissociation of Ro5-4864 from intact cultures of astrocytes (0.057/min) was observed by Bender and Hertz (1985), who found the dissociation of flunitrazepam or

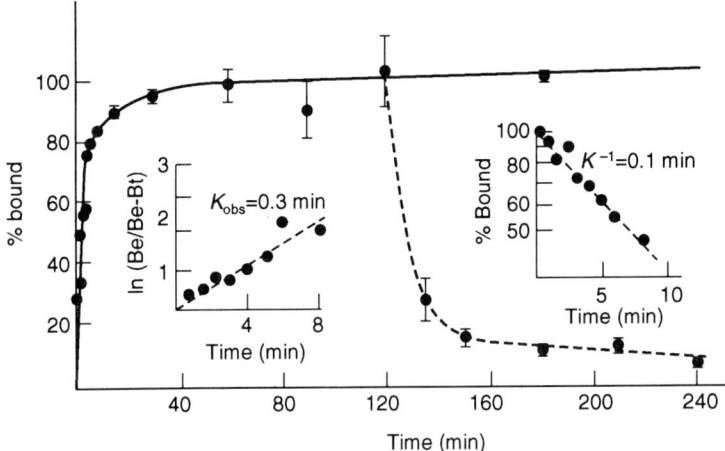

Figure 2 Time-course for [³H] PK 11195 binding to erythrocyte membranes incubated with 1 nM PK 11195. After 120 min of incubation 10 μM unlabelled PK 11195 was added to reverse binding (dashed line). From Olson *et al.* (1988).

diazepam to be somewhat faster ($K^{-1} = 0.2-0.3/\text{min}$) (Bender and Hertz, 1984, 1986a; Hertz and Bender, 1988); however Talwar and Sher (1987) found a much lower rate constant for diazepam dissociation. An especially rapid dissociation of flunitrazepam and diazepam might explain the finding of very little flunitrazepam binding to an astrocytic membrane fraction reported by MacCarthy and Harden (1981), and the low B_{max} values for flunitrazepam and diazepam in homogenates or membranes obtained from astrocytic cultures (Table 2; see also Tardy *et al.*, 1981, 1985; Bender and Hertz, 1984). More importantly, this will in all probability also apply to preparations obtained from the brain *in situ*. The less rapid dissociation of Ro5-4864 appears to be the most logical explanation of why displacement of Ro5-4864 binding by diazepam or flunitrazepam can be demonstrated in experiments using CNS membranes, whereas no binding of labelled diazepam or flunitrazepam has ever been demonstrated, which displays peripheral-type characteristics.

At 37 °C the rate constant for dissociation of diazepam from intact astrocytic cultures is increased to 2.6/min, i.e. the $t_{1/2}$ is reduced to 0.3 min (Hertz and Bender, 1988), confirming previous observations by Sher (1985) of a more rapid dissociation at higher temperatures. If this increase in dissociation constant at 37 °C applies also to binding of Ro5-4864 or PK 11195 to brain membranes, the dissociaton may be so fast that even binding of these ligands will be difficult or impossible to demonstrate using conventional techniques.

2.5 Pharmacological characteristics within the group of benzodiazepines

The ability of other benzodiazepines (or of isoquinolines like PK 11195 known to interact with the MBR) to displace benzodiazepines from their binding sites on intact cultures of astrocytes has been studied for at least three reasons: (1) to compare astrocytes in primary cultures with astrocytes *in situ*; (2) to compare pharmacological characteristics in intact and homogenized cells; and (3) to establish a detailed rank order of potency between different benzodiazepines in their binding to the MBR. Inhibition of binding of [^{3}H]Ro5-4864 to astrocytic cultures was potently inhibited (IC$_{50}$ vaues in the nanomolar range) by the benzodiazepines flunitrazepam, diazepam and Ro7-3351 (Bender and Hertz, 1985). Other benzodiazepines (alprazolam, clonazepam and chlordiazepoxide) inhibited [^{3}H]Ro5-4864 binding less potently, i.e. with IC$_{50}$ values in the micromolar range. The slope factors were significantly below unity or close to being significantly below unity. The possible effects of GABA were tested at 10, 30, 100, 300 and 1000 μM (at 1.8 nM Ro5-4864), but at none of these concentrations did this compound have any significant effect (Bender and Hertz, 1985).

The rank order of potency for the inhibition of [^{3}H]Ro5-4864 binding in cultured astrocytes is the same as that observed for displacement of [^{3}H]Ro5-4864 in brain membrane preparations (Marangos *et al.*, 1982a; Schoemaker *et al.*, 1983; Weissman *et al.*, 1984). Accordingly, it seems possible to use these cells as faithful models of their *in vivo* counterparts. They have the advantage that effects of benzodiazepines can be studied in a brain preparation expressing no CBR.

Pharmacological similarities and dissimilarities of [^{3}H]flunitrazepam binding sites in intact and homogenized astrocytic (and neuronal) cultures were studied using unlabelled Ro5-4864 (selectively high affinity for the MBR), diazepam (equal affinity for both receptors) and clonazepam (selectively high affinity for the CBR) (Bender and Hertz, 1985). Table 5 shows that Ro5-4864 inhibits [^{3}H]flunitrazepam binding in intact and homogenized astrocytes with identical K_i (and therefore also IC$_{50}$) values. In intact astrocytes, diazepam was only slightly (although significantly) less active, but its potency decreased four-fold after homogenization. Clonazepam had an extremely weak effect in astrocytic preparations, especially after homogenization, confirming the inability of clonazepam to displace Ro5-4864 in astrocytic membranes (Braestrup *et al.*, 1978). For comparison, diazepam was the most potent displacer of the binding in intact cerebral cortical, mainly GABAergic neurones, with the same potency as in astrocytes. It was followed by clonazepam, with a K_i value which was 20–25 times lower than in astrocytes. The least potent compound was Ro5-4864, which was almost inactive. The

Table 5 Inhibition of [^{3}H]flunitrazepam binding in intact and homogenized primary cultures of astrocytes and neurones.

Culture	Intact [K_i (nM)]	Homogenized [K_i (nM)]
Astrocytes		
Ro5-4864	3.7(3.2–4.2)[a]	3.9(3.7–4.2)
Diazepam	6.5(5.5–7.6)	27.7(27.0–28.3)
Clonazepam	597(557–642)	1947(1870–2025)
Neurones		
Diazepam	6.7(6.1–7.4)	102.0(71.2–147.0)
Clonazepam	26.0(21.0–33.0)	31.9(26.5–38.4)
Ro5-4864	>10000	ND[b]

Intact or homogenized astrocytes and neurones were incubated with 1.8 nM [^{3}H]flunitrazepam for 10 min at 0–4 °C in the presence of various inhibitory compounds. Non-specific binding was defined as the amount of [^{3}H]flunitrazepam bound in the presence of 50 μM unlabelled flunitrazepam. The K_i (inhibition constant) (with 95% confidence limits) was calculated using the Cheng–Prussoff equation: $K_i = IC_{50} (1 - [L]/K_D)$, where $[L]$ is the radioligand concentration, K_D the apparent dissociation constant of the ligand and IC_{50} the concentration of inhibitory compound producing 50% inhibition of binding. IC_{50} values (with 95% confidence limits) were calculated by log-probit analysis as described by Goldstein (1984).

[a] 95% confidence limit.
[b] Not determined.
From Bender and Hertz (1984).

order of potency was changed after homogenization, which did not alter the K_i value for clonazepam, whereas that for diazepam became 15 times higher.

In both astrocytes and neurones, the binding of [^{3}H]diazepam was inhibited by all tested benzodiazepines (a total of 14) and benzodiazepine antagonists (a total of 3), including the isoquinoline PK 11195 (Figure 3). Their IC_{50} values were in the nanomolar range, except for Ro5-4864, Ro5-2181 and PK 11195 in neurones, and clonazepam, flurazepam, Ro11-3624, Ro5-2181 and chlordiazepoxide in astrocytes, which had IC_{50} values in the low micromolar ranges (Bender and Hertz, 1987a). As can be seen from Figure 3, the complete rank order of potency for inhibition of [^{3}H]diazepam binding by benzodiazepines was distinctly different in astrocytes and in neurones.

Several benzodiazepines (i.e. clonazepam, flunitrazepam, midazolam, Ro5-4864, Ro11-3128(+), Ro11-3624(−) and chlordiazepoxide) and two benzodiazepine antagonists, Ro15-1788 (neuronal) and PK 11195 (mitochondrial) were also tested at 37 °C. In astrocytes, the rank order of

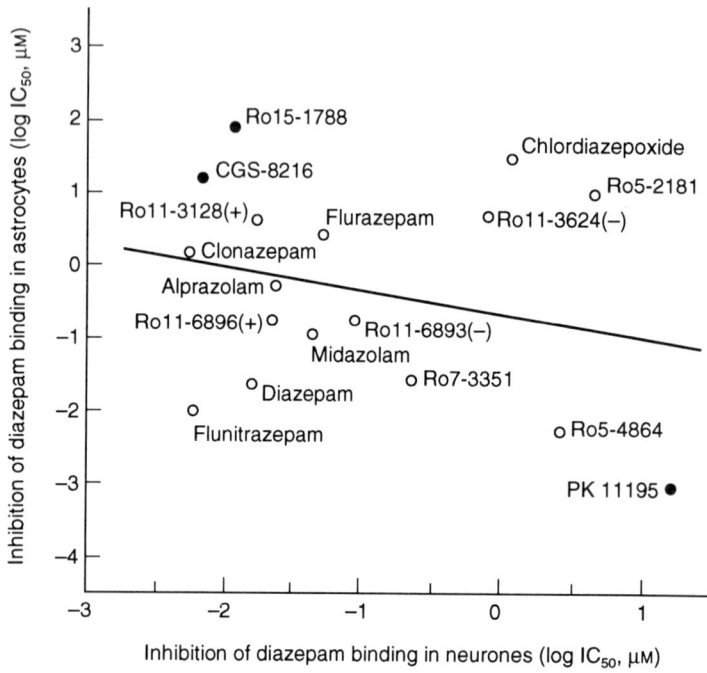

Figure 3 Correlation between IC_{50} values for inhibition of astrocytic and neuronal [³H]diazepam binding by benzodiazepine agonists (○) and benzodiazepine antagonists (●). Correlation coefficient, $r = -0.25$ (N.S.). From Bender and Hertz (1987a).

potency was not different from that observed at 0 °C, and little change occurred in IC_{50} values (Bender, 1988).

The only compounds, of the 17 tested, which have slope factors significantly lower than one in astrocytes were clonazepam (0.65) and PK 11195 (0.71), suggesting negative co-operativity or multiple binding sites (Bender and Hertz, 1987a). At 37 °C most slope factors for the seven drugs studied at this temperature (see above) decreased and, for some of the drugs (most notably Ro5-4864, 0.57, and chlordiazepoxide, 0.35), the difference was marked. The slope factors for clonazepam and PK 11195, which were <1 even at 0 °C, remained unaltered. Thus, most 'typical' benzodiazepines (which have relatively similar affinity for the neuronal and the mitochondrial benzodiazepine receptor) have Hill coefficients and slope factors close to one. This applies, e.g. to flunitrazepam, where neither the Hill coefficient, calculated from a saturation study (Bender and Hertz, 1984), nor the slope factor, calculated from its displacement of labelled diazepam (Bender and Hertz, 1987a), were significantly different from one. The exceptions are the drugs showing extreme differences between their affinity for the neuronal and the mitochondrial receptor: PK 11195 displaced diazepam with a slope factor < 1 both at 0 °C and at 37 °C and Ro5-4864 and chlordiazepoxide did so at 37 °C. Moreover, all the benzodiazepines tested displaced Ro5-4864 with slope factors below one (Bender and Hertz, 1985).

2.6 Effects of benzodiazepines on binding sites for other drugs and effects of other agents on the mitochondrial binding site

Benzodiazepines interacting with the MBR displace binding of some other types of drugs to their specific binding sites (especially transport proteins) or vice versa. The interaction between benzodiazepines and binding sites for dihydropyridines and other inhibitors of the L-channel for calcium (e.g. verapamil) has attracted considerable interest. At low nanomolar concentrations these drugs are specific inhibitors of the L-channel (Janis *et al.*, 1987). High-affinity nitrendipine binding has repeatedly been demonstrated to brain cortical homogenates and membranes, in general with K_D values between 0.2 and 1 nM and B_{max} values of 0.1–0.2 pmol/mg protein (e.g. Gould *et al.*, 1982; Ehlert *et al.*, 1982; Murphy and Snyder, 1982; Marangos *et al.*, 1982b; Janis *et al.*, 1987). At higher concentrations (low micromolar level), dihydropyridines bind to an anion channel located in the inner mitochondrial membrane (Zernig *et al.*, 1990) and presumably also to a multitude of other sites (Janis *et al.*, 1987). It has long been known that benzodiazepines are able to displace the binding of the dihydropyridine [^{3}H]nitrendipine, for example from guinea pig ileum (Bolger *et al.*, 1983; Luchowski *et al.*, 1984) and cardiac membranes (Holck and Osterrieder, 1985), but the affinity is low ($IC_{50} \approx 30$–40 μM). Conversely, at low micromolar concentrations, dihydropyridines are able to displace binding of Ro5-4864 from heart, kidney and brain membranes (Cantor *et al.*, 1984).

Both high- and low-affinity binding of dihydropyridines to cultured neuronal and glial cells has been observed (Litzinger and Brenneman, 1985; Litzinger *et al.*, 1986; Hertz and Bender, 1988; Hertz and Code, 1992). In cortical astrocytes, the high-affinity binding site (K_D 2.5 nM) is induced by treatment with dibutyryl cyclic AMP, presumably a differentiating agent (Hertz, 1990), which also is required for the expression of the L-channel itself (MacVicar, 1984; Barres *et al.*, 1990). The expression of low-affinity binding sites ($K_D > 200$ nM) is not dependent upon treatment with dibutyryl cyclic AMP and might be associated with the mitochondrial ion channel localized in the inner mitochondrial membrane, as previously discussed.

41

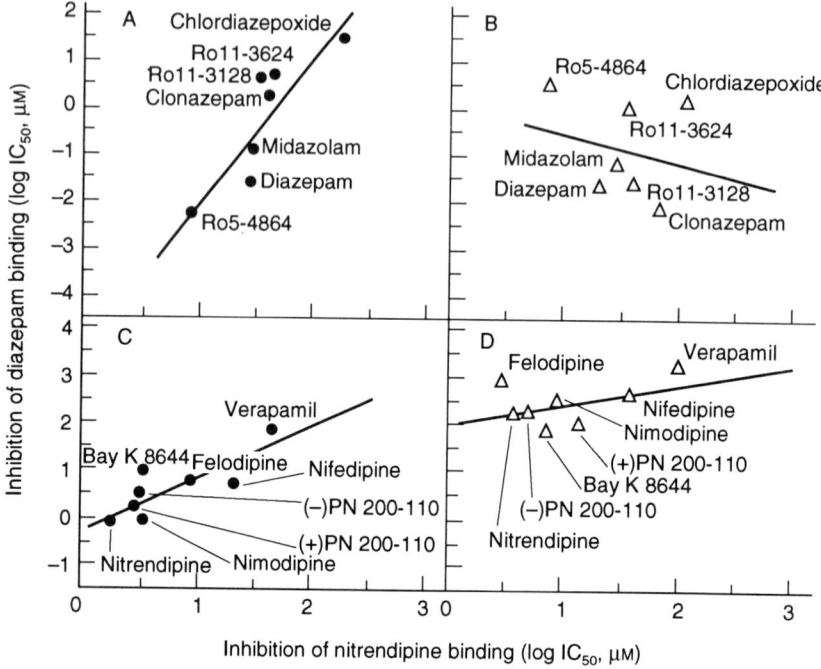

Figure 4 Correlation between IC$_{50}$ values for inhibition of [^{3}H]diazepam and [^{3}H]nitrendipine binding by various benzodiazepines (A and B) and calcium channel antagonists (C and D) in astrocytes (A and C) and neurones (B and D). (A) $r = 0.88$; $p < 0.01$; (B) $r = 0.24$ (N.S.); (C) $r = 0.83$; $p < 0.02$; (D) $r = 0.44$ (N.S.). From Hertz and Bender (1988).

The correlation between the potency of different benzodiazepines and calcium channel antagonists (dihydropyridines and verapamil) to displace diazepam and nitrendipine (both 1.8 nM) in primary cultures of astrocytes and of neurones is shown in Figure 4 (Bender, 1988; Hertz and Bender, 1988). In astrocytes (Figure 4A), the correlaton is significant for the displacement of the two ligands by benzodiazepines ($r = 0.88$; $p < 0.01$), whereas this is obviously not the case in neurones (Figure 4B). However, it should be noted that the concentrations of benzodiazepines required to displace nitrendepine binding are orders of magnitude larger than those required to displace diazepam, and similar to those inhibiting nitrendipine binding in guinea pig ileum and heart (see above). This obviously does not suggest a correlation between the mitochondrial benzodiazepine binding site and an L-channel, although it also does not prove that such an interaction cannot exist. Conversely, dihydropyridines and verapamil displace diazepam and

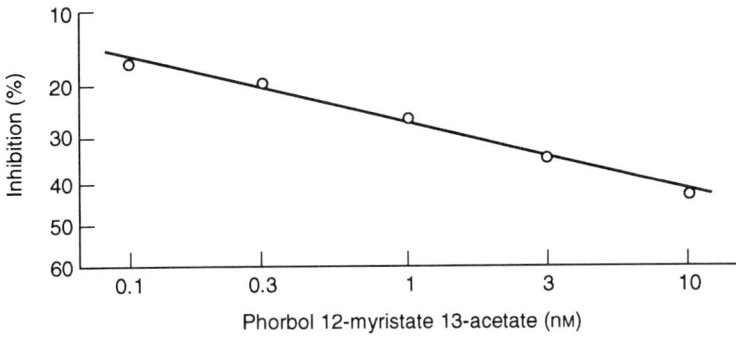

Figure 5 Dose–response curve for inhibition of [^{3}H]diazepam binding in primary cultures of astrocytes by phorbol 12-myristate 13-acetate. From Bender and Hertz (1988a).

nitrendipine with a similar rank order of potency ($r = 0.83$; $p < 0.02$) in astrocytes (Figure 4C), but not in neurones (Figure 4D). From the IC$_{50}$ value for nitrendipine's displacement of its own binding in astrocytes (1.7 μM), it is obvious that this is not a displacement from the high-affinity binding site (K_D 2.5 nM). Very high nanomolar or low micromolar concentrations of the dihydropyridines are also required for displacement of diazepam.

All three, the phorbolester, phorbol 12-myristate 13-acetate, an activator of protein kinase C (Bender and Hertz, 1988a), atrial natriuretic peptide (ANP) (Bender and Hertz, 1987b) and octadecaneuropeptide (ODN, 'anxiety peptide'), a constituent of diazepam-binding inhibitor (DBI) or endozepine (Bender and Hertz, 1986b), cause 35% inhibition of diazepam binding to intact cultures of astrocytes at 3–30 nM, but the slopes of the dose–response curves are very shallow (Figure 5); therefore much lower concentrations have some effect and the IC$_{50}$ values are considerably higher (Bender and Hertz, 1988a; Hertz and Bender, 1988). ODN displaces diazepam less potently from intact cultured neurones (Bender and Hertz, 1986b; Hertz and Bender, 1988) and it has little or no effect on Ro5-4864 binding to mitochondria obtained from rat adrenal cortex or membranes of cultured astrocytes (Guidotti et al., 1990). The convulsant agents pentylenetetrazole and picrotoxin potently inhibit binding of diazepam to primary cultures of astrocytes, but not to primary cultures of neurones, and several clinically used anticonvulsants, including carbamazepine and phenobarbital, displace diazepam binding from cultured astrocytes, but not from cultured neurones at pharmacologically relevant concentrations (Bender and Hertz, 1988b). PK 11195 has an anticonvulsant effect, and rats made tolerant to carbamazepine show cross-tolerance to PK 11195 but not to diazepam (Weiss and Post, 1991).

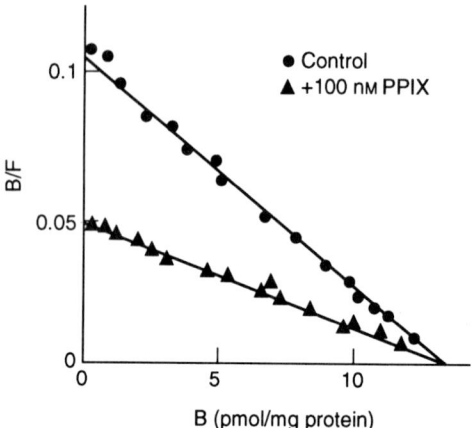

Figure 6 Scatchard analysis of [^{3}H]PK 11195 binding to rat kidney mitochondria in the presence and absence of protoporphyrin IX (PPIX). Rat kidney mitochondria were incubated for 90 min in the presence of 0.05–30 nM [^{3}H]PK 11195 with and without the addition of 100 nM protoporphyrin IX. Non-specific binding was less than 10% of total binding. From Verma and Snyder (1988).

Other physiologically occurring substances or drugs have also been found to interact with the mitochondrial benzodiazepine binding site. This applies to porphyrins, especially those with prominent physiological function, which at low nanomolar concentrations displace PK 11195 from adrenal gland or rat kidney mitochondria (Figure 6) in a competitive manner (Verma and Snyder, 1988, 1989). For this reason, porphyrins were suggested to be the endogenous ligand for the MBR. The ratio between IC$_{50}$ values for displacement of [^{3}H]PK 11195 by, respectively, PK 11195 and protoporphyrin IX, is well preserved across species (Verma and Snyder, 1988). However, there is a pronounced species difference with respect to IC$_{50}$ values for displacement of PK 11195 binding with Ro5-4864 (Table 6).

The enzyme pyruvate dehydrogenase is able to prevent binding of Ro5-4864 and PK 11195 to brain or kidney membranes, but not to displace the binding once it has taken place (Daval *et al.*, 1989). The action is specific for this enzyme, which is part of the pyruvate dehydrogenase complex. This multiple-enzyme complex oxidizes and decarboxylates cytosolic pyruvate and condenses it with coenzyme A to acetyl CoA, which subsequently enters the tricarboxylic acid cycle and reacts with oxalocetate to form citrate. Obviously a transport mechanism across the mitochondrial membrane is involved in this series of events.

Table 6 Displacement of [^{3}H] PK 11195 in membranes from tissues of various species.

Species/tissue	IC$_{50}$ (nM)		
	PK 11195	Protoporphyrin	Ro5-4864
Rat			
Brain	1.5 ± 0.1	25 ± 7.1	8.0 ± 0.5
Heart	2.0 ± 0.2	22 ± 10.1	10 ± 1.2
Kidney	1.5 ± 0.4	27 ± 5.6	6.0 ± 0.7
Guinea pig			
Brain	2.5 ± 0.7	30 ± 10	10 ± 1.5
Rabbit			
Kidney	2.1 ± 0.3	14 ± 5.4	483 ± 62
Bovine			
Cerebral cortex	1.7 ± 0.4	26 ± 14.3	800 ± 141
Heart	2.0 ± 0.4	17 ± 2.5	960 ± 230
Human			
Placenta	3.1 ± 1.5	30 ± 3.1	262 ± 98

Tissue membranes were incubated for 60 min at 4 °C with 1 nM [^{3}H] PK 11195 and at least ten different concentrations of inhibitor. Non-specific binding was <10% of the total binding in all preparations tested. Data are mean ± SEM. From Verma and Snyder (1988).

2.7 Effects of mitochondrial benzodiazepine ligands

The most important questions to be answered about mitochondrial benzo-diazepine ligands are: (1) which physiological response(s) do benzodiazepines evoke by stimulating mitochondrial receptors? (2) to what extent do the responses contribute to the known neuropharmacological effects of benzodiazepines? and (3) which other beneficial or adverse effects result from stimulation of MBR sites? In order to attempt to answer these questions some knowledge is needed of the pharmacologically relevant benzodiazepine concentrations. In serum, these will ordinarily not exceed 0.5–1.0 μM, but due to pronounced protein binding, the cerebrospinal concentration and the extracellular concentration in the CNS will normally be at most 100 nM (Kanto et al., 1975; Polc, 1988), i.e. concentrations compatible with those to be expected from the saturation kinetics.

Effects of nanomolar concentrations of benzodiazepines on the mitochondrial receptor include inhibition of release of hormones, especially steroid hormones. This seems of special importance on account of the high density of MBR in adrenal glands. Diazepam inhibits potassium-stimulated aldosterone secretion in a dose-dependent manner ($IC_{50} = 14$ nM), suggesting either that benzodiazepine receptors may play an active role in the regulation of aldosterone secretion or that the voltage-dependent calcium channel may be a possible site of action for diazepam (Shibata *et al.*, 1986). It is in support of a direct action of benzodiazepines on steroid synthesis that Wilkinson *et al.* (1980) demonstrated a stimulatory effect of diazepam on the production of testosterone *in vitro*, but the IC_{50} was extremely high (240 μM). More recently, Ritta *et al.* (1987) have described a stimulatory effect of Ro5-4864 (100 nM–10 μM) on testosterone secretion. It has also been observed that several benzodiazepines, including Ro5-4864, caused a dose-dependent stimulation of the conversion of cholesterol to pregnenolone in bovine adrenal mitochondria (Yanagibashi *et al.*, 1989). The rank order of potency was typical of the MBR and the stimulation was not inhibited by Ro15-1788, an antagonist of the CBR. Very interestingly, Knudsen and Nielsen (1990) observed that the amino acid sequence of diazepam binding inhibitor (DBI) is identical to that of a transmitochondrial acyl CoA carrier, needed for lipid synthesis. Analogously, Besman *et al.* (1989) discovered that transport of cholesterol from the outer to the inner mitochondrial membrane of the adrenal gland is stimulated by a protein which is identical to DBI, except for the two last residues, and that diazepam has a similar effect. However, low micromolar concentrations were required. Also, Mukhin *et al.* (1989) observed a correlation between the ability of different benzodiazepines to potently inhibit binding of PK 11195 in an adrenal tumour cell line (and in rat and bovine adrenocortical preparations) and to enhance the rate of progesterone synthesis. However, both the 'agonist' Ro5-4864 and the 'antagonist' PK 11195 stimulated progesterone synthesis to the same extent. This role of MBR is extensively reviewed in Chapter 4.

Another mitochondrial effect is the inhibition of ADP-stimulated respiration (state III respiration) by Ro5-4864 (25 μM) but not by clonazepam observed by Moreno-Sanchez *et al.* (1991b), who suggested that benzodiazepines might cause an unspecific inhibition within the respiratory chain and, in addition, exert a specific inhibition of succinate transport. An inhibition of state III respiration is in agreement with a previous observation by Hirsch *et al.* (1989a), using rat kidney mitochondria, but these authors also observed an increase in state IV respiration. The opposite effects on the rates of state III and IV respiration indicate a decrease in respiratory control ratio (RCR), which defines the degree of coupling between mitochondrial respiration and oxidative phosphorylation (Trupower and Haggerty, 1980). As can be seen in Figure 7, Ro5-4864 potently inhibited RCR ($IC_{50} \approx 20$ nM), whereas flunitrazepam had a very weak effect, and clonazepam was even less potent ($IC_{50} > 10$ μM; not shown in the figure). Several other compounds including

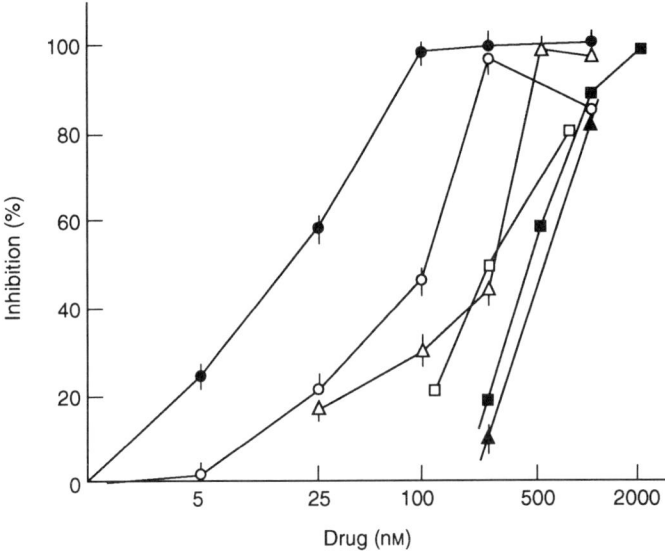

Figure 7 Inhibitory effects on respiratory control ratio in rat kidney mitochondria by the following ligands for the 'mitochondrial' benzodiazepine receptor: Ro5-4864 (●); PK 11195 (○); diazepam (△); deuteroporphyrin IX (□); mesoporphyrin (■); and flunitrazepam (▲). The effect elicited by 250 nM Ro5-4864 was taken as 100%. Values are means ± SEM. From Hirsch *et al.* (1989a).

porphyrins had a similar effect (Figure 7), and there was an excellent correlation between rank orders of potency for inhibition of mitochondrial binding of Ro5-4864 and of RCR. However, as also noted by the authors, the agonist Ro5-4864 and the antagonist PK 11195 had identical effects, with Ro5-4864 being the more potent of the two compounds. Moreover, only a small percentage of respiration was inhibited. The same authors (Hirsch *et al.*, 1989b) also observed that in intact mitochondria there were approximately the same number of binding sites for [³H]PK 11195 as for [³H]Ro5-4864, and their apparent K values were identical. In mitochondrial fragments, there were 80% more binding sites for [³H]Ro5-4864 than for [³H]PK 11195. An extremely potent inhibition of oxygen consumption in neuroblastoma cells (a supposedly neuronal cell line)* by Ro5-4864 and PK 11195 was observed by Larcher *et al.* (1989) but, again Ro5-4864 and PK 11195 exerted similar effects. Such an inhibition might be a result of the interaction with pyruvate dehydrogenase (which inhibits binding of these two ligands to brain and

---

* Such a 'mix-up' in neuronal and glial cell lines (but not in primary cultures) is quite common.

kidney membranes) but there is no direct inhibitory effect on pyruvate dehydrogenase activity by Ro5-4864 (Daval *et al.*, 1989).

A possible interaction between the MBR and calcium channels has repeatedly been suggested (see also Section 6). A potentiation of the inotropic effect of the calcium channel agonist Bay K 8644 by 500 nM of an Ro5-4864 analogue has recently been observed by Bolger *et al.* (1989) in heart muscle. At micromolar concentrations this enhancement was converted into an inhibition. Mestre *et al.* (1984) found Ro5-4864, even at low micromolar concentrations, to decrease action potential duration and contractility in a guinea pig heart preparation, an effect which was reversed by PK 11195. Calcium channel blockers caused the same effects as Ro5-4864 and they, too, could be reversed both by PK 11195 and by an increase of the extracellular calcium concentration (Mestre *et al.*, 1985).

In order to test the hypothesis of a correlation between calcium homeostasis and the mitochondrial, but not the neuronal, benzodiazepine receptor in brain, we have recently measured free intracellular calcium concentration in primary cultures of either astrocytes or neurones (Code *et al.*, 1991; Hertz and Code, 1992; Z. Zhao, W.E. Code and L. Hertz, unpub. obs.). In both cell types, a potassium-induced depolarization increased the free intracellular calcium concentration, which reached a maximum at ≈ 40 mM potassium. In neurones, the additional presence of a benzodiazepine did not alter the response, but in the astrocytic cultures, nanomolar concentrations of benzodiazepines (e.g. 10–20 nM midazolam or somewhat higher diazepam concentrations) enhanced the effect of the submaximum potassium depolarization evoked by, e.g. 20 mM potassium (Figure 8), but had no effect at normal or very high potassium concentrations. A further rise in the midazolam concentration greatly reduced the response (Z. Zhao, W.E. Code and L. Hertz, unpub. obs.). Such a decrease of the benzodiazepine effect at higher concentrations is identical to the observations in heart muscle by Bolger *et al.* (1989). The enhancing effect of the benzodiazepine on the potassium-induced increase in free intracellular calcium concentration was blocked by PK 11195, which did not change the potassium effect as such (Figure 8). In contrast, the dihydropyridine calcium channel blocker nifedipine annihilated both the effect of potassium and the additional effect of the benzodiazepine (Code *et al.*, 1991). Surprisingly, the neuronal benzodiazepine antagonist Ro15-1788 had a similar inhibitory effect with only slightly lower potency: The effect of 20 nM midazolam could be completely inhibited by 1.0 μM of either PK 11195 or Ro15-1788, but the inhibition was only partial at 0.3 μM of either drug.

Although it appears likely that the activation of the MBR enhanced calcium channel activity, it cannot be excluded that the additional amount of free intracellular calcium originated from intracellularly bound calcium. This is especially a possibility since mitochondrial benzodiazepines like Ro5-4864 can

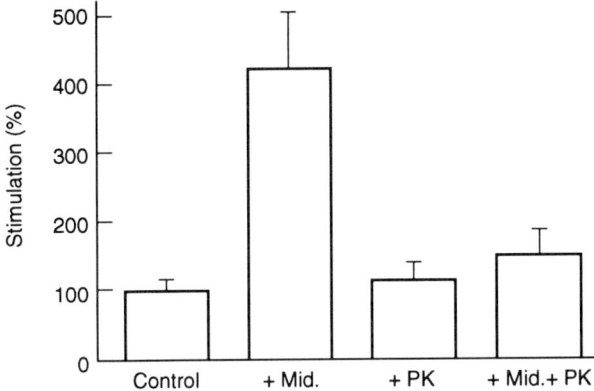

Figure 8 Effects of midazolam (Mid., 10 nM), of the mitochondrial benzodiazepine antagonist PK 11195 (PK, 1 μM), and of midazolam (10 nM) plus PK 11195 (1 μM) on the increase in free intracellular calcium concentration evoked by an increase in the extracellular potassium concentration to 20 mM. The increase induced by exposure to the elevated potassium concentration alone, i.e. the control value (about a doubling of the normal free intracellular calcium concentration), is indicated as 100%. From Code *et al.* (1991).

displace calcium from rat heart and kidney mitochondria, an effect which can be blocked by magnesium, and might suggest an action at the inner mitochondrial membrane (Moreno-Sanchez *et al.*, 1991a).

2.8 Concluding remarks

An attempt has been made to synthesize fragments of a pattern from some of the known characteristics and functions of the MBR. It seems safe to conclude that this receptor is not always exclusively located on mitochondrial membranes. There is also good evidence that several benzodiazepines, as well as PK 11195 and some physiologically important ligands like ANP, have slope factors well below one, indicating either negative co-operativity or more than one binding site. Among the roles of benzodiazepines at the mitochondrial binding site (s), the stimulation of steroid secretion has been investigated in most detail. By and large, the effects are of sufficient potency to suggest that they do, indeed, represent actions occurring by binding to the mitochondrial receptor. However, the distinct agonist effect by PK 11195 is at variance with the findings that this agent has an antagonistic action in brain and muscle.

The same applies to the effects of benzodiazepines and PK 11195 on energy metabolism, some of which were also very potent.

The observation that PK 11195 functions as a reliable antagonist at the MBR at some sites but is a potent agonist at other binding sites has at least two consequences. The first of these is that there must be at least two different mitochondrial receptor sites, maybe one located at the mitochondrial membrane (where PK 11195 is an agonist), and another at the cell membrane (where PK 11195 is an antagonist). In some tissues, one of these may dominate and in other tissues the other may be the only or the major MBR. The other, and important consequence, is that the lack of potent inhibition of effects by Ro5-4864 (and other benzodiazepines), by PK 11195 *in vivo* cannot be used as an indication that these effects are not evoked by stimulation of an MBR. Also, the ability of Ro15-1788 to counteract benzodiazepine effects on free intracellular calcium concentration in astrocytes almost as potently as PK 11195 shows that inhibition by this supposedly neuronal antagonist cannot be used as an unequivocal indication that the effect is, indeed, exerted on the CBR.

It appears that the MBR does play a prominent role in normal physiological functions, such as permeability of the mitochondrial membrane, production of steroids and regulation of calcium channels. One might therefore wonder why neurones have no need for this receptor site, but utilize a different CBR, which is generally believed to exert all its actions on chloride transport across the cell membrane. Maybe the answer to this question lies in the extensive metabolic and physiological interactions between neurones and astrocytes (e.g. Hertz, 1986, 1992; Hertz and Peng, 1992, Hertz *et al.*, 1992). As an even more intriguing possibility, the pronounced effects on CNS function of benzodiazepines might occur because most of these drugs at usual doses will act on both central and mitochondrial receptors. Thus, Majewska *et al.* (1988) have suggested that enhancement by classical, therapeutically used benzodiazepines (e.g. diazepam) of the synthesis of the steroid pregnelonone sulphate (an allosteric inhibitor of the neuronal $GABA_A$ receptor) at MBR sites (i.e. mainly on astrocytes) may counterbalance the enhanced inhibition evoked by direct stimulation of GABA currents by the same drug at the neuronal benzodiazepine binding site.

Langer and Arbilla (1988) have suggested a revised nomenclature for benzodiazepine receptors, i.e. omega 1, 2 and 3, with omega 1 and 2 corresponding to BZ_1 and BZ_2 receptors and omega 3 to the mitochondrial receptor. They indicate that most benzodiazepines are non-selective ligands for omega 1 and 2, while selectivity for the omega 1 receptor is present in several non-benzodiazepines, e.g. beta carbolines. On the basis of the presently reviewed material one should, perhaps, instead suggest four subtypes (e.g. omega 1, 2, 3 and 4 or BZ_{A1}, BZ_{A2}, BZ_{B1}, and BZ_{B2}), of which omega 1 and 2 (or the BZ_As) are similar to the corresponding subtypes in the Langer–Arbilla terminology, whereas omega 3 and 4 correspond to their omega 3 subtype (or the BZ_Bs) and can be distinguished from each other according to the behaviour of PK 11195 as an antagonist (omega 3 or BZ_{B1}) or agonist (omega 4 or BZ_{B2}). All 'typical' benzodiazepines (e.g. diazepam) are ligands for all four subtypes, but with differences in potency, which for a few of the benzodiazepines are pronounced. Omega 4 would correspond to the genuinely mitochondrial receptor (located on mitochondria and influencing respiratory control and steroid synthesis) and omega 3 to the mitochondrial receptor, regulating calcium homeostasis, which at least partly may be localized on non-mitochondrial elements. One common property of all of these sites (central and mitochondrial) is that they appear to be located on transport proteins, regulating ion channels at the cell membrane and/or the mitochondrial membrane and substrate carriers across the mitochondrial membrane.

Acknowledgements

I would like to express my sincere thanks to Mrs Maureen Shargool for typing this manuscript.

My own research has been supported by Medical Research Council of Canada.

References

Anholt, R.R., Murphy, K.M., Mack, G.E. and Snyder, S.H. (1984). Peripheral-type benzodiazepine receptors in the central nervous system: localization to olfactory nerves. *J. Neurosci.* **4**, 593–603.

Anholt, R.R., De Souza, E.B., Oster-Granite, M.L. and Snyder, S.H. (1985). Peripheral-type benzodiazepine receptors: autoradiographic localization in whole-body sections of neonatal rats. *J. Pharmacol. Exp. Ther.* **233**, 517–526.

Anholt, R.R., Pedersen, P.L., De Souza, E.B. and Snyder, S.H. (1986). The peripheral-type benzodiazepine receptor. Localization to the mitochondrial outer membrane. *J. Biol. Chem.* **261**, 576–583.

Antkiewicz-Michaluk, L., Guidotti, A. and Kruger, K.E. (1988). Molecular characterization and mitochondrial density of a recognition site for peripheral-type benzodiazepine ligands. *Mol. Pharmacol.* **34**, 272–278.

Barres, B.A., Chun, L.L.Y. and Corey, D.P. (1990). Ion channels in vertebrate glia. *Ann. Rev. Neurosci.* **13**, 441–474.

Basile, A.S. and Skolnick, P. (1986). Subcellular localization of 'peripheral-type' binding sites for benzodiazepines in rat brain. *J. Neurochem.* **46**, 305–308.

Benavides, J., Malgouris, C., Imbault, F., Begassat, F., Uzan, A., Renault, C., Dubroeucq, M.C., Gueremy, C. and LeFur, G. (1983a). 'Peripheral-type' benzodiazepine binding sites in rat adrenals: binding studies with [^{3}H]PK 11195 and autoradiographic localization. *Arch. Int. Pharmacodyn.* **266**, 38–49.

Benavides, J., Quarteronet, D., Imbault, F., Malgouris, C., Uzan, A., Renault, C., Dubroeucq, M.C., Gueremy, C. and LeFur, G. (1983b). Labeling of 'peripheral-type' benzodiazepine binding sites in the rat brain by using [^{3}H]PK 11195, an isoquinoline carboxamide derivative: kinetic studies and autoradiographic localization. *J. Neurochem.* **41**, 1744–1750.

Benavides, J., Quarteronet, D., Plouin, P.F., Imbault, F., Phan, T., Uzan, A., Renault, C., Dubroeucq, M.C., Gueremy, C. and Le Fur, G. (1984a). Characterization of peripheral type benzodiazepine binding sites in human and rat platelets by using [^{3}H]PK 11195. Studies in hypertensive patients. *Biochem. Pharmacol.* **33**, 2467–2472.

Benavides, J., Savaki, H.E., Malgouris, C., Laplace, C., Daniel, M., Begassat, F., Desban, M., Uzan, A., Dubroeucq, M.C. and Renault, C. (1984b). Autoradiographic localization of peripheral benzodiazepine binding sites in the cat brain with [^{3}H]PK 11195. *Brain Res. Bull.* **13**, 69–77.

Benavides, J., Dubois, A., Gotti, B., Bourdiol, F. and Scatton, B. (1990). Cellular distribution of omega 3 (peripheral type benzodiazepine) binding sites in the normal and ischaemic rat brain: an autoradiographic study with the photoaffinity ligand [^{3}H]PK 14105. *Neurosci. Lett.* **114**, 32–38.

Bender, A.S. (1988). *Characterization of Astrocytic and Neuronal Benzodiazepine Receptors.* Ph.D. Thesis, University of Saskatchewan.

Bender, A.S. and Hertz, L. (1984). Flunitrazepam binding to intact and homogenized astrocytes and neurons in primary cultures. *J. Neurochem.* **43**, 1319–1327.

Bender, A.S. and Hertz, L. (1985). Binding of [^{3}H]Ro5-4864 in primary cultures of astrocytes. *Brain Res.* **341**, 41–49.

Bender, A.S. and Hertz, L. (1986a). Benzodiazepine receptors in astrocytes. In *Dynamic Properties of Glial Cells, Cellular and Molecular Aspects* (T. Grisar, G. Franck, L. Hertz, W.T. Norton, M. Sensenbrenner and D.M. Woodbury, eds), pp. 405–414. Pergamon Press, Oxford.

Bender, A.S. and Hertz, L. (1986b). Octadecaneuropeptide (ODN; 'anxiety peptide') displaces diazepam more potently from astrocytic than from neuronal binding sites. *Eur. J. Pharmacol.* **132**, 335–336.

Bender, A.S. and Hertz, L. (1987a). Pharmacological characterization of diazepam receptors in neurons and astrocytes in primary cultures. *J. Neurosci. Res.* **18**, 366–372.

Bender, A.S. and Hertz, L. (1987b). Inhibition of ^{3}H diazepam binding in primary cultures of astrocytes by atrial natriuretic peptide and by a cyclic GMP analog. *Brain Res.* **436**, 189–192.

Bender, A.S. and Hertz, L. (1988a). Astrocytic benzodiazepine receptors: correlation with a calcium channel and involvement in anxiety and convulsions. In *Biochemical Pathology of Astrocytes* (M.D. Norenberg, L. Hertz and A. Schousboe, eds), pp. 599–610. Alan R. Liss, New York.

Bender, A.S. and Hertz, L. (1988b). Evidence for involvement of the astrocytic benzodiazepine receptor in the mechanism of action of convulsant and anti-convulsant drugs. *Life Sci.* **43**, 477–484.

Bennett, J.P. and Yamamura, H.I. (1985). Neurotransmitter, hormone or drug receptor binding methods. In *Neurotransmitter Receptor Binding* (H.I. Yamamura, S.J. Enna and M.J. Kuhar, eds), pp. 61–89. Raven Press, New York.

Besman, M.J., Yanagibashi, K., Lee, T.D., Kawamura, M., Hall, P.F. and Shively,

J.E. (1989). Identification of des- (Gly-Ile) -endozepine as an effector of corticotropin-dependent adrenal steroidogenesis: stimulation of cholesterol delivery is mediated by the peripheral benzodiazepine receptor. *Proc. Natl. Acad. Sci. USA* **86**, 4897–4901.

Bolger, G.T., Gengo, P., Klockowski, R., Luchowski, E., Siegel, H., Janis, R.A., Triggle, A.M. and Triggle, D.J. (1983). Characterization of binding of the Ca^{2+} channel antagonist, [^{3}H]nitrendipine, to guinea pig ileal smooth muscle. *J. Pharmacol. Exp. Ther.* **225**, 291–309.

Bolger, G.T., Newman, A.H., Rice, K.C., Lueddens, H.W., Basile, A.S. and Skolnick, P. (1989). Characterization of the effects of AHN 086, an irreversible ligand of 'peripheral' benzodiazepine receptors, on contraction in guinea-pig atrial and ileal longitudinal smooth muscle. *Can. J. Physiol. Pharmacol.* **67**, 126–134.

Bosmann, H.B., Case, K.R. and DiStefano, P. (1977). Diazepam receptor characterization: specific binding of a benzodiazepine to macromolecules in various areas of rat brain. *FEBS Lett.* **82**, 368–372.

Braestrup, C. and Nielsen, M. (1983). Benzodiazepine receptors. In *Biochemical Studies of CNS Receptors* (L. L. Iversen, S.D. Iversen and S.H. Snyder, eds), *Handbook of Psychopharmacology*, Vol. 17, pp. 285–384. Plenum Press, New York and London.

Braestrup, C. and Squires, R.F. (1977). Specific benzodiazepine receptors in rat brain characterized by high-affinity [^{3}H]diazepam binding. *Proc. Natl. Acad. Sci. USA* **74**, 3805–3809.

Braestrup, C., Nissen, C., Squires, R.F. and Schousboe, A. (1978). Lack of brain specific benzodiazepine receptors on mouse primary astroglial cultures. *Neurosci. Lett.* **9**, 45–49.

Cantor, E.H., Kenessey, A., Semenuk, G. and Spector, S. (1984). Interaction of calcium channel blockers with non-neuronal benzodiazepine binding sites. *Proc. Natl. Acad. Sci. USA* **81**, 1549–1552.

Code, W.D., White, H.S. and Hertz, L. (1991). The effect of midazolam on calcium signaling in astrocytes. *Ann. N.Y. Acad. Sci.* **625**, 430–432.

Daval, L., Post, R.M. and Marangos, P.J. (1989). Pyruvate dehydrogenase interactions with peripheral-type benzodiazepine receptors. *J. Neurochem.* **52**, 110–116.

Davies, L.P. and Huston, V. (1981). Peripheral benzodiazepine binding sites in heart and their interaction with dipyridamole. *Eur. J. Pharmacol.* **73**, 209–211.

De Souza, E.B., Anholt, R.R., Murphy, K.M., Snyder, S.H. and Kuhar, M.J. (1985). Peripheral-type benzodiazepine receptors in endocrine organs: autoradiographic localization in rat pituitary, adrenal, and testis. *Endocrinology* **116**, 567–573.

Del Zompo, M., Post, R.M. and Tallman, J.F. (1983). Properties of two benzodiazepine binding sites in spinal cord. *Neuropharmacology* **22**, 115–118.

Doble, A., Benavides, J., Ferris, O., Bertrand, P., Menager, J., Vaucher, N., Burgevin, M.C., Uzan, A., Gueremy, C. and Le Fur, G. (1985). Dihydropyridine and peripheral type benzodiazepine binding sites: subcellular distribution and molecular size determination. *Eur. J. Pharmacol.* **119**, 153–167.

Doble, A., Malgouris, C., Daniel, M., Daniel, N., Imbault, F., Basbaum, A., Uzan, A., Gueremy, C. and Le Fur, G. (1987). Labelling of peripheral-type benzodiazepine binding sites in human brain with [^{3}H]PK 11195: anatomical and subcellular distribution. *Brain Res. Bull.* **18**, 49–61.

Ehlert, F.J., Roeske, W.R., Itoga, E. and Yamamura, H.I. (1982). The binding of [^{3}H]nitrendipine to receptors for calcium channel antagonists on the heart, cerebral cortex and ileum of rats. *Life Sci.* **30**, 2191–2202.

Fares, F., Bar-Ami, S., Brandes, J.M. and Gavish, M. (1987). Gonadotropin- and estrogen-induced increase of peripheral-type benzodiazepine binding sites in the hypophyseal-genital axis of rats. *Eur. J. Pharmacol.* **133**, 97–102.

French, J.F. and Matlib, M.A. (1987). Peripheral-type benzodiazepine receptor exists in vascular smooth muscle. *J. Biophys.* **51**, 36a.

Gallagher, D.W., Mallorga, P., Oertel, W., Henneberry, R. and Tallman, J. (1981). [^{3}H] Diazepam binding in mammalian central nervous system: a pharmacological characterization. *J. Neurosci.* **1**, 218–225.

Goldstein, A. (1984). *Biostatistics.* Macmillan, New York.

Gould, R.J., Murphy, K.M.M. and Snyder, S.H. (1982). [^{3}H] Nitrendipine-labeled calcium channels discriminate inorganic calcium agonists and antagonists. *Proc. Natl. Acad. Sci. USA* **79**, 3656–3660.

Guidotti, A., Berkovich, A., Muhkin, A. and Costa, E. (1990). Diazepam-binding inhibitor: response to Knudsen & Nielsen. *Biochem. J.* **265**, 928–929.

Henn, F.A. and Henke, D.J. (1978). Cellular localization of [^{3}H] diazepam receptors. *Neuropharmacology* **17**, 985–988.

Hertz, L. (1986). Potassium transport in astrocytes and neurons in primary cultures. *Ann. N.Y. Acad. Sci.* **481** (H.F. Cserr, ed.), 318–333.

Hertz, L. (1990). Dibutyryl cyclic AMP treatment of astrocytes in primary cultures as a substitute for normal morphogenic and 'functiogenic' transmitter signals. In *Molecular Aspects of Development and Aging in the Nervous System* (A. Privat, E. Giacobini, P. Timiras and A. Vernadakis, eds), pp. 227–243. Plenum Press, New York.

Hertz, L. (1992). Autonomic control of neuronal-astrocytic interactions regulating metabolic activities and ion fluxes in the CNS. *Brain Res. Bull.*, in press.

Hertz, L. and Bender, A.S. (1988). Astrocytic benzodiazepine receptors: Their possible role in regulation of brain excitability. In *Glial Cell Receptors* (H.K. Kimelberg, ed.), pp. 159–181. Raven Press, New York.

Hertz, L. and Code, W.E. (1992). Regulation of calcium homeostasis in astrocytes. In *Calcium Antagonists: Pharmacology and Clinical Research* (R. Paoletti, T. Godfraind and P.M. Vankoullen, eds), in press.

Hertz, L. and Peng, L. (1992). Energy metabolism at the cellular level of the CNS. *Can. J. Physiol. Pharmacol.*, in press.

Hertz, L. and Mukerji, S. (1980). Diazepam receptors on mouse astrocytes in primary cultures: displacement by pharmacologically active concentrations of benzodiazepines or barbiturates. *Can. J. Physiol. Pharmacol.* **58**, 217–220.

Hertz, L., Mukerji, S. and Schousboe, A. (1980). Diazepam binding to astrocytes or neurons in primary cultures. *Trans. Am. Soc. Neurochem.* **11**, 248.

Hertz, L., Drejer, J. and Schousboe, A. (1988). Energy metabolism in glutamatergic neurons, GABAergic neurons and astrocytes in primary cultures. *Neurochem. Res.* **13**, 605–610.

Hertz, L., Peng, L., Westergaard, N., Yudkoff, M. and Schousboe, A. (1992). Neuronal-astrocytic interactions in metabolism of transmitter amino acids of the glutamate family. In *Alfred Benszon Symposium* (A. Schousboe and N. Diemer, eds). pp. 30–50, Munksgaard, Copenhagen.

Hirsch, J.D., Beyer, C.F., Malkowitz, L., Beer, B. and Blume, A.J. (1989a). Mitochondrial benzodiazepine receptors mediate inhibition of mitochondrial respiratory control. *Mol. Pharmacol.* **35**, 157–163.

Hirsch, J.D., Beyer, C.F., Malkowitz, L., Loullis, C.C. and Blume, A.J. (1989b). Characterization of ligand binding to mitochondrial benzodiazepine receptors. *Mol. Pharmacol.* **35**, 164–172.

Holck, M. and Osterrieder, W. (1985). The peripheral, high affinity benzodiazepine binding site is not coupled to the cardiac Ca^{2+} channel. *Eur. J. Pharmacol.* **118**,

293–301.

Hullihan, J.P., Spector, S., Taniguchi, T. and Wang, J.K. (1983). The binding of [^{3}H]diazepam to guinea-pig ileal longitudinal muscle and the in vitro inhibition of contraction by benzodiazepines. *Br. J. Pharmacol.* **78**, 321–327.

Ishii, K., Kano, T., Akutagawa, M., Makino, M., Tanaka, T. and Ando, J. (1982). Effects of flurazepam and diazepam in isolated guinea-pig taenia coli and longitudinal muscle. *Eur. J. Pharmacol.* **83**, 329–333.

Janis, R.A., Silver, P.J. and Triggle, D.J. (1987). Drug action and cellular calcium regulation. *Adv. Drug Res.* **16**, 309–591.

Kanto, J., Kangas, L. and Siirtola, T. (1975). Cerebrospinal-fluid concentrations of diazepam and its metabolites in man. *Acta Pharmacol. Toxicol.* **36**, 328–334.

Kendall, D.A. and Nahorski, S.R. (1985). Dihydropyridine calcium channel activators and antagonists influence depolarization-evoked inositol phospholipid hydrolysis in brain. *Eur. J. Pharmacol.* **115**, 31–36.

Knudsen, J. and Nielsen, M. (1990). Diazepam-binding inhibitor: a neuropeptide and/or an acyl-CoA ester binding protein? *Biochem. J.* **265**, 927–928.

Langer, S.Z. and Arbilla, S. (1988). Limitations of the benzodiazepine receptor nomenclature: a proposal for a pharmacological classification as omega receptor subtypes. *Fundam. Clin. Pharmacol.* **1**, 159–170.

Larcher, J.C., Vayssiere, J.L., Le Marquer, F.J., Cordeau, L.R., Keane, P.E., Bachy, A., Gros, F. and Croizat, B.P. (1989). Effects of peripheral benzodiazepines upon the O_2 consumption of neuroblastoma cells. *Eur. J. Pharmacol.* **161**, 197–202.

Le Fur, G., Guilloux, F., Rufat, P., Benavides, J., Uzan, A., Renault, C., Dubroeucq, M.C. and Gueremy, C. (1983a). Peripheral benzodiazepine binding sites: effect of PK 11195, 1-(2-chlorophenyl)-*N*-methyl-(1-methylpropyl)-3-isoquinoline-carboxamide. II. In vivo studies. *Life Sci.* **32**, 1849–1856.

Le Fur, G., Perrier, M.L., Vaucher, N., Imbault, F., Flamier, A., Benavides, J., Uzan, A., Renault, C., Dubroeucq, M.C. and Gueremy, C. (1983b). Peripheral benzodiazepine binding sites: effect of PK 11195, 1-(2-chlorophenyl)-*N*-methyl-*N*-(1-methylpropyl)-3-isoquinolinecarboxamide. I. In vitro studies. *Life Sci.* **32**, 1839–1847.

LeFur, G., Vaucher, N., Perrier, M.L., Flamier, A., Benavides, J., Renault, C., Dubroeucq, M.C., Gueremy, C. and Uzan, A. (1983c). Differentiation between two ligands for peripheral benzodiazepine binding sites, [^{3}H]Ro5-4864 and [^{3}H]PK 11195, by thermodynamic studies. *Life Sci.* **33**, 449–457.

Litzinger, M.J. and Brenneman, D.E. (1985). [^{3}H]Nitrendipine binding in developing dissociated fetal mouse spinal cord neurons. *Biochem. Biophys. Res. Commun.* **127**, 112–119.

Litzinger, M.J., Foster, G. and Brenneman, D.E. (1986). [^{3}H]Nitrendipine binding in non-neuronal cell cultures. *Biochem. Biophys. Res. Commun.* **136**, 783–788.

Luchowski, E.M., Yousif, F., Triggle, D.J., Maurer, S.C., Sarmiento, J.G. and Janis, R.A. (1984). Effects of metal cations and calmodulin antagonists on [^{3}H]nitrendipine binding in smooth and cardiac muscle. *J. Pharmacol. Exp. Ther.* **230**, 607–613.

MacVicar, B.A. (1984). Voltage-dependent calcium channels in glial cells. *Science* **226**, 1345–1347.

Majewska, M.D., Mienville, J.M. and Vicini, S. (1988). Neurosteroid pregnenolone sulfate antagonizes electrophysiological responses to GABA in neurons. *Neurosci. Lett.* **90**, 279–284.

Marangos, P.J., Patel, J., Boulenger, J.P. and Clark-Rosenberg, R. (1982a).

Characterization of peripheral-type benzodiazepine binding sites in brain using [^{3}H]Ro5-4864. *Mol. Pharmacol.* **22**, 26–32.

Marangos, P.J., Patel, J., Miller, C. and Martino, A.M. (1982b). Specific calcium antagonist binding sites in brain. *Life Sci.* **31**, 1575–1585.

McCarthy, K.D. and Harden, T.K. (1981). Identification of two benzodiazepine binding sites on cells cultured from rat cerebral cortex. *J. Pharmacol. Exp. Ther.* **216**, 183–191.

Mestre, M., Carriot, T., Belin, C., Uzan, A., Renault, C., Dubroeucq, M.C., Gueremy, C. and LeFur, G. (1984). Electrophysiological and pharmacological characterization of peripheral benzodiazepine receptors in a guinea pig heart preparation. *Life Sci.* **35**, 953–962.

Mestre, M., Carriot, T., Belin, C., Uzan, A., Renault, C., Dubroeucq, M.C. Gueremy, C., Doble, A. and Le Fur, G. (1985). Electrophysiological and pharmacological evidence that peripheral type benzodiazepine receptors are coupled to calcium channels in the heart. *Life Sci.* **36**, 391–400.

Mohler, H. and Okada, T. (1977). Properties of ^{3}H-diazepam binding to benzodiazepine receptors in rat cerebral cortex. *Life Sci.* **20**, 2101–2110.

Moreno-Sanchez, R., Bravo, C., Gutierrez, J., Newman, A.H. and Chiang, P.K. (1991a). Release of Ca^{2+} from heart and kidney mitochondria by peripheral-type benzodiazepine receptor ligands. *Int. J. Biochem.* **23**, 207–213.

Moreno-Sanchez, R., Hogue, B.A., Bravo, C., Newman, A.H., Basile, A.S. and Chiang, P.K. (1991b). Inhibition of substrate oxidation in mitochondria by the peripheral-type benzodiazepine receptor ligand AHN 086. *Biochem. Pharmacol.* **41**, 1479–1484.

Mukherjee, S. and Das, S.K. (1989). Subcellular distribution of 'peripheral type' binding sites for [^{3}H]Ro5-4864 in guinea pig lung. Localization to the mitochondrial inner membrane. *J. Biol. Chem.* **264**, 16 713–16 718.

Mukhin, A.G., Papadopoulos, V., Costa, E. and Krueger, K.E. (1989). Mitochondrial benzodiazepine receptors regulate steroid biosynthesis. *Proc. Natl. Acad. Sci. USA* **86**, 9813–9816.

Murphy, K.M.M. and Snyder, S.H. (1982). Calcium antagonist receptor binding sites labeled with [^{3}H]nitrendipine. *Eur. J. Pharmacol.* **77**, 201–202.

Olson, J.M., Ciliax, B.J., Mancini, W.R. and Young, A.B. (1988). Presence of peripheral-type benzodiazepine binding sites on human erythrocyte membranes. *Eur. J. Pharmacol.* **152**, 47–53.

Polc, P. (1988). Electrophophysiology of benzodiazepine receptor ligands, multiple mechanisms and sites of action. *Progr. Neurobiol.* **31**, 349–423.

Ritta, M.N., Campos, M.B. and Calandra, R.S. (1987). Effect of GABA and benzodiazepines on testicular androgen production. *Life Sci.* **40**, 791–798.

Saano, V. (1986). Affinity of various compounds for benzodiazepine binding sites in rat brain, heart and kidneys in vitro. *Acta Pharmacol. Toxicol.* **58**, 333–338.

Schoemaker, H., Bliss, M. and Yamamura, H.I. (1981). Specific high-affinity saturable binding of [^{3}H]Ro5-4864 to benzodiazepine binding sites in the rat cerebral cortex. *Eur. J. Pharmacol.* **71**, 173–175.

Schoemaker, H., Boles, R.G., Horst, W.D. and Yamamura, H.I. (1983). Specific high affinity binding sites for [^{3}H]Ro5-4864 in rat brain and kidney. *J. Pharmacol. Exp. Ther.* **225**, 61–69.

Sher, P.K. (1985). Characteristics of benzodiazepine receptor binding on living cultures of mouse cerebral cortex at physiological temperature. *Dev. Brain Res.* **21**, 133–136.

Sher, P.K. and Machen, V.L. (1984). Properties of [^{3}H]diazepam binding sites on cultured murine glia and neurons. *Dev. Brain Res.* **14**, 1–6.

Shibata, H., Kojima, I. and Ogata, E. (1986). Diazepam inhibits potassium-induced aldosterone secretion in adrenal glomerulosa cell. *Biochem. Biophys. Res. Commun.* **135**, 994–999.

Skerritt, J.H., Chen-Chow, S., Johnston, G.A.R. and Davies, L.P. (1982). Purines interact with 'central' but not 'peripheral' benzodiazepine binding sites. *Neurosci. Lett.* **34**, 63–68.

Talwar, D. and Sher, P.K. (1987). Benzodiazepine receptor development in murine glial cultures. *Dev. Neurosci.* **9**, 183–189.

Tardy, M., Costa, M.F., Rolland, B., Fages, C. and Gonnard, P. (1981). Benzodiazepine receptors on primary cultures of mouse astrocytes. *J. Neurochem.* **36**, 1587–1589.

Tardy, M., Fages, C., Dupre, G., Costa, M.F., Bardakdjian, J. and Rolland, B. (1985). Further characterization of [^{3}H]flunitrazepam binding sites on cultured mouse astroglia. *Neurochem. Res.* **10**, 809–817.

Trumpower, B.L. and Haggerty, J.G. (1980). Inhibition of electron transport in the cytochrome bc1 segment of the mitochondrial respiratory chain by a synthetic analogue of ubiquinone. *J. Bioenergy Biomembr.* **12**, 151–164.

Verma, A. and Snyder, S.H. (1988). Characterization of porphyrin interactions with peripheral type benzodiazepine receptors. *Mol. Pharmacol.* **34**, 800–805.

Verma, A. and Snyder, S.H. (1989). Peripheral type benzodiazepine receptors. *Ann. Rev. Pharmacol. Toxicol.* **29**, 307–322.

Villinger, J.W. (1984). Specific [^{3}H]Ro5-4864 binding to rat spinal cord membranes: Evidence for peripheral type benzodiazepine recognition sites. *Neurosci. Lett.* **46**, 267–270.

Villinger, J.W. (1985). Characterization of peripheral-type benzodiazepine recognition sites in the rat spinal cord. *Neuropharmacology* **24**, 95–98.

Watanabe, Y., Shibuya, T., Khatami, S. and Salafsky, B. (1986). Comparison of typical and atypical benzodiazepines on the central and peripheral benzodiazepine receptors. *Jpn. J. Pharmacol.* **42**, 189–197.

Weiss, S.R. and Post, R.M. (1991). Contingent tolerance to carbamazepine: a peripheral-type benzodiazepine mechanism. *Eur. J. Pharmacol.* **193**, 159–163.

Weissman, B.A., Bolger, G.T., Isaac, L., Paul, S.M. and Skolnick, P. (1984). Characterization of the binding of [^{3}H]Ro5-4864, a convulsant benzodiazepine, to guinea pig brain. *J. Neurochem.* **42**, 969–975.

Wilkinson, M., Moger, W.H. and Grovestine, D. (1980). Chronic treatment with Valium (diazepam) fails to affect the reproductive system of the male rat. *Life Sci.* **27**, 2285–2291.

Yanagibashi, K., Ohno, Y., Nakamichi, N., Matsui, T., Hayashida, K., Takamura, M., Yamada, K., Tou, S. and Kawamura, M. (1989). Peripheral-type benzodiazepine receptors are involved in the regulation of cholesterol side chain cleavage in adrenocortical mitochondria. *J. Biochem.* **106**, 1026–1029.

Zernig, G., Graziadei, I., Moshammer, T., Zech, C., Reider, N. and Glossmann, H. (1990). Mitochondrial Ca^{2+} antagonist binding sites are associated with an inner mitochondrial membrane anion channel. *Mol. Pharmacol.* **38**, 362–369.

CHAPTER 3

ENDOGENOUS AND SYNTHETIC LIGANDS OF MITOCHONDRIAL BENZODIAZEPINE RECEPTORS: STRUCTURE–AFFINITY RELATIONSHIPS

Jean-Jacques Bourguignon

Laboratoire de Pharmacochimie Moléculaire, Centre de Neurochimie CNRS, 5 rue Blaise Pascal, 67084 Strasbourg Cedex, France

Table of Contents

3.1 Introduction

The so-called benzodiazepine (BZ) receptors present at least two physically and pharmacologically distinct classes (Braestrup and Squires, 1977; Mohler and Okada, 1977; Doble and Martin, 1992; Langer and Arbilla, 1988a,b).

Mitochondrial benzodiazepine receptors (MBR) are present in both peripheral tissues and the central nervous system (Braestrup and Squires, 1977), possibly on neuronal (Anholt *et al.*, 1984; Benavides *et al.*, 1984) and glial cells (Schoemaker *et al.*, 1982; Basile and Skolnick, 1986). However, unlike their central counterpart, they are located in the mitochondrial outer membrane (Anholt, 1986; Anholt *et al.*, 1986; Hirsch *et al.*, 1988), and differ in several respects from the well-known neuronal BZ receptor, which is a part of the $GABA_A/Cl^-$ channel complex receptor. In particular, MBR are specifically labelled by certain tritiated ligands.

The biochemical and physiological properties of MBR have been reviewed (Saano *et al.*, 1989; Verma and Snyder, 1989), but there has not been an exhaustive discussion of compounds that bind significantly to these receptors ($IC_{50} \leqslant 1 \, \mu M$).

This chapter focuses on the affinities of different chemical classes of ligands of MBR with regard to their structural specificities. Their selectivity for the central BZ receptors (CBR) will be introduced here as another characteristic of such compounds.

3.2 Potential endogenous ligands of MBR

Insight into the physiological role of the MBR would be greatly facilitated by the identification of the endogenous ligands for these sites (De Robertis *et al.*, 1988). Two hypotheses are cited here. The earlier hypothesis resulted from the isolation and characterization of the diazepam binding inhibitor (DBI) (Costa *et al.*, 1987; Berkovich *et al.*, 1990) or endozepine, an endogenous 11 kDa polypeptide of 86 amino acids (Besman *et al.*, 1989) that competitively displaces benzodiazepine binding from rat brain membrane preparations with rather low affinity ($K_i = 5 \, \mu M$). It is not selective, since it has the same affinity (μM range) for both CBR and MBR. However, DBI is present in high concentration ($10-25 \, \mu M$) in rat brain. It has also been demonstrated that in peripheral tissues where high levels of DBI have been found, high densities of MBR-containing membranes are also present.

Trikontatetraneuropeptide (TTN), a shorter fragment of DBI (fragment 17–50) with the same affinity for MBR as DBI, is more selective for MBR. ODN, an octadecaneuropeptide tryptic fragment of DBI (fragment 33–50), has been suspected to contain the binding site domain of DBI for binding to CBR (same IC_{50} of 5 μM as that of DBI or CBR) and was found to be selective for CBR. However, the same ODN was highly potent on cultured astrocytes, which have MBR receptors.

The C-terminal region of TTN (lysine 50), which is also present in short DBI fragments (fragments 22–50, 33–50, 42–50), is essential for binding of

Table 1 Affinity of DBI and DBI fragments for CBR and MBR.

Name	Fragment	IC$_{50}$ or K_i (μM)	
		CBR	MBR
DBI	1–86	5[a]	5[c]
TTN	17–50	>100[b]	6[d]
	22–50	>50[b]	7[d]
ODN	33–50	5[a]	>100[d]
ODN-NH$_2$	33–50	>100[a,b]	>100[c,d]
	42–50	30[a]	15[d]

[a] [^{3}H] BCCM, K_i.
[b] [^{3}H] Ro15-1788, IC$_{50}$.
[c] [^{3}H] PK 11195, K_i.
[d] [^{3}H] Ro5-4864, IC$_{50}$.

DBI to MBR. The carboxyl group of the C-terminal lysine 50 residue has a critical effect on the affinity of these peptides for BZ receptors, as the ODN-amide is devoid of affinity for both CBR and MBR (see Table 1).

Despite this apparent selectivity, it is likely that the role of this endogenous peptide, which is present in almost all tissues, probably involves biological targets other than BZ receptors with higher affinity for DBI. Knudsen *et al.* (1989) have revealed sequence identity of DBI and acyl-CoA binding protein and therefore suggested a role in termination of fatty acid synthesis.

More recently, the physical and biochemical characteristics of MBR have been hypothesized to correspond to those of porin, a mitochondrial pore protein (Anholt, 1986). The fully planar porphyrins are thought to be MBR endogenous ligands (Verma and Snyder, 1988). These compounds have been found to be potent (nanomolar range affinities) and selective ligands for MBR in binding experiments (Hirsch *et al.*, 1988). Structure–affinity data revealed protoporphyrin IX and mesoporphyrin IX as the most potent derivatives within this series. Their potencies (IC$_{50}$ of about 10–20 nM) are associated with the presence of lipophilic groups surrounding the four cyclic tetramer pyrrole rings within their structures.

3.3 Benzodiazepines and structurally related compounds

Historically, benzodiazepines (compounds **1a–i**) have been used to label CBR and MBR (Braestrup and Squires, 1977). A large series of such compounds has therefore been tested for their capacity to inhibit binding to the CBR

Table 2 Selectivity of benzodiazepine reference compounds.

Name	R	7	2'	4'	IC$_{50}$^{a} (nM)	
					CBR	MBR
Flunitrazepam	Me	NO$_2$	F	H	4	350
Diazepam	Me	Cl	H	H	6	80
Ro5-4864	Me	Cl	H	Cl	5000	6
Clonazepam	H	NO$_2$	Cl	H	3	>10 000

^{a} Data from Wang *et al.* (1984).

([^{3}H]flunitrazepam binding in brain: Czernik *et al.*, 1982) or the MBR ([^{3}H]Ro5-4864 binding in kidney: Schoemaker *et al.*, 1983). Some of these, used as reference compounds, are presented in Table 2 (Wang *et al.*, 1984; Hirsch *et al.*, 1988).

Diazepam and flunitrazepam are potent and fairly selective MBR ligands. However, it was found that suppressing the *N*-methyl group in their structure (i.e. clonazepam) led to a dramatic and selective loss of affinity for MBR, whereas shifting the chlorine from the 2' position to the 4' position produced the first selective probe for these receptors (Ro5-4864).

More detailed and specific structural criteria can be obtained from the data in Table 3. Binding experiments were performed on intact PC 12 cells, which have Ro5-4864 binding sites similar to MBR in other tissues (Morgan *et al.*, 1985).

All the *N*-unsubstituted benzodiazepines, including clonazepam and **1a** (see Table 3), are poor MBR ligands with micromolar affinities, where the replacement of the methyl group by larger alkyl substituents (**1c, d**) is tolerated (Wang *et al.*, 1984; Newman *et al.*, 1987). The chlorine atoms in the 7 and 4' positions in the structure of Ro5-4864 are both necessary, as the unchlorinated derivative **1b** is about two orders of magnitude less active than the reference compound, the chlorine atoms being of equal importance (consider IC$_{50}$ values of diazepam and **1g**). The substitution of the chlorine in the 7 position

Table 3 Affinity of benzodiazepines for MBR.

No.	Name Ro[a]	R	R_1	R_2	$IC_{50}{}^{b}$ (nM)
	5-4864	Me	Cl	Cl	3
1a	5-2752	H	Cl	Cl	10 000
1b	5-3464	Me	H	H	700
1c	5-6993	Et	Cl	Cl	2
1d	5-6945	$CH_2CH=CH_2$	Cl	Cl	3
	Diazepam	Me	Cl	H	72
1e	7-3351	Me	Cl	OH	46
1f	5-6669	Me	Cl	OMe	8
1g	5-5115	Me	H	Cl	54
1h	5-6531	Me	F	Cl	13
1i	5-5120	Me	NO_2	Cl	34

[a] BZDs are indicated by their Roche identification code.
[b] [^{3}H] Ro5-4864 binding on intact PC_{12} cells.

by a fluorine (**1h**) or nitro group (**1i**), or substitution of the chlorine in the 4′ position by an OH group (**1e**) leads to less active compounds, but replacing the chlorine in the same position by a methoxy group is tolerated (**1f**).

Other compounds with a benzodiazepine ring in their structure that have been tested as MBR ligands are shown in Figure 1. Their N-amide function is incorporated into either a triazole (triazolam: Lueddens and Skolnick, 1987) or an imidazole ring (midazolam). In brotizolam and KW 1937, the benzodiazepine ring has been replaced by the isosteric thienyl group (Weissman et al., 1985). Affinities of the 4,5 (**1j**) and 2,2 (medazepam) dihydrobenzo-diazepines are also given in Figure 1 (Morgan et al., 1985; Bond et al., 1985). Since these data are reported from different papers, the affinity of each compound has been evaluated by the ratio of its IC_{50} values, with Ro5-4864 being used as the reference compound.

The imine nitrogen of Ro5-4864 plays an important role in binding to MBR, as the compound with the corresponding sp^3 nitrogen (4,5-dihydro derivative)

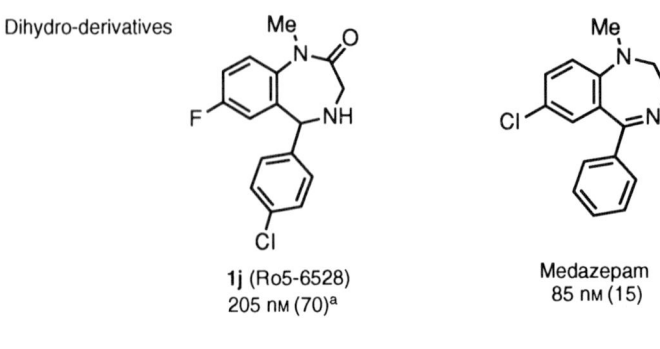

Dihydro-derivatives

1j (Ro5-6528)
205 nM (70)[a]

Medazepam
85 nM (15)

Triazolo [4,3a] benzodiazepines

Triazolam
41 nM (25)

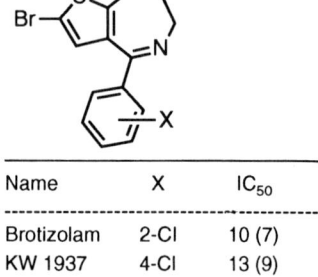

Name	X	IC$_{50}$
Brotizolam	2-Cl	10 (7)
KW 1937	4-Cl	13 (9)

Reference compounds

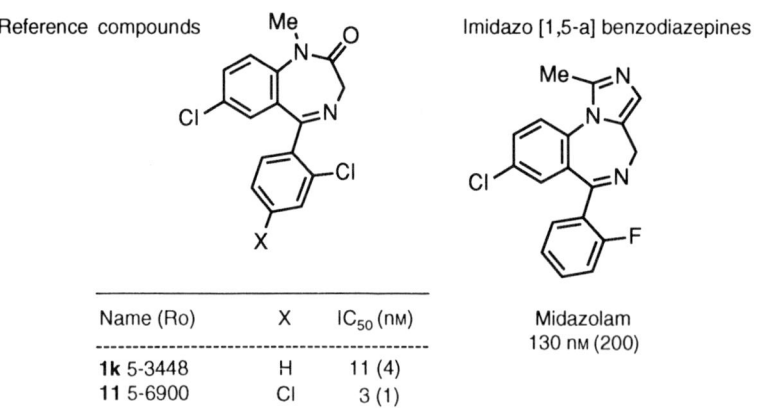

Name (Ro)	X	IC$_{50}$ (nM)
1k 5-3448	H	11 (4)
11 5-6900	Cl	3 (1)

Imidazo [1,5-a] benzodiazepines

Midazolam
130 nM (200)

Figure 1 Affinity of other benzodiazepine derivatives for MBR. [a] The ratio between the IC$_{50}$ value of a given MBR ligand and that of Ro5-4864 determined in the same experiment is given in parentheses.

is significantly less active (compare **1j** and **1h** in Table 3). This sp^3 nitrogen may be involved in a typical hydrogen bond with the receptor. Incorporating the amide of benzodiazepines in a triazole ring leads to a slight decrease of affinity (compare **1k** and triazolam). As expected from data on benzodiazepines, the positional thieno-isomers brotizolam and KW 1937 show about the same affinity for MBR (5–10 times less than Ro5-4864).

Replacement of the triazole ring by an imidazole ring (as in midazolam) is detrimental to activity, as this latter compound is about 200 times less active than the reference compound.

The quinazolinone-2 derivative proquazone (Figure 2) fulfills the main criteria characteristic of benzodiazepine derivatives: (1) a generally flat

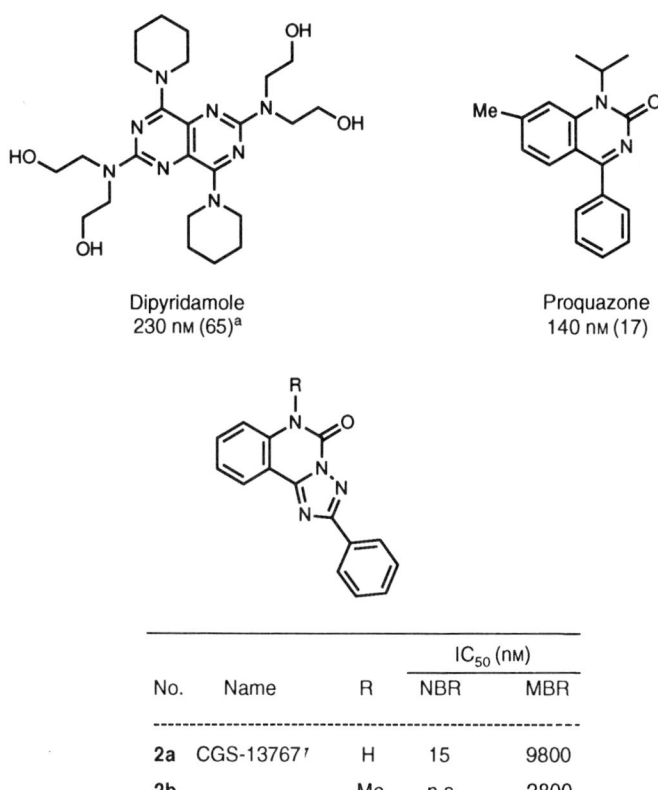

Dipyridamole
230 nM (65)[a]

Proquazone
140 nM (17)

No.	Name	R	IC$_{50}$ (nM)	
			NBR	MBR
2a	CGS-13767[ʹ]	H	15	9800
2b		Me	n.s.	2800

Figure 2 Structurally related benzodiazepine-like compounds [a] The ratio between the IC$_{50}$ value of a given MBR ligand and that of Ro5-4864 determined in the same experiment is given in parentheses.

molecule (mainly a bicyclic heterocycle); and (2) an N-substituted, lipophilic amide close to an sp^2 nitrogen of a phenyl imine moiety.

Even though this compound does not possess chlorine atoms, which confer to benzodiazepines a significant increase in affinity (see Table 3), proquazone is only 17 times less active than Ro5-4864 (Saano, 1986). Dipyridamole shows in its symmetrical structure some similarities with benzodiazepines and proquazone, and has an sp^2 nitrogen with a lipophilic group (non-ionized piperidine), which may correspond to the phenyl ring in proquazone and Ro5-4864. As observed for most of the chemical classes of BZ receptor ligands, dipyridamole is not selective, as its affinity for CBR is only ten times lower than that for MBR (2.3 μM versus 0.2 μM) (Hirsch et al., 1988). In other preparations, dipyridamole has been found to be more potent (10 times less than Ro5-4864) (Hirsch et al., 1988; Davies and Huston, 1981).

Recently, a series of [1,2,4]triazolo[1,5-c]quinazoline-5(6H)-ones (TZQs) has been reported as CBR ligands (compounds **2a, b**; Francis et al., 1991a). The compound CGS 13767 (**2a**, Figure 2) has a nanomolar IC_{50} value for these receptors.

The potential maps (Brandau et al., 1991) and geometrical parameters (Figure 6) of TZQs suggest that these compounds are structurally equivalent to classical benzodiazepine ligands. Hence some of them would be expected to present significant affinity for the MBR. It has been found that the reference compound CGS 13767 does show a reasonable affinity for MBR (10 μM).

In addition, as found earlier for the benzodiazepine series, introducing a methyl group on the amide nitrogen in the structure of the CGS reference compound leads to another TZQ compound with a μM range IC_{50} value for MBR (compare the TZQ **2b** and its corresponding benzodiazepine **1b** in Table 3). Interestingly, the same compound is not recognized by CBR.

3.4 2-Phenyl quinoline carboxamides (PK 11195 series) and related compounds

The first series of MBR ligands chemically unrelated to benzodiazepines was developed by Pharmuka Laboratories (PK series, compounds **3a–c**, see Figure 3) (Renault, 1986). The 2-phenyl quinoline **3a** was a good and fairly selective MBR ligand. Replacing the N,N-diethylamide in **3a** by a more lipophilic group (N-methyl-N-isobutyl, **3b**) led to a significant loss of affinity for CBR. The imine nitrogen in **3b** is not necessary for MBR binding, as shown by the greater affinity of the isoquinoline **3c**. Further investigations led to another series of isoquinolines (PK 11195 series), in which the benzene-fused ring is shifted towards the freely rotating aromatic ring (compare **3c** and PK 11195).

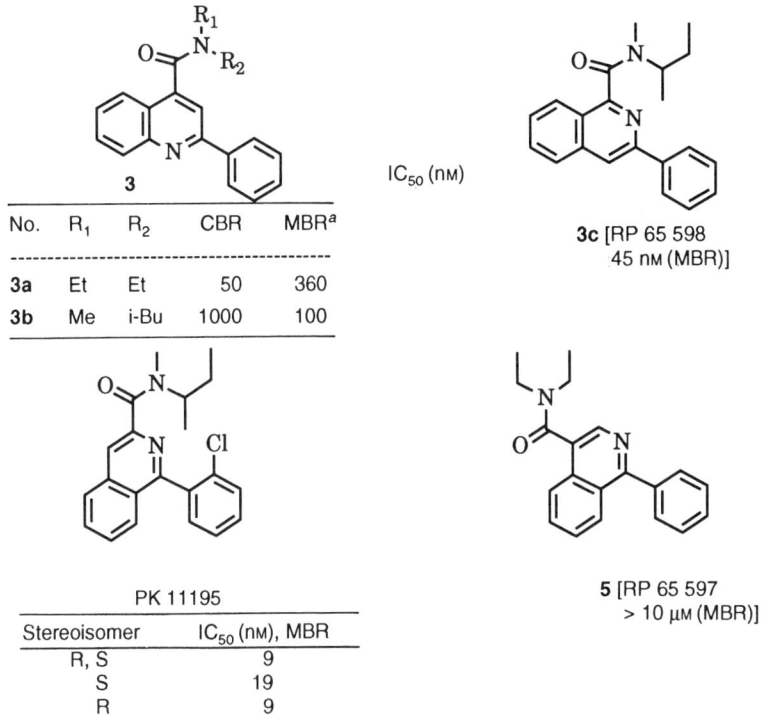

IC$_{50}$ (nM)

3

No.	R$_1$	R$_2$	CBR	MBR[a]
3a	Et	Et	50	360
3b	Me	i-Bu	1000	100

3c [RP 65 598
45 nM (MBR)]

PK 11195

Stereoisomer	IC$_{50}$ (nM), MBR
R, S	9
S	19
R	9

5 [RP 65 597
> 10 μM (MBR)]

6

Name	X	R	R$_1$	R$_2$	IC$_{50}$ (nM) CBR	IC$_{50}$ (nM) MBR
6a	O	H	Morpholino		8	
−Q1 (*l*-isomer)	CH$_2$	CH$_3$	Et	Et		5
+Q1 (*d*-isomer)	CH$_2$	CH$_3$	Et	Et		3500
+Q1 (*d, l*)	CH$_2$	CH$_3$	Et	Et		15

Figure 3 Affinity of PK 11195 and related compounds. [a] [^{3}H]Diazepam binding in kidney.

The well-known *o*-chlorophenyl derivative PK 11195 is a potent and selective MBR ligand (Le Fur *et al.*, 1983), and its tritiated derivative is used as an efficient tool for labelling MBR in binding experiments (Riond *et al.*, 1989). There is a good correlation between IC_{50} values determined with $[^3H]Ro5$-4864 and $[^3H]PK$ 11195 in mitochondria (Hirsch *et al.*, 1988).

Data obtained with the aim of establishing a structure–activity relationship for PK 11195 derivatives are listed in Table 4. In this discussion the compound *N*,*N*-diethyl amide (**4a**) has been chosen as the reference compound for the PK 11195 series (B = CH, A = N, compounds **4a–f**), and the following conclusions can be drawn (Dubroeucq *et al.*, 1983):

(1) Introducing a chlorine atom on the freely rotating aromatic ring in *ortho* (**4d**) or *meta* (**4g**) is favourable for affinity, the latter derivative being about 20 times more active than the reference compound **4a**.
(2) Replacing the *N*,*N*-diethyl group in **4a** by the lipophilic *N*,*N*-dibutyl residue significantly increased the affinity (compare **4a** and **4b**); the optimum is obtained with the *N*-methyl-*N*-isobutyl carboxamide **4c**

Table 4 Affinity of PK 11195 derivatives for MBR.

No.	A	B	X	R_1	R_2	K_i^a (nM)
4a	N	CH	H	Et	Et	150
4b	N	CH	H	Bu	Bu	21
4c	N	CH	H	Me	i-Bu	2
4d	N	CH	2-Cl	Et	Et	17
4e	N	CH	2-Cl	Pr	Pr	18
PK 11195	N	CH	2-Cl	Me	i-Bu	8
4f	N	CH	2-Cl	Me	Bn	117
4g	N	CH	3-Cl	Et	Et	9
4h	CH	N	H	Et	Et	27
4i	CH	N	2-F	Me	i-Bu	2
4j	N	N	H	Et	Et	27
4k	CH	CH	H	Et	Et	5

^{a} $[^3H]$ Diazepam binding in rat kidney.

($IC_{50} = 2$ nM). The amide substitution has similar effects in the 2-chlorophenyl series (**4d**–**e**); the compound PK 11195 showed optimal affinity, while replacement of the isobutyl by a benzyl group led to a significant decrease in affinity (compare PK 11195 and **4f**).

(3) It is noteworthy that, when the optimized N-methyl,N-isobutyl carboxamide is present in the structure of **4**, the compound exhibits a nanomolar IC_{50} value and no further increase in activity is obtained by introducing a chlorine atom on the benzene ring (compare PK 11195 and **4c**).

(4) The quinazoline **4j** (B = A = N) and quinoline **4h** (B = N, A = CH) with an IC_{50} of 27 nM are more potent derivatives than the corresponding isoquinoline **4a** (B = CH, A = N). However, the most active N,N-diethyl carboxamide derivative was found in the naphthalene series (A = B = CH, compound **4k**) with an IC_{50} of 5 nM.

(5) As the isoquinoline PK 11195 possesses an asymmetric carbon in its structure, both R and S enantiomers have been tested and were found to be nearly equipotent (Figure 3).

Two other structurally related PK 11195 compounds are shown in Figure 3. The isoquinoline **5** (shift of the carboxamide from position 3 to position 4, compare **4a** and **5**) is completely inactive. The 2-phenyl, quinoline 4-propionamides **6** (X = CH_2), represented in Figure 3 by Q1, can be considered as the carbaisostere analogues of another series of 2-phenyl, quinoline 4-oxoacetamide (**6a**, X = O) described earlier as potent CBR ligands (Benavides et al., 1987). Enantiomers +Q1 and −Q1 significantly differ in their IC_{50} values, the *l*-enantiomer being 700-fold more active than the *d*-isomer, and with an affinity equivalent to that of PK 11195 (Dubroeucq et al., 1986). These data illustrate the stereochemical characteristics of the recognition pattern of MBR ligands.

3.5   Imidazo[1,2-a]pyridine-3-acetamides (Alpidem series)

These compounds were first described by Almirante as potent anticonvulsant drugs (Almirante, 1969). More recently, this series of imidazo[1,2-a]pyridine-3-acetamides has been described as ligands of BZ receptors, but few data are available. Two compounds of the series have shown high affinity for CBR and have been developed by Synthelabo's researchers as a sedative (zolpidem) and an anxiolytic (alpidem) (Langer and Arbilla, 1988a,b; Langer et al., 1990). Interestingly, the latter compound showed similar nanomolar IC_{50}

Table 5 Imidazo[1-2a]pyridine-3 acetamides.

Name	X	R	CBR[a]	MBR[b]
			\multicolumn{2}{c}{K_i (nM)}	
Zolpidem	Me	Me	26	1040
Alpidem	Cl	n-Pr	7	1

[a] [^{3}H] Diazepam in cerebellum.
[b] [^{3}H] Ro5-4864 in kidney.

values for both CBR and MBR (Table 5). Moreover, in accordance with our observations in other MBR ligand series, the replacement of the N,N-dimethylamide in zolpidem by the more lipophilic N,N-dipropyl group appears to be closely linked with the striking increase in affinity of alpidem for MBR.

3.6 3-Acyl-4-aminoacetamide quinolines and other related SR compounds

This series of MBR ligands (compounds **6, 6a**) recently patented by the Sanofi group, merits attention, as some compounds belonging to these two classes are both potent (IC_{50} less than 1 nM), and selective (IC_{50} for CBR generally more than 10 μM) (Mendes *et al.*, 1989). Main data are listed in Table 6. All the compounds are characterized by the presence of a lipophilic side-chain attached at the exo nitrogen of the 4-amino quinoline system, and bear a tertiary amide. Thus, compounds **7a–q** (SR series) resemble the structure of the quinolines (compound **6**, X = O or CH$_2$). However, their potencies and selectivities towards MBR may be associated with the presence in their structure of a typical N-arylamide.

Table 6 Quinoline-4-aminoacetamides (SR compounds).

No.	Name SR[a]	A	B	C	X	R	R_1	R_2	R_3	R_4	IC_{50}[b]
7a	26485	CH	CH	CH	Cl	Ph	H	H	Et	Et	40
7b	26290	CH	CH	CH	Cl	Ph	H	H	Pr	Pr	10
7c	26386	CH	CH	CH	Cl	Ph	H	H	Bu	Bu	16
7d	26412	CH	CH	CH	Cl	Ph	H	H	Me	p-ClPh	0.4
7e	26397	CH	CH	CH	Cl	Ph	H	H	Pr	p-ClPh	4
7f	26450	CH	CH	CH	Cl	Ph	Me	H	Me	p-ClPh	2
7g	26841	CH	CH	CH	Cl	Ph	H	Me	Me	p-ClPh	17
7h	27196	CH	CH	CH	Cl	Ph	H	H	Me	p-MeOPh	0.8
7i	26778	CH	CH	CH	Cl	Ph	H	H	Me	p-CF$_3$Ph	27
7j	26310	CH	CH	CH	Cl	OEt	Me	H	Pr	Pr	1
7k	26423	CH	CH	CH	Cl	OEt	Me	H	Me	p-ClPh	1
7l	26830	N	CH	CH	Cl	Ph	H	H	Me	p-ClPh	0.1
7m	26276	N	CH	CH	H	Ph	H	H	Me	p-ClPh	0.3
7n	26651	N	CH	CH	H	Ph	H	H	Me	o-ClPh	0.9
7o	26803	N	CH	CH	H	Ph	H	H	Me	Ph	7
7p	26409	CH	N	CH	H	Ph	H	H	Me	p.ClPh	84
7q	26919	CH	CH	N	H	Ph	H	H	Me	p-ClPh	52

[a] Compounds are named by their Sanofi Recherche (SR) identification code.
[b] [^{3}H] PK 11195 displacement.

The introduction of a methyl group on the N-exo nitrogen ($R_1 = $ Me), or in the position α to the N(exo) ($R_2 = $ Me), led to less active compounds, particularly with the former derivative (compare respectively **7f**, **g** and **7d**, the reference SR 26412). The decrease in affinity of **7f** (40 times less active) associated with the loss of the NH group in its structure may result from the loss of an H-donor group involved in an inter- (with the receptor) or intramolcular interaction with the oxygen of the adjacent carbonyl group.

The tertiary amide is highly lipophilic. When compared with the N,N-diethyl carboxamide **7a** which possesses an IC_{50} value of 40 nM, introducing larger, more lipophilic substituents (dibutyl or dipropyl) on the carboxamide was well tolerated. Moreover, in the compound **7b**, replacing one of the two propyl chains by a lipophilic aromatic group (**7e**) led to a compound about 10 times more active than the N,N-diethyl derivative **7a**. The compound **7e** presents some steric hindrance in the vicinity of the carboxamide side-chain, which may reduce its ability to bind. This suggestion is supported by the high potency of the N-methyl- compared to the N-propyl-, N-aryl carboxamide **7d**, which has an IC_{50} value less than 1 nM.

All these compounds have in common in their structure an acyl group (ester or benzoyl group) in the *ortho* position to the exo nitrogen of the quinoline ring. These acyl groups seem to be equivalent for binding, as illustrated by the similar IC_{50} values measured for **7f** and **7k**, and **7b** and **7j**, respectively.

The benzoyl derivatives of the SR series have been developed more extensively, probably because of their better *in vivo* activities.

Other structure–affinity relationships for compounds in this series are as follows:

(1) The chlorine atom on the terminal N-phenyl carboxamide of SR 26412 can be replaced satisfactorily by a methoxy group (**7h**), but replacing it by a CF_3 group led to a compound which was about 70 times less active (**7i**). The chlorine atom can be shifted into the *ortho* position without significant change of affinity (compare in the naphtyridine series **7m** and **7n**).

(2) The chlorine in the 7 position is optimal, since replacing it by a fluorine or by a trifluoromethyl group led to markedly less active compounds (data not shown); its presence is however not crucial (compare **7l** and **7m** in the naphtyridine series).

(3) Azaisosteres (structures A, B, C and D) have been tested and only compounds **7l–o** in the naphtyridine series (A = N) show similar affinity to the reference compound SR 26412. The other isomers **7p** and **7q** were several orders of magnitude less potent (Table 6).

3.7 Imidazo[1,5-c]quinazolinone-6-acetamides (NCS 1000 series)

In order to support our hypothesis that triazoloquinazolinones (TZQs) (compounds **2** in Figure 2) are structurally equivalent to benzodiazepinones, we prepared the deaza-3 derivatives of the triazolo-quinazolinones (NCS 1000 series) (Bourguignon *et al.*, 1991) and tested them as CBR ligands. The N-unsubstituted imidazo-quinazolinone **8a** (Table 7) was about two orders of magnitude less active than its corresponding TZQ analogue **2a**, suggesting

Table 7 Imidazo[1,5-c]quinazolinones.

No.	NCS	R_1	R_2	R	MBR^b IC_{50} (nM)
8a	1005	Ph	H	H	4500^a
8b	1027	Ph	H	Me	n.s.
8c	1010	Ph	H	CH_2CO_2Et	n.s.
8d	1008	Ph	H	$CH_2CON(Et)_2$	400
8e	1026	Ph	H	$CH_2CON(Pr)_2$	11
8f	1037	Ph	H	$CH_2CON\langle\rangle$	n.s.
8g	1039	Ph	H	$CH_2CONMePh$	1.9
8h	1038	Ph	H	$CH_2CONHPh$	n.s.
8i	1012	p-ClPh	H	$CH_2CON(Et)_2$	1500
8j	1016	m-ClPh	H	$CH_2CON(Et)_2$	58
8k	1028	o-ClPh	H	$CH_2CON(Et)_2$	280
8l	1014	p-MeOPh	H	$CH_2CON(Et)_2$	140
8m	1015	p-MePh	H	$CH_2CON(Et)_2$	210
8n	1033	t-Bu	H	$CH_2CON(Et)_2$	130
8o	1030	m-ClPh	H	$CH_2CON(Pr)_2$	2.4
8p	1044	m-ClPh	H	$CH_2CONMePh$	1.9
8q	1017	Ph	Me	$CH_2CON(Et)_2$	30
8r	1032	Ph	Br	$CH_2CON(Et)_2$	1.8
8s	1044	m-ClPh	Br	$CH_2CONMePh$	0.7

n.s., not significant.
^{a} [^{3}H]Ro15-1788 (rat brain), $IC_{50} = 350$ nM; other compounds were found inactive ($IC_{50} > 10\ \mu$M).
^{b} [^{3}H]PK 11195, rat heart.

that the N-3 sp^2 nitrogen of the triazole ring in TZQ may be involved in a hydrogen bond with the receptor. However, this compound did not show any significant affinity for MBR. Introducing a methyl group on the N-amide nitrogen of **8a** led to an inactive compound (**8b**) for both CBR and MBR. However, differently functionalized side-chains have been introduced on the N-amide of imidazo quinazolinones of type **8** (Table 7). All these N-substituted compounds have been found totally inactive on MBR (data not shown), except when the compounds were bearing a tertiary lipophilic acetamide group. In particular, N,N-diethyl acetamide **8d** (NCS 1008), with an IC_{50} value of 400 nM on MBR, was chosen as a lead compound for our studies. Increasing the lipophilicity of the amide was favourable for the affinity. Thus the N,N-dipropylamide (**8e**) and the N-methyl N-phenyl amide (**8g**) were found to be 10 times and 100 times more active, respectively, than the lead compound NCS 1008. Unexpectedly the relatively bulky piperidino-amide **8f** was found to be inactive. The loss of affinity observed with the NH-phenylamide **8h** and other tested secondary amides may result from the possibility that an internal hydrogen bond is established in their structure with the carbonyl of the quinazolinone ring.

When considering other data on structure–activity relationships obtained with other MBR-ligand classes cited above, the effects of substitution on the terminal amide in our series were very similar to those found with the SR compounds, alpidem and PK 11195.

In further work, the effects of substituents at the imidazole ring have been evaluated (see compounds **8i**–**r**). The introduction of different substituents (Cl, OMe, Me) on the 2-phenyl ring generally led to more active compounds; the m-chloro derivative **8j** was found to be about 10 times more active than NCS 1008. Interestingly, the phenyl group can be advantageously replaced by aliphatic, bulky groups such as a tertiobutyl group (**8n**). These data suggest that the m-Cl phenyl ring in **8j** is probably involved in a hydrophobic rather than an electronic interaction with the receptor. More generally, increasing the lipophilicity in the vicinity of positions 2 and 3 of the imidazole ring significantly increased the affinity (**8q**–**s**). When the 3 position of the imidazole ring is substituted by a bromine (**8r**, **s**), the compound had nanomolar affinity. Moreover, combining in the same compound the structural criteria required for optimized binding to MBR led to the highly potent and selective MBR ligand (**8s**) (NCS 1044) with an IC_{50} value less than 1 nM. Systematic studies within the NCS 1000 series led us to consider the importance of the benzene fused ring for binding. Suppressing it led to an inactive compound (compare **9** in Figure 4 with **8d**), clearly showing that the benzene fused ring in our NCS compounds, and probably also in TZQ compounds and benzodiazepinones, is crucial for MBR binding. In support of our hypothesis that benzodiazepinones and triazoloquinazolinones TZQ are structurally equivalent, structure–affinity data from our NCS 1000 series (deaza-3 TZQ compounds) suggested the synthesis of the TZQ acetamide **2c** (R = CH_2 CON(Et)$_2$, NCS 1056 in Figure 4), which shows an IC_{50} value of 60 nM for MBR, and is more potent than its 3-deaza analogue (NCS 1008, IC_{50} = 400 nM). However, this compound is less selective, as it presents a significant affinity for CBR (1.7 μM), whereas the corresponding 3-deaza derviatives are selective (IC_{50}s of other NCS compounds for NBR > 100 μM).

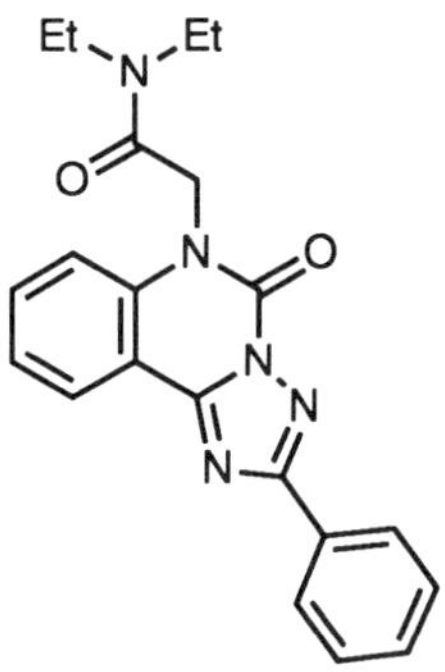

2c (NCS 1056) 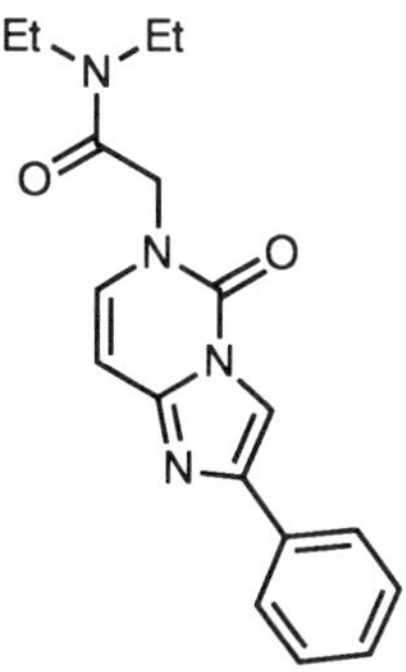   9 (NCS 1040)

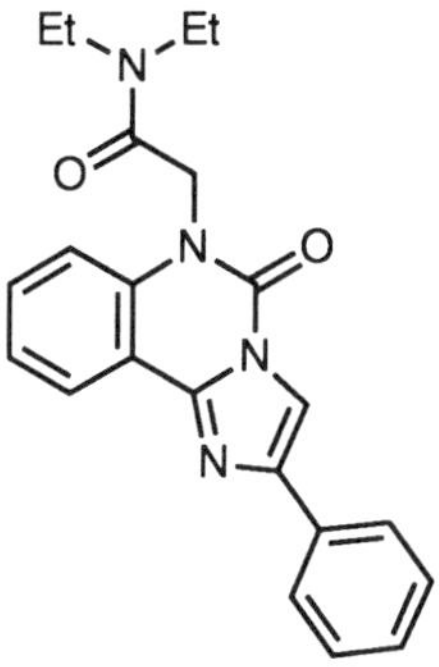

8d (NCS 1008)

| No. | (NCS) | IC$_{50}$ (nM) | |
		CBR^a	MBR^a
8d	1008	$> 10^5$	400
2c	1056	1700	60
9	1040	$> 10^5$	$> 10^5$

Figure 4   NCS 1000 compounds. ^{a} See Table 7.

3.8 Pyridazine derivatives

Another class of BZ ligands is represented by the 6-aryl triazolo[4,3-b]pyridazine CL 218 872 (Figure 5), which only has micromolar affinity for CBR (Albright *et al.*, 1981), and no affinity for MBR (Schoemaker *et al.*, 1983). However their original pharmacological properties prompted us to develop novel 7-substituted TZP (Bourguignon *et al.*, 1985) (compounds **10**), which were tested on both CBR and MBR (Table 8). IC$_{50}$ values for the most potent triazolopyridazines (TZP) reached 100 nM for CBR (i.e. **10c**),

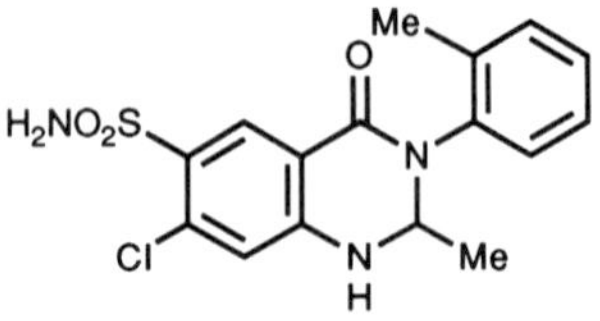

Metolazone
1 µM

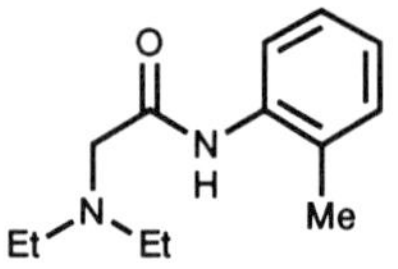

Lidocaine
0.3 µM

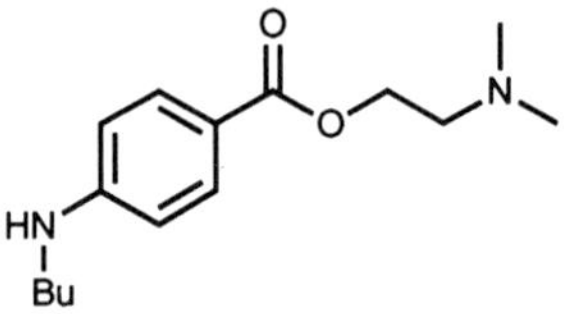

Tetracaine
300 µM

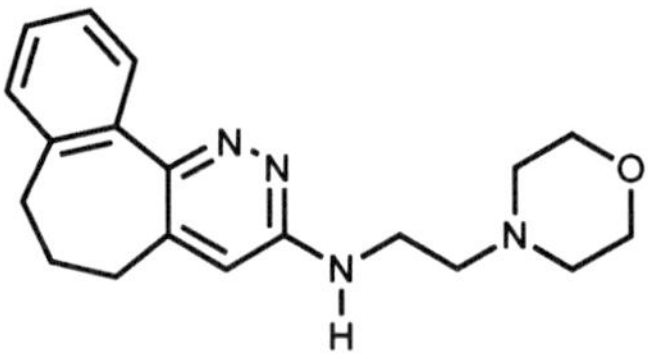

SR 95639
3 µM

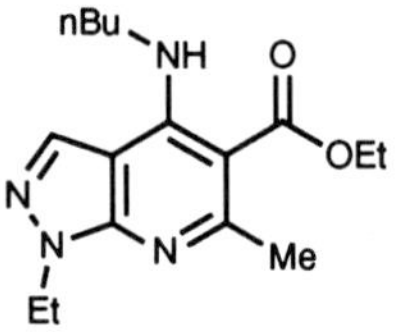

Tracazolate
0.6 µM

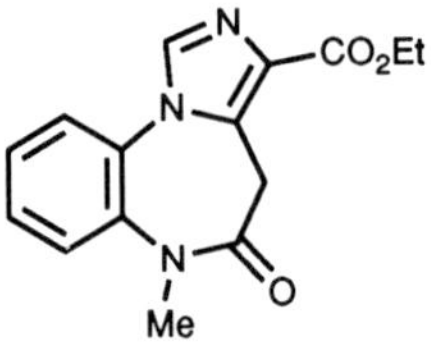

Ro15-1788
400 µM

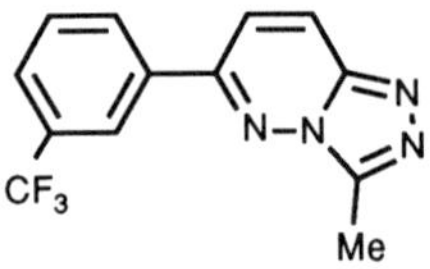

CL 218 879
> 100 µM

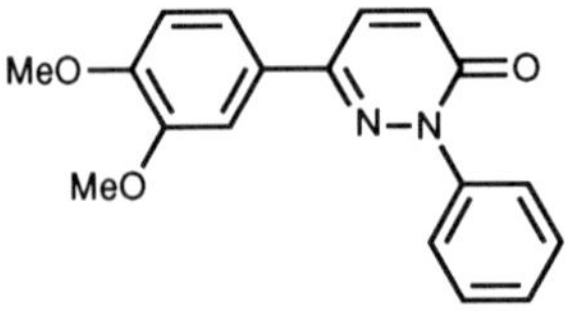

AAL 198
0.77 µM

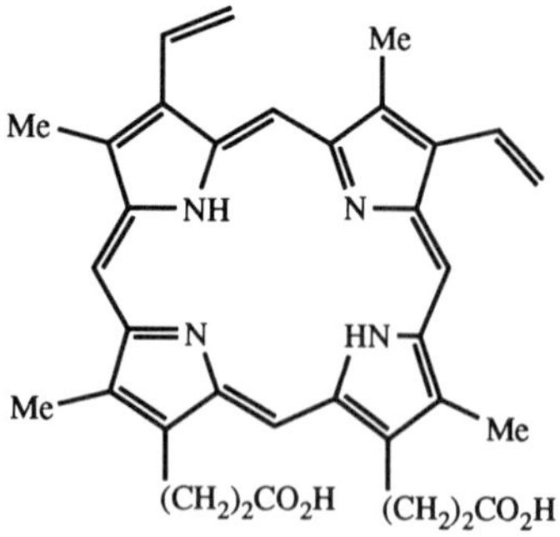

Protoporphyrin IX
20 nM

Figure 5 Other pharmacological agents tested as MBR ligands.

Table 8 Affinity of triazolo[4,3-b]pyridazines.

No.	R_1	R_2	IC$_{50}$ (μM)	
			CBR	MBR
10a	Me	Ph	n.s.	n.s.
10b	Me	Bn	27	0.9
10c	Ph	Ph	0.1	7.5
10d	o-ClPh	Ph	1.7	2.5
10e	m-ClPh	Ph	100	0.6
10f	p-ClPh	Ph	100	1.7
10g	o-MeOPh	Ph	7	13
10h	m-MeOPh	Ph	5	0.4
10i	p-MePh	Ph	1.2	1.2

and 400 nM for MBR (i.e. **10h**). Structure–affinity relationship data for MBR binding clearly show: (1) the importance of the aromatic ring attached at the triazole ring (compare **10a** and **10c**); (2) the presence of a substituent (Cl or OMe) at the aromatic ring generally increases the affinity, particularly in the *meta* position. Thus, the O-chloro derivative (**10d**) has micromolar affinity for both CBR and MBR, whereas shifting the chlorine atom to the *meta* or *para* position led to a dramatic loss of affinity for CBR with no significant change of affinity for MBR (compare **10d** and **10e, f**). The effects of substituting on the triazole phenyl ring in the TZP series are similar to those found earlier on the phenyl imine group in the chemical class of benzodiazepines.

In the same class of pyridazine derivatives, a diaryl 2,6 pyridazone-3 has been described in the patent literature as a potent ligand of CBR (Tahara *et al.*, 1986). Another diaryl 2,6 pyridazone prepared by our group as a potent cyclic AMP phosphodiesterase inhibitor (Griebel *et al.*, 1991) (AAL-198, Figure 5) had significant and selective affinity for MBR (IC$_{50}$ = 0.8 μM). The 6-phenyl acyl imidazo[1,2,b]pyridazines (compounds **11**, Table 9) constitute another series of novel pyridazine derivatives developed by us. The 2-acylimidazo-pyradazines **11a–c** show micromolar affinities for both CBR and MBR, but introducing a methyl group in the 8 position of the pyridazine

Table 9 Acyl-imidazo[1,2-b]pyridazines.

No.	R	R_1	R_2	IC$_{50}$ (μM)	
				CBR	MBR
11a	H	CO_2Et	H	0.4	11
11b	Me	CO_2Et	H	>10	3.9
11c	H	COPh	H	n.t.	1.5
11d	H	H	CO_2Et	2	5.5
11e	H	H	COPh	1.7	0.2
11f	Me	H	COPh	n.s.	0.14

n.t., not tested.
n.s., >100 μM.

ring ($R = Me$) led to a dramatic loss of affinity for CBR (compare **11a** and **11b**). Whereas the 3-imidazo ester **11d** was not selective, replacing the 3-ester moiety in its structure by a benzoyl group led to potent ligands of MBR (IC$_{50}$ of 200 nM for **11e**). As observed with other compounds within this series, the presence of a methyl group in the 8 position led to a potent and selective MBR ligand (**11f**, IC$_{50}$ = 140 nM). Further investigations starting from the structure of **11f** should bring about selective compounds with IC$_{50}$ values for MBR in the nanomolar range.

3.9 Other pharmacological agents

As the physiological role of MBR remains unknown, different pharmacological agents have been tested as MBR ligands. Beyond different classes of diuretics, only metolazone (Lukeman and Fanestil, 1987) shows some affinity. Anaesthetics have also been evaluated. Lidocaine (Clark and Post, 1990) appeared to be a fairly potent (IC$_{50}$ of 300 nM) and water-soluble MBR ligand, whereas tetracaine, another anaesthetic was found to be inactive. The 3-morpholino ethylaminopyridazine derivative SR 95639 (Figure 5), which has structural similarities with tetracaine, has been described as a muscarinic agonist (Schumacher *et al.*, 1989). Analysis of its selectivity profile revealed a micromolar IC_{50} value for MBR. It is noteworthy that both metolazone and lidocaine have within their structures the same highly lipophilic *N-ortho*-methylphenyl amide moiety. Calcium channel blockers, particularly dihydro-pyridines, have been considered in this context, but only nitrendipine showed an IC_{50} of $2\,\mu\text{M}$ (Cantor *et al.*, 1984; Rampe and Triggle, 1986). Some convulsant pyretroids like deltamethrin inhibit [^{3}H]Ro5-4864 binding, but they are not potent ($IC_{50} = 40\,\mu\text{M}$) and they probably bind to an allosteric recognition site (Devaud and Murray, 1988).

The anxiolytic tracazolate (Figure 5), a pyrazolo pyridine for which the mechanism of action remains unknown, has an IC_{50} value in the micromolar range for MBR (Schoemaker *et al.*, 1983). Its structure resembles the SR compound (butyl side-chain of tracazolate corresponding to the lipophilic acetamide in SR compounds). The lack of a typical carboxamide in the tracazolate structure may be associated with its fair affinity.

3.10   Discussion

Starting from this survey of the different chemical classes of synthetic MBR ligands that have significant affinities ($IC_{50} < 1\,\mu\text{M}$), efforts were made to characterize the main structural features of these ligands in terms of geometrical and electronic parameters.

As several MBR ligands are not selective towards CBR (benzodiazepines, pyridazine derivatives, some quinolines or imidazopyridines like alpidem), we proceeded by re-examining our earlier pharmacophoric model of CBR ligands (Tebib *et al.*, 1987). Several characteristics, illustrated with the benzodiazepine Ro5-4864, can be deduced:

(1)   the presence of two electronegative regions named $\delta2$ and $\delta2$;
(2)   the existence of two regions containing aromatic rings.

One of the aromatic ring regions constitutes the reference plane of the molecule (the benzo ring in benzodiazepines) which generally belongs to a fused bicyclic π-aromatic (PAR) system (quinoline, isoquinoline, imidazo-pyridine); this confers to the molecule a relatively planar overall geometry.

One of the distinguishing features of the PAR region is that it contains both $\delta1$ and $\delta2$ regions, whereas the second region, generally a freely rotating aromatic ring (FRA), may come out of the mean-plane of the compound.

The first series of ligands we have considered in Figure 6 includes

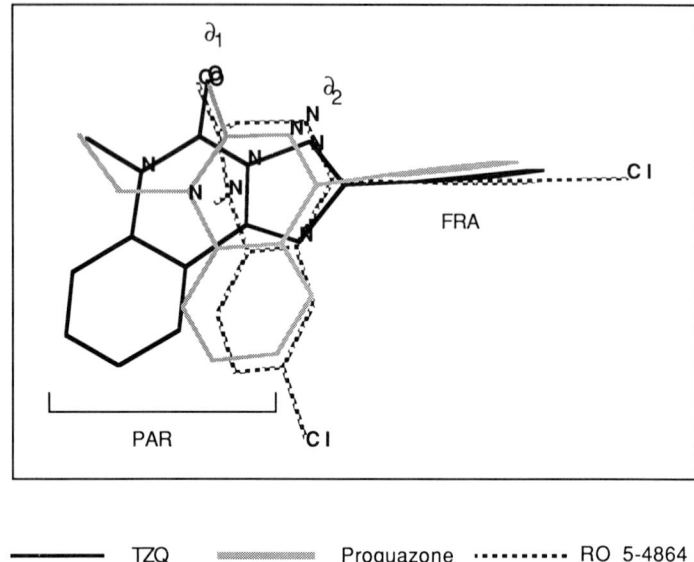

—————— TZQ ▨▨▨▨ Proquazone ·········· RO 5-4864

Figure 6 Structural similarities between the benzodiazepinone Ro5-4864, proquazone and TZQ.

benzodiazepine Ro5-4864, TZQ (CGS 13767) and proquazone. These compounds have similar features, and are probably recognized by the MBR as more or less chemically equivalent compounds.

When compared to these compounds, a second series of MBR ligands is characterized by the presence in their structure of a tertiary lipophilic amide, which significantly increases the global lipophilicity of such compounds, when referred to CBR ligands. Moreover, as substitutions on the amides PK 11195, alpidem, Q1, SR 26412, and NCS 1056 present identical effects, they have been superposed in a common region, which constitutes a third typical electronegative zone surrounded by lipophilic groups (LA). Thus, the pharmacophoric model (Plate 1) reveals the presence of a large zone containing dipoles $\delta 1$, $\delta 2$ and LA, probably involved in hydrogen bonding or dipolar interactions with the receptor (upper part of the pharmacophore). This relatively hydrophilic area is surrounded by large hydrophobic regions containing aromatic systems designated FRA and PAR. However, rings in both PAR and FRA regions seem to be involved in hydrophobic rather than electronic interactions, as supported by structure–affinity relationship data from our work and from the literature. It has been found that introducing lipophilic chlorine atoms on the aromatic rings of MBR ligands or replacing either the benzene fused ring in PAR (Williams *et al.*, 1989; Francis *et al.*, 1991b), or the phenyl ring in FRA, by cyclic or branched alkyl groups (our work on the NCS series; Bourguignon *et al.*, 1992) is not detrimental to MBR binding.

Our model suggests a large overlap in the pharmacophore of both the Ro5-4864 benzodiazepine and the quinoline carboxamide PK 11195 binding sites; this is supported by the existence of a good correlation between IC_{50} values determined with $[^3H]$Ro5-4864 and $[^3H]$PK 11195 (Newman *et al.*, 1987; Hirsch *et al.*, 1988). However, the MBR contains a histidine residue, which slightly differentiates its binding sites (Benavides *et al.*, 1984). In addition, other compounds such as arachidonate, diethylpyrocarbonate and thiol reagents have been used for the characterization of differences in binding sites for PK 11195 and Ro5-4864 (Skowronski *et al.*, 1987). (For the structure of the MBR refer to Chapter 1.)

When compared with the previous structural requirements for both CBR and MBR ligands, the characterization of a third amide, called LA, surrounded by lipophilic substitutents, constitutes a novel region of the pharmacophore which may be associated with the selectivity of several MBR ligands.

Several pharmacophoric models have been described for MBR ligands (Tebib *et al.*, 1987; Fryer, 1990). However, to our knowledge, this work constitutes the first systematic analysis of MBR ligands known to date (Georges *et al.*, 1990), and the model we are proposing here may aid the design of novel potent and selective MBR ligands with increased water-solubility. These compounds may better elucidate the role of MBR and the pharmacological properties of specific MBR ligands for therapeutic purposes than the compounds that have so far been available.

Acknowledgements

I would like to thank Mustapha Abarghaz, who contributed to the development of the NCS series in our laboratory, with the help of a fellowship from Servier. Many thanks to Thierry Langer and Rémy Hoffmann for their work in the graphic computer processing, and to the pharmacologists who were respectively responsible for binding experiments with NCS compounds (Jean-François Renaud de la Faverie and Alain Lombet) and TZP analogues (Peter Keane and André Bachy from Sanofi Recherche). I am also indebted to Françoise Herth for excellent secretarial assistance.

References

Albright, J.D., Moran, D.B., Wright, W.B. Jr., Collins, J.B., Beer, B., Lippa, A.S. and Greenblatt, E.N. (1981). Synthesis and anxiolytic activity of 6-(substituted-phenyl)-1,2,4-triazolo[4,3-b]pyridazines. *J. Med. Chem.* **24**, 592–600.

Almirante, L., Mugnaini, A., Rugarli, P., Gamba, A., Zefelippo, E., De Toma, N. and Murmann, W. (1969). Derivatives of imidazole. III. Synthesis and pharmacological activities of nitriles, amides, and carboxylic acid derivatives of imidazo[1,2-a]pyridine. *J. Med. Chem.* **12**, 122–126.

Anholt, R.R.H., Murphy, K.M.M., Mack, G.E. and Snyder, S.H. (1984). Peripheral-type receptors in the central nervous system: localization to olfactory nerves. *J. Neurosci.* **4**, 593–603.

Anholt, R.R.H. (1986). Mitochondrial benzodiazepine receptors as potential modulators of intermediary metabolism. *Trends Pharmacol. Sci.* **12**, 506–511.

Anholt, R.R.H., Pedersen, P.L., De Souza, E.B. and Snyder, S.H. (1986). The peripheral-type benzodiazepine receptor. Localization of the mitochondrial outer membrane. *J. Biol. Chem.* **261**, 576–583.

Basile, A.S. and Skolnick, P. (1986). Subcellular localization of 'peripheral-type' binding sites for benzodiazepines in rat brain. *J. Neurochem.* **46**, 305–309.

Benavides, J., Begassat, F., Phan, T., Tur, C., Uzan, A., Renault, C., Dubroeucq, M. C., Geúrémy, C. and Le Fur, G. (1984). Histidine modification with diethylpyro-carbonate induces a decrease in the binding of an antagonist, PK 11195, but not of an agonist, Ro5-4864, of the peripheral benzodiazepine receptors. *Life Sci.* **35**, 1249–1256.

Benavides, J., Dubroeucq, M.C., Le Fur, G. and Renault, C. (1987). Preparation of (quinolyoxy)acetamides as anxiolytics. *Eur. Pat. Appl.*, EP 205 375; *Chem. Abstr.* **107**, 175 897c.

Berkovich, A., McPhie, P., Campagnone, M., Giidotti, A. and Hensley, P. (1996). A natural processing product of rat diazepam binding inhibitor, triakontatetraneupeptide (diazepam binding inhibitor 17–50) contains an α-helix, which allows discrimination between benzodiazepine binding site subtypes. *Mol. Pharmacol.* **37**, 164–172.

Besman, M.J., Yanagibashi, K., Lee, T.D., Lawamura, M., Hall, P.F. and Shivley, J.E. (1989). Identification of des(Gly-Ile)-endozepine as an effector of corticotropin-dependent adrenal steroiodogenesis: stimulation of cholesterol delivery is mediated by the peripheral benzodiazepine receptor. *Proc. Natl. Acad. Sci. USA* **86**, 4897–4901.

Bond, P.A., Cundall, R.L. and Rolfe, B. (1985). [^{3}H]diazepam binding to human granulocytes. *Life Sci.* **37**, 11–16.

Bourguignon, J.J., Chambon, J.P. and Wermuth, C.G. (1985). Triazolo[4,3,b]-pyridazines, procédé pour leur préparation et compositions pharmaceutiques les contenant. *Eur. Pat.*, EP 156 734.

Bourguignon, J.J., Wermuth, C.G., Renaud de la Faverie, J.F., Thollon, C. and Lombet, A. (1991). Preparation of imidazo[1,2-c]quinazolines as smooth muscle relaxants. *Eur. Pat. Appl.* EP446 141 11th Sept. 1991, *Chem. Abstr.* **116**, 21068g.

Braestrup, C. and Squires, R.F. (1977). Specific benzodiazepine receptors in rat brain characterized by high-affinity [^{3}H]diazepam binding. *Proc. Natl. Acad. Sci. USA* **74**, 3805–3809.

Brandau, B., Bourguignon, J.J. and Wermuth, C.G. (1991). Ligands of the central benzodiazepine receptors: computer-aided characterization of their specific interactions. *Pharmacochem. Libr.* **16**, 253–256.

Cantor, E.H., Kenessey, A., Semenuk, G. and Spector, S. (1984). Interaction of calcium channel blockers with non-neuronal benzodiazepine binding sites. *Proc. Natl. Acad. Sci. USA* **81**, 1549–1552.

Clark, M. and Post, R.M. (1990). Lidocaine binds with high affinity to peripheral-type benzodiazepine receptors. *Eur. J. Pharmacol.* **179**, 473–475.

Costa, E., Berkovich, A. and Guidotti, A. (1987). The regulation of GABAergic receptors by a novel family of endogenous neuropeptides. *Life Sci.* **41**, 799–803.

Czernik, A.J., Petrack, B., Kalinsky, H.J., Psychoyos, S., Cash, W.D., Tsai, C., Rinehart, R.K., Granat, F.R. and Lovell, R.A. (1982). CGS 8216: receptor binding characteristics of a potent benzodiazepine antagonist. *Life Sci.* **30**, 363–372.

Davies, L.P. and Huston, V. (1981). Peripheral benzodiazepine binding sites in heart and their interaction with dipyridamole. *Eur. J. Pharmacol.* **73**, 209–211.

De Robertis, E., Pena, C., Paladini, A.C. and Medina, J.H. (1988). New developments on the search for the endogenous ligand(s) of central benzodiazepine receptors. *Neurochem. Int.* **13**, 1–11.

Devaud, L.L. and Murray, T.F. (1988). Involvement of peripheral-type benzodiazepine receptors in the proconvulsant actions of pirethroid insecticides. *J. Pharmacol. Exp. Ther.* **247**, 14–22.

Doble, A. and Martin, I.L. (1982). Multiple benzodiazepine receptors: no reason for anxiety. *Trends Pharmacol. Sci.* **13**, 76–81.

Doble, A., Malgouris, C., Daniel, M., Daniel, N., Imbault, F., Basbaum, A., Uzan, A., Guérémy, C. and Le Fur, G. (1987). Labelling of peripheral-type benzodiazepine binding sites in human brain with [3H]PK 11195: anatomical and subcellular distribution. *Brain Res. Bull.* **18**, 49–61.

Dubroeucq, M.C., Renault, C. and Le Fur, G. (1983). Nouveaux dérivés d'arène et d'hétéroarène carboxamides, leur procédé de préparation et médicaments les contenant. *Eur. Pat.*, EP 94 271.

Dubroeucq, M.C., Benavides, J., Doble, A., Guilloux, F., Allam, D., Vaucher, N., Bertrand, P., Guérémy, C., Renault, C., Uzan, A. and Le Fur, G. (1986). Stereoselective inhibition of the binding of [3H]PK 11195 to peripheral-type benzodiazepine binding sites by a quinoline propanamide derivative. *Eur. J. Pharmacol.* **128**, 269–272.

Francis, J.E., Bennett, D.A., Hyun, J.L., Rovinski, S.L., Amrick, C.L., Loo, P.S., Murphy, D., Neale, R.F. and Wilson, D.E. (1991a). Anxiolytic properties of certain annelated[1,2,4]triazolo[1,5-c]pyrimidin-5(6H)-ones. *J. Med. Chem.* **34**, 2899–2906.

Francis, J.E., Cash, W.D., Barbaz, B.S., Bernard, P.S., Lovell, R.A., Mazzenga, G.C., Friedmann, R.C., Hyun, J.L., Braunwalder, A.F., Loo, P.S. and Bennett, D.A. (1991b). Synthesis and benzodiazepine binding activity of a series of novel [1,2,4]triazolo[1,5-c]quinazolin-5(6H) ones. *J. Med. Chem.* **34**, 281–290.

Fryer, R.I. (1990). Ligand interactions at the benzodiazepine receptor. In *Comprehensive Medicinal Chemistry* (ed. Emmet, J.C.), Vol. 3, pp. 539–566. Pergamon Press, New York.

Georges, G., Vercauteren, D.P., Vanderveken, D.J., Horion, R., Evrard, G., Fripiat, J.G., André, J.M. and Durant, F. (1990). Structural and electronic analysis of peripheral benzodiazepine ligands: description of the pharmacophoric elements for their receptors. *Int. J. Quant. Chem. Quant. Biol. Symp.* **17**, 1–25.

Griebel, G., Misslin, R., Vogel, E. and Bourguignon, J.J. (1991). Behavioural effects of rolipram and structurally related compounds in mice: behavioural sedation of cAMP phosphodiesterase inhibitors. *Pharmacol. Biochem. Behav.* **39**, 321–323.

Hirsch, J.D., Beyer, C.F., Malkowitz, L., Loullis, C.C. and Blume, A.J. (1988). Characterization of ligand binding to mitochondrial benzodiazepine receptors. *Mol. Pharmacol.* **34**, 164–172.

Knudsen, J., Jojrup, P., Hansen, H.O., Hansen, H.F. and Roepstorff, P. (1989). Acyl-CoA-binding protein in the rat. Purification, binding characteristics, tissue concentrations and amino acid sequence. *Biochem. J.* **262**, 513–519.

Langer, S.Z. and Arbilla, S. (1988a). Imidazopyridines as a tool for the characterization of benzodiazepine receptors: a proposal for a pharmacological classification as Omega receptor subtypes. *Pharmacol. Biochem. Behav.* **29**, 763–766.

Langer, S.Z. and Arbilla, S. (1988b). Limitations of the benzodiazepine receptor nomenclature: a proposal for a pharmacological classification as omega receptor subtypes. *Fundam. Clin. Pharmacol.* **2**, 159–170.

Langer, S.Z., Arbilla, S., Tan, S., Lloyd, K.G., George, P., Allen, J. and Wick, A.E. (1990). Selectivity for omega-receptor subtypes as a strategy for the development of anxiolytic drugs. *Pharmacopsychiatry* **23**, 103–107 (Suppl.).

Le Fur, G., Guilloux, F., Rufat, P., Benavides, J., Uzan, A., Renaul, C., Dubroeucq, M.C. and Gurérémy, C. (1983). Peripheral benzodiazepine binding sites: effect of PK 11195, 1-(2-chlorophenyl)-*N*-methyl-(1-methylpropyl)-3 isoquinoline carboxamide. *Life Sci.* **32**, 1849–1856.

Le Fur, G., Perrier, M.L., Vaucher, N., Imbault, F., Flamier, A., Benavides, J., Uzan, A., Renault, C., Dubroeucq, M.C. and Guérémy, C. (1983). Peripheral benzodiazepine binding sites: effect of PK 11195, 1-(2-chlorophenyl)-N-methyl-N-(1-methylpropyl)-3-isoquinoline carboxamide. I. In vitro studies. *Life Sci.* **32**, 1839–1847.

Lueddens, H.M.W. and Skolnick, P. (1987). 'Peripheral-type' benzodiazepine receptors in the kidney: regulation of radioligand binding by anions and DIDS. *Eur. J. Pharmacol.* **133**, 205–214.

Lukeman, D.S. and Fanestil, D.D. (1987). Interactions of diuretics with the peripheral-type benzodiazepine receptor in rat kidney. *J. Pharmacol. Exp. Ther.* **241**, 950–955.

Mendes, E., Verneires, J.C., Keane, P.E. and Bachy, A. (1989). Amino-4 quinoléines et naphtyridines, leur procédé de préparation et leur application comme médicaments. *Eur. Pat.*, EP 346 208.

Morgan, J.I., Johnson, M.D., Wang, J.K.T., Sonnenfeld, K.H. and Spector, S. (1985). Peripheral-type benzodiazepines influence ornithine decarboxylase levels and neurite outgrowth in PC12 cells. *Proc. Natl. Acad. Sci. USA* **82**, 5223–5226.

Newman, A.H., Lueddens, H.W.M., Skolnick, P. and Rice, K.C. (1987). Novel irreversible ligands specific for 'peripheral' type benzodiazepine receptors: (±)-, (+)-, and (−)-1-(2-chlorophenyl)*N*-(1-methylpropyl)*N*-(2-isothiocyanatoethyl)-3-isoquinolinecarboxamide and 1-(2-isothiocyanatoethyl)-7-chloro-1,3-dihydro-5-(4-chlorophenyl)2*H*-1,4-benzodiazepin-2-one. *J. Med. Chem.* **30**, 1901–1905.

Rampe, D. and Triggle, D.J. (1986). Benzodiazepines and calcium channel function. *Trends Pharmacol. Sci.* **12**, 461–464.

Renault, C. (1986). Current developments in pharmaceutical research on Quinquina alkaloids. *Actual. Chim. Ther.* **13**, 131–151.

Riond, J., Vita, N., Le Fur, G. and Ferrara, P. (1989). Characterization of a peripheral-type benzodiazepine-binding site in the mitochondria of chinese hamster ovary cells. *FEBS Let.* **245**, 238–244.

Saano, V. (1986). Affinity of various compounds for benzodiazepine binding sites in rat brain, heart and kidney *in vitro*. *Acta Pharmacol. Toxicol.* **58**, 333–338.

Saano, V., Rägo, L. and Räty, M. (1989). Peripheral benzodiazepine binding sites. *Pharmacol. Ther.* **41**, 503–514.

Schoemaker, H., Boles, R.G., Horst, W.D. and Yamamura, H.I. (1983). Specific high-affinity binding sites for [^{3}H]Ro5-4864 in rat brain and kidney. *J. Pharmacol. Exp. Ther.* **225**, 61–69.

Schumacher, C., Steinberg, R., Kan, J.P., Michaud, J.C., Bourguignon, J.J., Wermuth, C.G., Feltz, P., Worms, P. and Bizière, K. (1989). Pharmacological characterization of the aminopyridazine SR 95639A, a selective M1 muscarinic agonist. *Eur. J. Pharmacol.* **166**, 139–147.

Skowronski, R., Beaumont, K., and Fanestil, D.D. (1987). Modification of the peripheral-type benzodiazepine receptor by arachidonate, diethylpyrocarbonate and thiol reagents. *Eur. J. Pharmacol.* **143**, 305–314.

Tahara, T., Kawakami, M., Takehara, S. and Sakamori, M. (1986). Benzo(h)cinnoline compounds and medicinal composition containing them. PCT Int. Appl. WO 86 01, 506, 13 March 1986, *Chem. Abstr.* **105**, 42 831 p.

Tebib, S., Bourguignon, J.J. and Wermuth, C.G. (1987). The active analog approach applied to the pharmacophore identification of benzodiazepine receptor ligands. *J. Comput. Aided Mol. Design* **1**, 153–170.

Verma, A. and Snyder, H. (1988). Characterization of porphyrin interactions with peripheral type benzodiazepine receptors. *Mol. Pharmacol.* **34**, 800–805.

Verma, A. and Snyder, S.H. (1989). Peripheral type benzodiazepine receptors. *Ann. Rev. Pharmacol. Toxicol.* **29**, 307–322.

Wang, J.K.T., Taniguchi, T. and Spector, S. (1984). Structural requirements for the binding of benzodiazepines to their peripheral-type sites. *Mol. Pharmacol.* **25**, 349–351.

Weisman, B.A., Cott, J., Jackson, J.A., Bolger, G.T., Weber, K.H., Horst, W.D., Paul, S.M. and Skolnick, P. (1985). 'Peripheral-type' binding sites for benzodiazepines in brain: relationship to the convulsant actions of Ro5-4864. *J. Neurochem.* **44**, 1494–1499.

Williams, M., Bennett, D.A., Loop, P.S., Braunwalder, A.F., Amrick, C.L., Wilson, D.E., Thompson, T.N., Schmutz, M., Yokoyoma, N. and Wasley, J.W.F. (1989). CGS 20625, a novel pyrazolopyridine anxiolytic. *J. Pharmacol. Exp. Ther.* **248**, 89–96.

PHARMACOLOGICAL EFFECTS OF PERIPHERALLY ACTING BENZODIAZEPINES

CHAPTER 4

THE ROLE OF MITOCHONDRIAL BENZODIAZEPINE RECEPTORS IN STEROIDOGENESIS

Karl E. Krueger[1,2] and Vassilios Papadopoulos[2]

[1] *Fidia-Georgetown Institute for the Neurosciences*
and the
[2] *Department of Anatomy and Cell Biology, Georgetown University School of Medicine, 3900 Reservoir Rd, Washington, DC 20007, USA*

Table of Contents

4.1 Introduction

The function of peripheral-type benzodiazepine receptors has been a topic of considerable interest because the pharmacological implications of these recognition sites are not known, despite the extensive therapeutic use of benzodiazepines such as diazepam which potently bind to these sites. An

PERIPHERAL BENZODIAZEPINE RECEPTORS
ISBN 0-12-282630-2

important criterion for disclosing the roles of these recognition sites must take into consideration their tissue and subcellular localization. Peripheral-type benzodiazepine recognition sites have been characterized as being predominantly associated with mitochondria, specifically on the outer mitochondrial membrane (Anholt et al., 1986; Basile and Skolnick, 1986; Hirsch et al., 1989b). This suggests a mitochondrial function for this class of binding sites, which have been given the name of mitochondrial benzodiazepine receptors (MBR). This chapter therefore focuses on the growing body of research pointing to the involvement of MBR in the regulation of steroid biosynthesis as it occurs at the mitochondrial level.

4.2 Functional cues gained from tissue distribution

Radioligand binding studies showed that MBR exhibit a very specific pattern of expression in different tissues (Anholt et al., 1985b). The adrenal cortex displays the greatest density of MBR, whereas in the testis these receptors are highly concentrated in Leydig cells (DeSouza et al., 1985). Secretory tissues such as in the kidney, lung, sweat glands and choroid plexus also show high levels (Anholt et al., 1985b) but steroidogenic cells contain a density which is substantially higher than that of these other tissues (Anholt, 1986).

In mitochondrial preparations from different tissues the density of MBR shows no correlation with the specific activity of the mitochondrial marker enzymes cytochrome c oxidase or succinate dehydrogenase (Antkiewicz-Michaluk et al., 1988a). The fact that the mitochondrial density of MBR between different tissues varies by more than 100-fold suggests that they are not ubiquitous to all mitochondria, but function in a process which is differentially utilized by various cells. Because the density of MBR in rat adrenocortical mitochondria is over 100 pmol/mg, the MBR protein accounts for about 0.2% of the total mitochondrial protein assuming a molecular weight of 18 000 (Doble et al., 1987; Antkiewicz-Michaluk, 1988b). Clearly, this unusually high mitochondrial density suggests that MBR participate in a function elaborated by steroidogenic cells.

Other evidence linking MBR with steroid synthesis has been derived from analysis of MBR expression following various manipulations of the endocrine system in the rat (see Chapter 12). Hypophysectomy causes a drastic reduction of MBR levels in the adrenal cortex and testis, and this effect is reversed by administration of the appropriate pituitary hormones which maintain trophic support for these endocrine tissues and act as physiological stimulators of steroidogenesis (Anholt et al., 1985a; Fares et al., 1987). It was also found that the density of MBR in ovaries fluctuates during the oestrous cycle (Fares *et al.*, 1988) with other tissues showing other patterns of steroid dependence with respect to regulation of MBR (Gavish *et al.*, 1986). The possible connections of MBR with steroidogenesis therefore become ostensible when considering these different paths of leading evidence.

4.3 Biochemical and cell biological basis of steroid biosynthesis

Before discussing how MBR participate in steroidogenesis it is helpful to first summarize the wealth of research which has been devoted to elucidating the mechanisms which regulate this biosynthetic pathway. All steroids are derived from cholesterol where the first metabolic step is an oxidative cleavage of the sterol side-chain to yield pregnenolone. This side-chain cleavage reaction is therefore the initial and controlling step in the synthesis of all steroids. Consequently, the rate at which steroid synthesis occurs is dependent on the rate at which pregnenolone is formed.

The enzyme catalysing the conversion of cholesterol to pregnenolone is referred to as the C_{27}-cholesterol side-chain cleavage cytochrome P-450 ($P-450_{scc}$). This enzyme is located on inner mitochondrial membranes and its activity depends upon an electron transport system comprised of adrenodoxin and adrenodoxin reductase. Extensive studies have shown that the catalytic activity of the $P-450_{scc}$ enzyme *per se* is not rate limiting in steroidogenesis (Hall, 1984). The basis for the regulation of this pathway instead is founded on the principal that the cell regulates the transport of cholesterol to $P-450_{scc}$ as the means by which the rate of pregnenolone formation is determined, i.e. the rate of steroid biosynthesis is governed by controlling the accessibility of the substrate cholesterol to the $P-450_{scc}$.

Physiological activators of steroidogenesis in peripheral endocrine organs, such as adrenocorticotropin (ACTH) and gonadotropins, acutely stimulate steroid synthesis in their target tissues by triggering an increase of cholesterol delivery to inner mitochondrial membranes for the $P-450_{scc}$ (Hall, 1984). These tropic hormones initially bind to a cell surface receptor coupled to the activation of cAMP synthesis. The increased intracellular levels of this second messenger then promote a number of events, all of which lead to facilitate cholesterol transport to inner mitochondrial membranes.

The first phase in this complex process is that cholesterol is recruited from different sites within cell and made available for intracellular transport. The second phase is characterized by movement of cholesterol to mitochondria and its incorporation into or near outer mitochondrial membranes. The third phase, which has been shown to be distinct from the second, is the translocation of cholesterol from outer to inner mitochondrial membranes and is termed intramitochondrial cholesterol transport. The final phase of this scheme is where cholesterol within inner mitochondrial membranes then becomes available to enter the steroidogenic pathway by having access to P-450$_{scc}$ to be metabolized to pregnenolone. Following this conversion, pregnenolone is further metabolized at smooth endoplasmic reticulum, ultimately giving rise to all final steroid products produced by the various steroidogenic cells.

With regard to elucidating the rate-limiting step in this entire process, the protein synthesis inhibitor cycloheximide will completely block the stimulation of steroidogenesis by pituitary tropic hormones. The block by cycloheximide was determined to be specifically at a step transpiring within the mitochondrion (Simpson et al., 1978). It was found that cholesterol accumulated at outer mitochondrial membranes, but its subsequent transfer to the inner membranes was prevented (Privalle et al., 1983). Therefore, intramitochondrial transport of cholesterol was identified as being the principal rate-determining step in steroid biosynthesis and pituitary tropic hormones activate this translocation process by a mechanism which is sensitive to cycloheximide.

Despite being recognized as the central point of control in steroidogenesis, the molecular mechanisms which operate during intramitochondrial cholesterol transport and identification of its functional molecular components have eluded scientists for nearly two decades. Several proteins have been proposed as being related to the cycloheximide-sensitive factor generated following tropic hormone stimulation (Pederson and Brownie, 1987; Yanagibashi et al., 1988; Alberta et al., 1989). It is clear though that a great deficit exists in our understanding of this important physiological process setting the stage for examining the role of MBR in steroid synthesis.

4.4   Regulation of steroidogenesis mediated by MBR

4.4.1   Effects of MBR ligands on steroidogenesis

With the background information given in the previous section it becomes much more apparent that MBR might provide a missing link in the regulation of steroid biosynthesis. Early reports on *in vivo* studies claimed that diazepam administration in rats elevated plasma glucocorticoid (Marc and Morselli, 1969) and testosterone levels (Arguelles and Rosner, 1975). A probable cause for these effects may be accounted for by an alteration of GABAergic mechanisms in the hypothalamic regulation of the pituitary. A more recent study by Calogero et al. (1990) upholds this possibility demonstrating that

PK 11195 and Ro5-4864 affect release of corticotropin-releasing factor and ACTH in the rat. It was also pointed out, however, that PK 11195, at doses lower than those required to affect ACTH release, still elevated plasma corticosterone levels. This latter finding suggests that PK 11195 may directly stimulate steroid synthesis in the adrenal cortex *in vivo*. More definitive evidence supporting this proposal will be provided by the following *in vitro* studies.

The first paper reporting an effect of MBR-selective ligands on steroidogenesis had described studies using crude testicular Leydig cell preparations and the benzodiazepine Ro5-4864 (Ritta *et al.*, 1987). This was followed by another study in somewhat more depth showing that both PK 11195 and Ro5-4864 stimulated testosterone secretion in crude Leydig cell preparations (Ritta and Calandra, 1989); however, because Sertoli cells are known to influence Leydig cell steroid secretion (Sharpe, 1986; Papadopoulos, 1991) the precise locus of action of the drugs could not be differentiated by these studies.

Within the same period of time several other laboratories reported that ligands with MBR specificity stimulated steroidogenesis. The most definitive studies provided a detailed pharmacological characterization to demonstrate unequivocally that the steroidogenic effects of these ligands were mediated specifically via their binding to MBR (Mukhin *et al.*, 1989; Papadopoulos *et al.*, 1990).

Figure 1 highlights the results from one of these studies in which the mouse Y-1 adrenocortical tumour cell line was used. Dose–response curves are shown for a series of ligands, covering over four orders of magnitude in their affinities for MBR and including members from several different classes of compounds. Those ligands with submicromolar affinities stimulated the secretion of 20α-hydroxyprogesterone, the major steroid product of Y-1 cells. Furthermore, those ligands with the highest affinities also exhibited the greatest efficacy (two-fold) in this stimulation (Figure 1A). In contrast, clonazepam and flumazenil, which are $GABA_A$ receptor-selective benzodiazepines, failed to stimulate steroidogenesis.

When the steroidogenic potencies of these ligands are compared with their affinities for MBR a highly significant correlation ($r = 0.98$) is found (Figure 1B). It should be noted that this correlation holds for the enantiomeric ligands (−)PK 14067 and (+)PK 14068, showing that the steroidogenic response exhibits stereospecificity characteristic of MBR (Dubroeucq *et al.*, 1986). These findings provide strong evidence that these ligands stimulate steroid synthesis by virtue of their binding to MBR. Qualitatively identical results were obtained with the mouse MA-10 Leydig tumour cell line (Papadopoulos *et al.*, 1990).

The effects of these ligands are not unique to cultured steroidogenic cell lines. Similar results are obtained with rat and bovine adrenocortical cell preparations (Mukhin *et al.*, 1989), as well as with purified rat Leydig cell

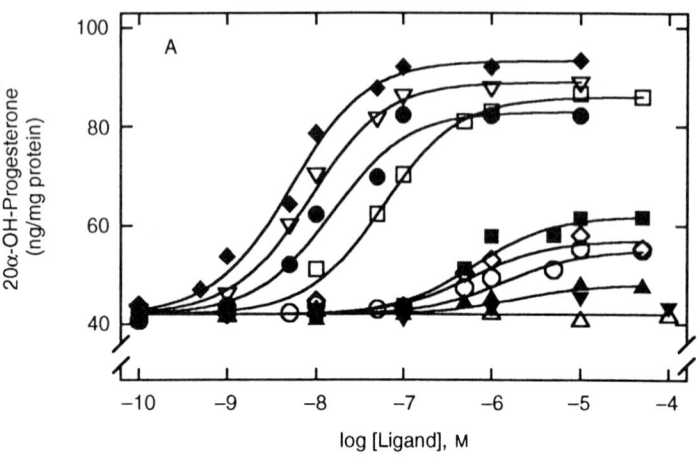

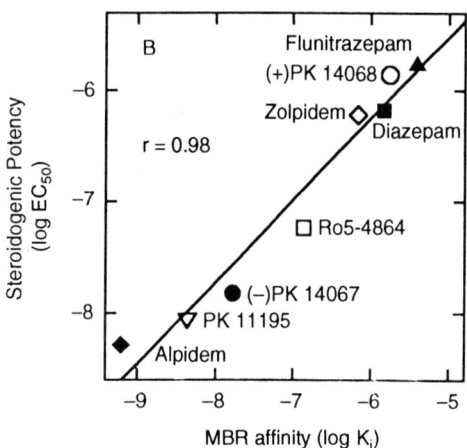

Figure 1 High-affinity MBR ligands stimulate steroidogenesis. (A) Y-1 adrenocortical cells were incubated with the designated concentrations of each ligand for 4 h, after which 20α-hydroxyprogesterone released into the medium was measured. Note the interruption of the abscissa. Symbols representing each active ligand are indicated in (B); the inactive benzodiazepines clonazepam (△) and flumazenil (▼) are also displayed. (B) The steroidogenic potency of each ligand, as determined by non-linear regression analysis shown in (A), is compared with the respective inhibitory constants determined by competition of specific [³H]PK 11195 binding in Y-1 cells. Adapted from Mukhin et al. (1989).

preparations (Papadopoulos *et al.*, 1990). In fact, further support for the involvement of MBR has been obtained by comparing rat and bovine adrenocortical preparations. MBR in rodent species exhibit a nanomolar affinity for Ro5-4864 whereas MBR of bovine origin are much lower in their affinity for this benzodiazepine (Awad and Gavish, 1987; Parola and Laird, 1991). Consistent with this species-specific difference, Ro5-4864 was found to be nearly 100-fold less potent at stimulating glucocorticoid synthesis in bovine adrenocortical cell preparations in comparison with preparations from rat adrenals (Mukhin *et al.*, 1989).

Investigations by other laboratories confirm these results, although similar detailed characterizations on the specificity by MBR ligands were not performed. These studies included bovine adrenocortical cell preparations (Yanagibashi *et al.*, 1989b; Besman *et al.*, 1989), human placental tissue (Barnea *et al.*, 1989), and ovarian granulosa cells (Amsterdam and Suh, 1991). It is therefore widely accepted that the effects of these ligands on steroidogenesis in a variety of systems are mediated by MBR.

Further studies were aimed at characterizing the steroidogenic stimulation by MBR ligands in more depth. These ligands stimulate pregnenolone formation in mitochondrial fractions from Leydig (Papadopoulos *et al.*, 1990) and adrenocortical cell lines (Krueger and Papadopoulos, 1990), further supporting the role of MBR in this action.

To examine if MBR ligands directly affect the activity of P-450$_{scc}$ the compound 22(R)-hydroxycholesterol was used. This cholesterol analogue is produced as an intermediate in the side-chain cleavage reaction of P-450$_{scc}$ and thus it can be metabolized by this enzyme to form pregnenolone. It also apparently has the ability to diffuse freely across membrane bilayers and therefore a transport process is not required for its entry to inner mitochondrial membranes. MBR ligands such as Ro5-4864 or PK 11195 show no effect on pregnenolone formation with either saturating or submaximal concentrations of 22(R)-hydroxycholesterol (Papadopoulos *et al.*, 1990). This finding suggests that the stimulation of MBR ligands does not result from a modulation of P-450$_{scc}$ enzymatic activity.

To provide more direct evidence that MBR mediate intramitochondrial cholesterol transport, subfractionation of outer and inner mitochondrial membranes has been performed to analyse cholesterol contents associated with each membrane fraction. In Y-1 cell mitochondrial preparations incubated in the presence of a P-450$_{scc}$ inhibitor (to prevent cholesterol conversion to pregnenolone), PK 11195 was found to facilitate a decrease of cholesterol in outer mitochondrial membranes accompanied by a concomitant increase of cholesterol associated with the corresponding inner mitochondrial membrane fractions (Krueger and Papadopoulos, 1990). These observations suggest that MBR play a role in the translocation of cholesterol from outer to inner mitochondrial membranes, the step which had previously been characterized as being the rate-limiting process in steroidogenesis.

The precise function of MBR in intramitochondrial cholesterol transport is still not known. Three possibilities can account for the role of MBR: (1) MBR may be an integral subunit of the intramitochondrial cholesterol transport apparatus; (2) MBR may function as a regulatory protein which modulates the transport apparatus; (3) MBR may function in an ancillary process which is required for intramitochondrial transport to occur. Obviously, further studies are necessary to elucidate more definitively the molecular aspects of MBR function in cholesterol transport.

4.4.2 Coupling of MBR to the activation of steroidogenesis by tropic hormones

An important concept related to the physiological significance of MBR concerns whether these receptors are directly involved in the steroid biosynthetic mechanisms stimulated by pituitary tropic hormones. Many of the ligands which bind to MBR show no effect on steroidogenesis stimulated with ACTH or human choriogonadotropin in adrenocortical (Krueger and Papadopoulos, 1990; Yanagibashi et al., 1989a) and Leydig (Papadopoulos et al., 1990) cell systems, respectively. The finding that the stimulation by MBR ligands is non-additive with that of pituitary tropic hormones is consistent with the possibility that these hormones activate a mechanism in which MBR participate.

The benzodiazepine flunitrazepam shows a different profile in its action on hormone-stimulated steroidogenesis. This ligand, which binds with moderately high affinity to MBR ($K_d \approx 200$ nM), antagonizes the steroidogenic efficacy of ACTH and human choriogonadotropin in Y-1 cells (see Figure 2A) and MA-10 cells, respectively, with an IC_{50} of about 500 nM (Papadopoulos et al., 1991b). The inhibition by flunitrazepam is only 40–60%, however, and the reason for this incomplete antagonism is not fully understood. However, flunitrazepam by itself acts as a partial agonist and therefore it stimulates steroidogenesis to a level slightly higher than the basal activity (Figures 1A and 2A).

Evidence to support the mediation of the antagonist action of flunitrazepam via its binding to MBR is demonstrated by the fact that other ligands recognized by MBR, which show no effect on tropic hormone stimulation of steroid biosynthesis, compete against the inhibitory action of flunitrazepam (Figure 2B). The series of ligands in this paradigm were the same as those shown in Figure 1, having been characterized for their effects on steroidogenesis in the absence of hormones. These ligands, having a 10 000-fold range in their affinities for MBR, exhibit an equivalent rank order of potency in their ability to prevent the antagonism by flunitrazepam ($r = 0.99$). The stereospecificity

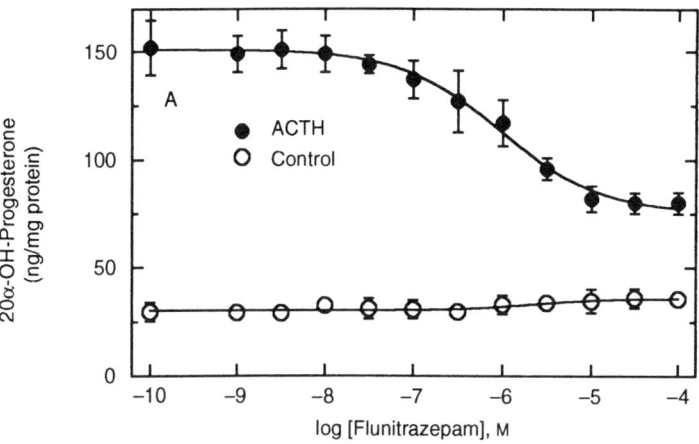

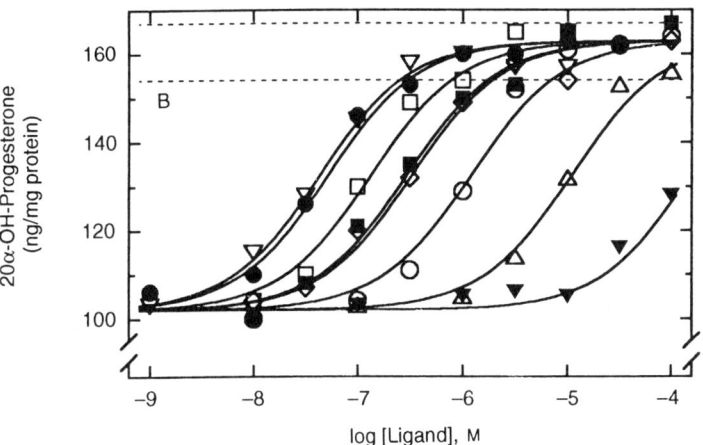

Figure 2   Flunitrazepam inhibits the stimulation of steroidogenesis by tropic hormones. (A) Y-1 adrenocortical cells in the presence (●) or absence (○) of 10 nM ACTH were incubated with the indicated concentrations of flunitrazepam for 2 h, after which 20α-hydroxyprogesterone in the medium was measured. Means ± SD of four cultures are shown. (B) Y-1 cells were incubated with 10 nM ACTH, 10 μM flunitrazepam, and the indicated concentrations of each competing ligand for 2 h before quantitation of 20α-hydroxyprogesterone in the medium. The range of steroid levels in medium from cells treated with 10 nM ACTH alone is shown by dashed lines. Note the interruption of the abscissa. Symbols representing each competing ligand are given in Figure 1. Adapted from Papadopoulos *et al.* (1991b).

characteristic of MBR is also manifest in this experimental model as (—) PK 14067 is over 100-fold more potent than (+) PK 14068 at competing against the inhibition of flunitrazepam. The intrinsic steroidogenic action of these competing ligands does not account for this prevention because the magnitude of this effect observed in the presence of the tropic hormone plus flunitrazepam is much greater than the maximal stimulation attained with even the most efficacious ligands in the absence of these other agents. Furthermore, note that clonazepam, which is ineffective at stimulating steroidogenesis (Figure 1A), at concentrations sufficient to displace flunitrazepam from MBR completely blocks the antagonism by flunitrazepam (Figure 2B). A similar scenario is observed with the lower affinity ligand flumazenil. These noteworthy results can be accounted for by the ability of the competing MBR ligands to displace flunitrazepam from MBR, thereby preventing the antagonism from occurring. These findings provide rather strong evidence that the inhibitory action of flunitrazepam on tropic hormone stimulation of steroidogenesis is mediated as a consequence of its binding to MBR.

Other studies were performed to better characterize this antagonism by flunitrazepam (Papadopoulos et al., 1991b). Although flunitrazepam is a good photoaffinity probe for $GABA_A$ receptors, its inhibitory effects are not due to covalent modification of a protein because the antagonism is reversible and is not dependent on photoactivation. Flunitrazepam also inhibits steroid synthesis induced by dibutyryl cAMP but it does not affect conversion of 22(R)-hydroxycholesterol to pregnenolone. Therefore the action by flunitrazepam is subsequent to the activation of adenylyl cyclase but prior to the reaction catalysed by P-450$_{scc}$. Subfractionation studies on mitochondrial membranes suggest that flunitrazepam partially inhibits the activation of cholesterol translocation from outer to inner mitochondrial membranes consistent with the proposal that this specific step involves the participation of MBR.

The significance of these studies with flunitrazepam is that they provide evidence that MBR are coupled to the physiological mechanisms by which steroidogenesis is activated in endocrine tissues. The reason why flunitrazepam is unique in its intrinsic activity to antagonize the stimulation by tropic hormones is still a mystery which requires a much better understanding of all the molecular components and their interactions functioning in intra-mitochondrial cholesterol transport.

Another important aspect concerning the relationship of MBR function with activation by tropic hormones is their differing sensitivities to cycloheximide. Whereas cycloheximide blocks the steroidogenic activity by tropic hormones, it has no effect on the stimulation by MBR ligands (Papadopoulos et al., 1990; Krueger and Papadopoulos, 1990). Because cycloheximide had previously been shown to block the tropic hormone action specifically at the stage of cholesterol translocation from outer to inner mitochondrial membranes (Privalle *et al.*, 1983), it is implied that the MBR play a role subsequent to the blockade by cycloheximide. In fact, when Y-1 cells are pre-treated for 2 h with ACTH and cycloheximide simultaneously, thereby permitting cellular transport of cholesterol to mitochondria without subsequent intramitochondrial transport, MBR ligands display a greater steroidogenic efficacy (Krueger and Papadopoulos, 1990). However, even with these manipulations, the maximal stimulation attained by MBR ligands is still considerably lower than the level of steroid biosynthesis achieved following tropic hormone treatment. This discrepancy implies that occupation of MBR with these pharmacological agents merely incompletely mimics the cellular mechanisms which regulate intramitochondrial cholesterol transport.

4.4.3 Diazepam binding inhibitor as an endogenous modulator of MBR activity

Independently from all the work on MBR discussed up to this point, another laboratory had purified a protein from bovine adrenal cortex that stimulated pregnenolone synthesis in adrenocortical mitochondrial preparations (Yanagibashi *et al.*, 1988). Sequencing of this 10 kDa steroidogenic protein (Besman *et al.*, 1989) revealed its identity as being a polypeptide termed diazepam binding inhibitor (DBI). [Editor's note: DBI was recently shown to be identical in sequence to acyl-coenzyme A binding protein, a protein which is induced during insulin-induced pre-adipocyte differentiation (Andersen *et al.*, 1991); see also Chapter 10).] This polypeptide was first isolated several years earlier from rat brain based on its ability to inhibit diazepam binding in rat brain membrane preparations (Guidotti *et al.*, 1983). DBI was suggested to modulate $GABA_A$ receptors allosterically by the report that it inhibits chloride channel conductance induced by GABA in primary neuronal cultures (Bormann *et al.*, 1985) and displaces $GABA_A$ receptor-selective ligands from brain membranes (Ferrero *et al.*, 1986). Later studies have shown that DBI also displaces ligands from MBR (Guidotti *et al.*, 1988).

Investigations focused on endocrine systems have subsequently shown that DBI is highly expressed in steroidogenic cells (Bovolin *et al.*, 1990; Rheaume *et al.*, 1990). The converging fields of study on DBI and MBR suggested that there may be an interplay between these two proteins in regulating steroidogenesis. Despite the fact that DBI displaces ligands from MBR at micromolar concentrations, it displays a steroidogenic potency at nanomolar levels in mitochondrial fractions (Papadopoulos *et al.*, 1991a). The stimulation of pregnenolone synthesis by PK 11195 is non-additive with that of DBI whereas flunitrazepam inhibits by 70% the stimulation by DBI. This pharmacological behaviour is reminiscent of the parallel effects found with these drugs on the stimulation of steroid synthesis in cells by pituitary tropic hormones. Therefore, it is currently hypothesized that tropic hormones may trigger an intracellular mechanism controlling the interaction of DBI with MBR to regulate steroidogenesis.

Current indications suggest that tropic hormones do not increase DBI levels within the 5–10 min period during which steroidogenesis is first activated (Massotti et al., 1991; Brown et al., 1992). If DBI is a mediator of the hormone action a more complex molecular scheme must be involved. Different proteolytic processing fragments of DBI have been identified in tissues and these may help account for an alternative mechanism involving DBI. Among these peptide fragments an octadecaneuropeptide (ODN) shows selectivity for GABA_A receptors, whereas a trikontatetraneuropeptide (TTN) is selective for MBR (Slobodyansky et al., 1989; Berkovich et al., 1990). In accordance with this specificity TTN potently stimulates pregnenolone synthesis in mitochondria, whereas ODN is only active at much higher concentrations (Papadopoulos et al., 1991a).

Other lines of research have shown that porphyrins at submicromolar concentrations interact with MBR (Verma et al., 1987; Verma and Snyder, 1988), leading to the supposition that porphyrins may be endogenous ligands for MBR. With regard specifically to steroid biosynthesis, porphyrins do not appear to be regulators of this pathway because protoporphyrin IX elicits a maximal efficacy which is 5- to 10-fold lower than that of PK 11195 or DBI (V. Papadopoulos and K.E. Krueger, unpubl. data).

4.4.4   Summarizing the role of MBR in steroidogenesis

The intracellular events that delineate the regulation of steroidogenesis are depicted in Figure 3, which shows how MBR and their modulation by DBI may participate in this process. The studies described here suggest that MBR mediate the rate-determining step of steroidogenesis and that their function is also linked to the activation of this process during tropic hormone stimulation.

Although this role for MBR is perhaps the most explicit function among those presently proposed, it should be stressed that the actual biochemical utility of MBR in this process is still not defined. Significant pieces have to be added to this puzzle before we can more substantially understand how intramitochondrial cholesterol transport is regulated and accomplished at the molecular level. Nevertheless, it is apparent that MBR are an important part of this mystery which has eluded scientists studying the regulation of steroid biosynthesis.

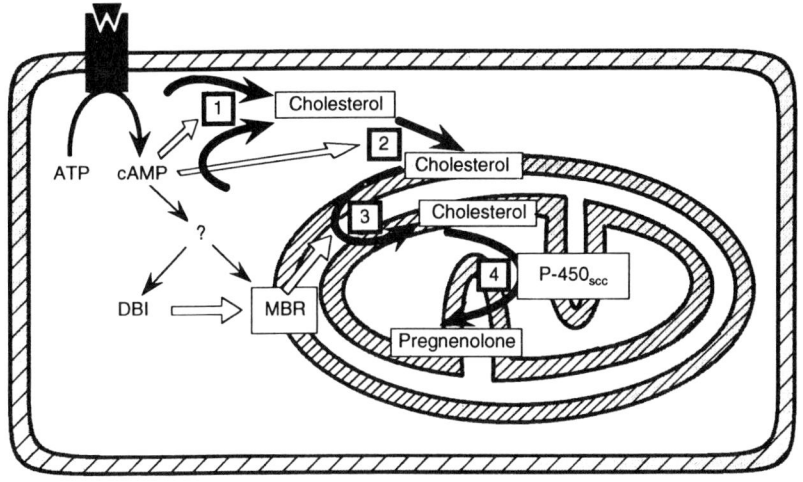

Figure 3 The participation of MBR in the stimulation of steroidogenesis by tropic hormones. This scheme depicts the chain of cellular mechanisms activated by tropic hormones during the stimulation of steroid biosynthesis. Binding of the hormone to its cell surface receptor stimulates adenylyl cyclase. The increased intracellular levels of cAMP trigger a series of events which serve to facilitate cholesterol transport to inner mitochondrial membranes where cholesterol conversion to pregnenolone is catalysed by the side-chain cleavage enzyme ($P\text{-}450_{scc}$). Cholesterol transport occurs via four principal stages indicated by the bold arrows. Elevation of cellular cAMP promotes the first three transport stages (indicated by open arrows). The first stage is represented by the recruitment of cholesterol from intracellular reserves such as cholesteryl ester hydrolysis or withdrawal from membranes. The second step is the transport of cholesterol to mitochondria, and incorporation into or around outer mitochondrial membranes. The third step, which is rate-limiting in this entire pathway, is the translocation of cholesterol from outer to inner mitochondrial membranes. The fourth step then follows where cholesterol now gains access to $P\text{-}450_{scc}$ to be converted to pregnenolone. As indicated, MBR participate in the third and rate-determining step of this complex process. Furthermore, MBR seem to be coupled to the mechanisms by which cAMP mediates the stimulation of steroidogenesis. Although still only vaguely understood, it appears that diazepam binding inhibitor (DBI) acts to regulate MBR function during tropic hormone activation. Experimental evidence has shown that activation of Step 3 is completely inhibited by cycloheximide downstream of cAMP but prior to MBR involvement, whereas the binding of flunitrazepam to MBR will partially antagonize the tropic hormone and DBI-mediated stimulation of Step 3. Taken from Krueger and Papadopoulos (1992).

4.5 Relationship of MBR structure to steroidogenesis

An 18 kDa protein comprising the ligand-binding domains of MBR has been completely sequenced as deduced from the corresponding cDNAs cloned from rat (Sprengel *et al.*, 1989), human (Riond *et al.*, 1991), and bovine (Parola *et al.*, 1991) species. This protein, although quite small, contains five potential transmembrane-spanning segments and therefore its chemical character is very lipophilic (see Chapter 1 for greater detail). This property falls within reason for a protein which participates in cholesterol transport between membranes.

There is only one other protein currently included in sequence databases which shows salient homology with the MBR protein. The crtK gene product of the carotenoid biosynthesis gene cluster in the photosynthetic bacterium *Rhodobacter capsulatus* (Armstrong *et al.*, 1989) shares many properties with MBR in its sequence similarity, molecular size, and hydrophobic character (Chapter 1; Krueger and Papadopoulos, 1992). The precise function of crtK is under investigation but it is clear that this protein participates in a biosynthetic process related to that of steroids. It might therefore be presumed that MBR and crtK fulfil similar functions in their respective isoprenoid metabolizing pathways. The explicit similarities of the MBR protein with crtK clearly reinforces the proposal, over many of the other more poorly defined functions, that MBR play an integral role in steroidogenesis.

4.6 Pharmacological significance in the brain

In the light of the demonstrations that MBR ligands regulate steroid biosynthesis, consideration should be given to the possibility that this pharmacological action may occur in the central nervous system. During recent years it has become apparent that *de novo* steroid synthesis occurs in the brain. Immunochemical studies of rat brain have demonstrated the presence of a protein recognized by antibodies directed against P-450$_{scc}$ of the adrenal cortex (Le Goascogne *et al.*, 1987; Warner *et al.*, 1989). Because much of the immunoreactivity was found in white matter, mitochondrial fractions were prepared from isolated oligodendrocytes and were found to synthesize pregnenolone (Hu *et al.*, 1987). Subsequently primary glial cultures from neonatal rat cortex were demonstrated to synthesize pregnenolone, progesterone, and some of its reduced metabolites from radiolabelled mevalonate (Jung-Testas *et al.*, 1989).

These pioneering studies suggest that certain glial cells have the capability to synthesize steroids from cholesterol. The C6 glioma cell line has recently been shown to contain P-450$_{scc}$ and exhibit steroidogenic activity (Papadopoulos *et al.*, 1992). Furthermore, pregnenolone synthesis in mitochondrial fractions from C6 cells is stimulated by MBR ligands and by DBI in qualitatively the same manner as has been observed in peripheral endocrine systems. These studies begin to point to the possibility that MBR ligands may elicit steroidogenic effects in the central nervous system and thus may have certain neuropsychotropic actions.

One area of active research concerning steroid effects on neuronal activity has been the modulation of $GABA_A$ receptor function by specific reduced metabolites of progesterone (Majewska *et al.*, 1986). These steroid derivatives, much like benzodiazepines and barbiturates, potentiate the action of GABA in chloride channel opening (Morrow *et al.*, 1990; Puia *et al.*, 1990) whereas pregnenolone sulphate antagonizes this activity (Mienville and Vicini, 1989). The ability of certain steroids to potently modulate $GABA_A$ receptors presents us with an unusual situation where the two predominant classes of benzodiazepine recognition sites, MBR and $GABA_A$ receptors, have a direct functional relationship with steroids. Therefore, it is not unlikely that a subtle contribution of the benzodiazepine pharmacological profile may arise from MBR. In addition, $GABA_A$ receptors seem to be setting a precedent where other neuronal components are now being identified as putative targets for steroid action (Chadwick and Widdows, 1990), further expanding the implications that steroidogenesis in the brain may have a more widespread significance in the modulation of synaptic transmission.

4.7 Possible correlations with other cellular activities

This chapter has provided extensive discussion giving strong support for the role of MBR in steroidogenesis; however, it is apparent that MBR are present in many tissues which do not synthesize steroids. As alluded to earlier, it is possible that steroidogenic cells amplify the expression of MBR to elaborate on a specific function common to many cells, a function which seems to be related to cholesterol transport in mitochondria. This final section will raise the question on whether MBR participation in cholesterol transport is related to other functions attributed to MBR.

Two laboratories independently reported that MBR ligands alter mitochondrial respiration (Hirsch *et al.*, 1989a; Larcher *et al.*, 1989). More recently another group has observed that mitochondrial morphology is changed and their replication is stimulated by nanomolar concentrations of PK 11195 or Ro5-4864 in different cell lines (Shiraishi *et al.*, 1990). Other evidence has indicated that PK 11195 and Ro5-4864, again at nanomolar levels, stimulate cell proliferation (Ikezaki and Black, 1990) and potentiate the mitogenic response to prolactin in Nb 2 lymphoma cells (Laird *et al.*, 1989). It is possible that all of these effects are secondary to changes in mitochondrial physiology which in turn may be dependent on cholesterol distribution in mitochondrial membranes. Alternatively, the specific function of MBR may directly influence mitochondrial processes in such a way as to produce these apparent effects.

There are some *in vivo* studies which may be pertinent to the relationship of MBR to sterol metabolism. Chronic diazepam administration was found to lower plasma cholesterol in several animal species given an atherogenic diet (Feszt *et al.*, 1977; Wong *et al.*, 1980). A similar effect was observed with dipyridamole, another drug which can bind to MBR at high concentrations (Zafra *et al.*, 1991). In contrast, PK 11195 increases total plasma cholesterol and triglyceride levels in rats (Horak *et al.*, 1990). Furthermore, PK 11195 was also found to antagonize the antiatherogenic effect of diazepam (Cuparencu *et al.*, 1990, 1991). At this point these observations are still incomplete in terms of identifying whether MBR function is involved with these atherogenic paradigms. The finding that PK 11195 antagonizes the action of diazepam on plasma cholesterol content is inconsistent with the agonistic properties both of these compounds have in steroidogenesis mediated through MBR. More definitive pharmacological studies are still needed to clarify this issue of the atherogenic activity of MBR ligands and whether this has any relationship to steroid biosynthesis or cholesterol transport.

Another interesting point of speculation which may be of relevance to these functional studies concerns some recent findings on the enzyme sterol 27-hydroxylase. This enzyme was first identified as being involved in bile acid synthesis (Wikvall, 1984), but recent molecular biological studies have revealed that this enzyme is also present in many tissues which do not produce bile acids (Andersson *et al.*, 1989). Sterol 27-hydroxylase is a cytochrome P-450 located at inner mitochondrial membranes utilizing the same electron transport system which serves P-450$_{scc}$. Because cholesterol and other sterol derivatives can serve as substrates for this enzyme, these developments have illuminated the possibility that cholesterol may enter an unknown metabolic scheme whose products and biological roles remain to be identified. Taking into account the very close similarities with the P-450$_{scc}$ system of steroidogenic cells, it seems quite plausible that MBR may play an analogous, if not identical, role in the accessibility of cholesterol for sterol 27-hydroxylase. Further advances in understanding this newly discovered metabolic pathway for cholesterol may thus provide novel insights into many of the biological effects attributed to MBR.

Acknowledgement

The authors were supported by NIH grants MH44284 and DK43358 and NSF grant DCB-9017752

References

Alberta, J.A., Epstein, L.F., Pon, L.A. and Orme-Johnson, N.R. (1989). Mitochondrial localization of a phosphoprotein that rapidly accumulates in adrenal cortex cells exposed to adrenocorticotropic hormone or to cAMP. *J. Biol. Chem.* **264**, 2368–2372.

Amsterdam, A. and Suh, B.S. (1991). An inducible functional peripheral benzodiazepine receptor in mitochondria of steroidogenic granulosa cells. *Endocrinology* **128**, 503–510.

Andersen, K.V., Ludvigsen, S., Mandrup, S., Knudsen, J. and Poulsen, F.M.L. (1991). The secondary structure in solution of acyl-coenzyme A binding protein from bovine liver using 1 H nuclear magnetic resonance spectroscopy. *Biochemistry* **30**, 10 654–10 663.

Andersson, S., Davis, D.L., Dahlbäck, H., Jörnvall, H. and Russel, D.W. (1989). Cloning, structure, and expression of the mitochondrial cytochrome P-450 sterol 26-hydroxylase, a bile acid biosynthetic enzyme. *J. Biol. Chem.* **264**, 8222–8229.

Anholt, R.R.H. (1986). Mitochondrial benzodiazepine receptors as potential modulators of intermediary metabolism. *Trends Pharmacol. Sci.* **7**, 506–511.

Anholt, R.R.H., DeSouza, E.B., Kuhar, M.J. and Snyder, S.H. (1985a). Depletion of peripheral-type benzodiazepine receptors after hypophysectomy in rat adrenal gland and testis. *Eur. J. Pharmacol.* **110**, 41–46.

Anholt, R.R.H., DeSouza, E.B., Oster-Granite, M.L. and Snyder, S.H. (1985b). Peripheral-type benzodiazepine receptors: Autoradiographic localization in whole-body sections of neonatal rats. *J. Pharmacol. Exp. Ther.* **233**, 517–526.

Anholt, R.R.H., Pedersen, P.L., De Souza, E.B. and Snyder, S.H. (1986). The peripheral-type benzodiazepine receptors: localization to the mitochondrial outer membrane. *J. Biol. Chem.* **261**, 576–583.

Antkiewicz-Michaluk, L., Guidotti, A. and Krueger, K.E. (1988a). Molecular characterization and mitochondrial density of a recognition site for peripheral-type benzodiazepine ligands. *Mol. Pharmacol.* **34**, 272–278.

Antkiewicz-Michaluk, L., Mukhin, A.G., Guidotti, A. and Krueger, K.E. (1988b). Purification and characterization of a protein associated with peripheral-type benzodiazepine binding sites. *J. Biol. Chem.* **263**, 17 317–17 321.

Arguelles, A.E. and Rosner, J. (1975). Diazepam and plasma testosterone levels. *Lancet* **ii**, 607.

Armstrong, G.A., Alberti, M., Leach, F. and Hearst, J.E. (1989). Nucleotide sequence, organization, and nature of the protein products of the carotenoid biosynthesis gene cluster of *Rhodobacter capsulatus*. *Mol. Gen. Genet.* **216**, 254–268.

Awad, M. and Gavish, M. (1987). Binding of [^{3}H]Ro5-4864 and [^{3}H]PK 11195 to cerebral cortex and peripheral tissues of various species: species differences and heterogeneity in peripheral benzodiazepine binding sites. *J. Neurochem.* **49**, 1407–1414.

Barnea, E.R., Fares, F. and Gavish, M. (1989). Modulatory action of benzodiazepines on human term placental steroidogenesis *in vitro*. *Mol. Cell. Endocrinol.* **64**, 155–159.

Basile, A.S. and Skolnick, P. (1986). Subcellular localization of 'peripheral-type' binding sites for benzodiazepines in rat brain. *J. Neurochem.* **46**, 305–308.

Berkovich, A., McPhie, P., Campagnone, M., Guidotti, A. and Hensley, P. (1990). A natural processing product of rat diazepam binding inhibitor, triakontatetra-neuropeptide (diazepam binding inhibitor 17–50) contains an α-helix, which allows discrimination between benzodiazepine binding site subtypes. *Mol. Pharmacol.* **37**, 164–172.

Besman, M.J., Yanagibashi, K., Lee, T.D., Kawamura, M., Hall, P.F. and Shively, J.E. (1989). Identification of des-(Gly-Ile) endozepine as an effector of corticotropin-dependent adrenal steroidogenesis: stimulation of cholesterol delivery is mediated by the peripheral benzodiazepine receptor. *Proc. Natl. Acad. Sci. USA* **86**, 4897–4901.

Bormann, J., Ferrero, P., Guidotti, A. and Costa, E. (1985). Neuropeptide modulation of GABA receptor Cl channels. *Regul. Peptides* **264**, 33–38.

Bovolin, P., Schlichting, J., Miyata, J., Ferrarese, C., Guidotti, A. and Alho, H. (1990). Distribution and characterization of diazepam binding inhibitor (DBI) in peripheral tissues of rat. *Regul. Peptides* **29**, 267–281.

Brown, A.S., Hall, P.F., Shoyab, M. and Papadopoulos, V. (1992). Endozepine/diazepam binding inhibitor in adrenocortical and Leydig cell lines: absence of hormonal regulation. *Mol. Cell. Endocrinol.* **83**, 1–9.

Calogero, A.E., Kamilaris, T.C., Bernardini, R., Johnson, E.O., Chrousos, G.P. and Gold, P.W. (1990). Effects of peripheral benzodiazepine receptor ligands on hypothalamic–pituitary–adrenal axis function in the rat. *J. Pharmacol. Exp. Ther.* **253**, 729–737.

Chadwick, D. and Widdows, K. (eds) (1990). *Steroids and Neuronal Activity*, Ciba Foundation Symposium 153, 284 pp. John Wiley & Sons, New York.

Cuparencu, B., Horak, J., De Santis, D., Losasso, C., Nicolella Budetta, S. and Marmo, E. (1990). The influence of peripheral-type benzodiazepine receptors angatonist PK 11195 on the high-density lipid-cholesterol level in hyperglycemic and hyper-lipidemic as well as in normoglycemic and normolipidemic rats. *Curr. Ther. Res.* **48**, 749–756.

Cuparencu, B., Horak, J., Marmo, E., De Santis, D., Lampa, E., Losasso, C. and Rossi, F. (1991). The influence of the peripheral-type benzodiazepine receptor antagonist PK 11195 on blood glucose and serum lipid levels in rats. Interactions with diazepam. *Curr. Ther. Res.* **49**, 409–414.

De Souza, E.B., Anholt, R.R.H., Murphy, K.M.M., Snyder, S.H. and Kuhar, M.S. (1985). Peripheral-type benzodiazepine receptors in endocrine organs: auto-radioraphic localization in rat pituitary, adrenal and testis. *Endocrinology* **116**, 567–573.

Doble, A., Ferris, O., Burgevin, M.C., Menager, J., Uzan, A., Dubroeucq, M.C., Renault, C., Gueremy, C. and Le Fur, G. (1987). Photoaffinity labeling of peripheral-type benzodiazepine binding sites. *Mol. Pharmacol.* **31**, 42–49.

Dubroeucq, M.C., Benavides, J., Doble, A., Guilloux, F., Allam, D., Vaucher, N., Bertrand, P., Gueremy, C., Renault, C., Uzan, A. and Le Fur, G. (1986). Stereoselective inhibition of the binding of [^{3}H]PK 11195 to peripheral-type benzodiazepine binding sites by a quinolinepropanamide derivative. *Eur. J. Pharmacol.* **128**, 269–272.

Fares, F., Bar-Ami, S., Brandes, J.M. and Gavish, M. (1987). Gonadotropin- and estrogen-induced increase of peripheral-type benzodiazepine binding sites in the hypophyseal-genital axis of rats. *Eur. J. Pharmacol.* **133**, 97–102.

Fares, F., Bar-Ami, S., Brandes, J.M. and Gavish, M. (1988). Changes in the density of peripheral benzodiazepine binding sites in genital organs of the female rat during the oestrous cycle. *J. Reprod. Fert.* **83**, 619–625.

Ferrero, P., Santi, M.R., Conti-Tronconi, B., Costa, E. and Guidotti, A. (1986). Study of an octadecaneuropeptide derived from diazepam binding inhibitor (DBI): biological activity and presence in rat brain. *Proc. Natl. Acad. Sci. USA* **83**, 827–831.

Feszt, T., Buska, C., Cuparencu, B. and Horak, J. (1977). Influence of some benzodiazepines on serum lipids levels in hyperlipidaemic rabbits. *Agressologie* **18**, 265–267.

Gavish, M., Okun, F., Weizman, A. and Youdim, M.B.H. (1986). Modulation of peripheral benzodiazepine binding sites following chronic estradiol treatment. *Eur. J. Pharmacol.* **127**, 147–151.

Guidotti, A., Forchetti, C.M., Corda, M.G., Konkel, D., Bennett, C.D. and Costa, E. (1983). Isolation, characterization, and purification to homogeneity of an endogenous polypeptide with agonistic action on benzodiazepine receptors. *Proc. Natl. Acad. Sci. USA* **80**, 3531–3535.

Guidotti, A., Berkovich, A., Ferrarese, C., Santi, M.R. and Costa, E. (1988). Neuronal–glial differential processing of DBI to yield ligands to central and peripheral benzodiazepine recognition sites. In L.E.R.S. Monograph Series, Vol. 6: *Imidazopyridines in Sleep Disorders* (eds Sauvant, J.P., Langer, S.Z. and Morselli, P.L.), pp. 25–38. Raven Press, New York.

Hall, P.F. (1984). Cellular organization for steroidogenesis. *Int. Rev. Cytol.* **86**, 53–95.

Hirsch, J.D., Beyer, C.F., Malkowitz, L., Beer, B. and Blume, A.J. (1989a). Mitochondrial benzodiazepine receptors mediate inhibition of mitochondrial respiratory control. *Mol. Pharmacol.* **35**, 157–163.

Hirsch, J.D., Beyer, C.F., Malkowitz, L., Loullis, C.C. and Blume, A.J. (1989b). Characterization of ligand binding to mitochondrial benzodiazepine receptors. *Mol. Pharmacol.* **35**, 164–172.

Horak, J., Cuparencu, B., De Santis, D., Losasso, C., Nicolella Budetta, S. and Marmo, E. (1990). The influence of the benzodiazepine receptors antagonist PK 11195 on blood glucose and serum lipids in rats: interactions with diazepam. *Curr. Ther. Res.* **48**, 739–743.

Hu, Z.Y., Bourreau, E., Jung-Testas, I., Ribel, P. and Baulieu, E.-E. (1987). Neurosteroids: Oligodendrocyte mitochondria convert cholesterol to pregnenolone. *Proc. Natl. Acad. Sci. USA* **84**, 8215–8219.

Ikezaki, K. and Black, K.L. (1990). Stimulation of cell growth and DNA synthesis by peripheral benzodiazepine. *Cancer Lett.* **49**, 115–120.

Jung-Testas, I., Hu, Z.Y., Baulieu, E.-E. and Robel, P. (1989). Neurosteroids: Biosynthesis of pregnenolone and progesterone in primary cultures of rat glial cells. *Encodrinology* **125**, 2083–2091.

Krueger, K.E. and Papadopoulos, V. (1990). Peripheral-type benzodiazepine receptors mediate translocation of cholesterol from outer to inner mitochondrial membranes in adrenocortical cells. *J. Biol. Chem.* **265**, 15015–15022.

Krueger, K.E. and Papadopoulos, V. (1992). Mitochondrial benzodiazepine receptors and the regulation of steroid biosynthesis. *Ann. Rev. Pharmacol. Toxicol.* **32**, 211–237.

Laird, II, H.E., Gerrish, K.E., Duerson, K.C., Putnam, C.W. and Russell, D.H. (1989). Peripheral benzodiazepine binding sites in Nb 2 node lymphoma cells: effects on prolactin-stimulated proliferation and ornithine decarboxylase activity. *Eur. J. Pharmacol.* **171**, 25–35.

Larcher. J.-P., Vayssiere, J.-L., Le Marquer, F.J., Cordeau, L.R., Keane, P.E., Bachy. A., Gros, F. and Croizat, B.P. (1989). Effects of peripheral benzodiazepines upon the O_2 consumption of neuroblastoma cells. *Eur. J. Pharmacol.* **161**, 197–202.

Le Goascogne, C., Robel, P., Gouezou, M., Sananes, N., Baulieu, E.-E. and Waterman, M. (1987). Neurosteroids: cytochrome P-450$_{scc}$ in rat brain. *Science* **237**, 1212–1215.

Majewska, M.D., Harrison, N.L., Schwartz, R.D., Barker, J.L. and Paul, S.M. (1986). Steroid hormone metabolites are barbiturate-like modulators of the GABA receptor. *Science* **232**, 1004–1007.

Marc, V. and Morselli, P.L. (1969). Effect of diazepam on plasma corticosterone levels in the rat. *J. Pharm. Pharmacol.* **21**, 784–786.

Massotti, M., Slobodyansky, E., Konkel, D., Costa, E. and Guidotti, A. (1991).

Regulation of diazepam binding inhibitor in rat adrenal gland by adrenocorticotropin. *Endocrinology* **129**, 591–596.

Mienville, J.-M. and Vicini, S. (1989). Pregnenolone sulfate antagonizes GABA$_A$ receptor-mediated currents via a reduction of channel opening frequency. *Brain Res.* **489**, 190–194.

Morrow, A.L., Pace, J.R., Purdy, R.H. and Paul, S.M. (1990). Characterization of steroid interactions with γ-aminobutyric acid receptor-gated chloride ion channels: evidence for multiple steroid recognition sites. *Mol. Pharmacol.* **37**, 263–270.

Mukhin, A.G., Papadoloulos, V., Costa, E. and Krueger, K.E. (1989). Mitochondrial benzodiazepine receptors regulate steroid biosynthesis. *Proc. Natl. Acad. Sci. USA* **86**, 9813–9816.

Papadopoulos, V. (1991). Identification and purification of a human Sertoli cell-secreted protein (hSCSP-80) stimulating Leydig cell steroid biosynthesis. *J. Clin. Endocrinol. Metab.* **72**, 1332–1339.

Papadopoulos, V., Mukhin, A.G., Costa, E. and Krueger, K.E. (1990). The peripheral-type benzodiazepine receptor is functionally linked to Leydig cell steroidogenesis. *J. Biol. Chem.* **265**, 3772–3779.

Papadopoulos, V., Berkovich, A., Krueger, K.E., Costa, E. and Guidotti, A. (1991a). Diazepam binding inhibitor and its processing products stimulate mitochondrial steroid biosynthesis via an interaction with mitochondrial benzodiazepine receptors. *Endocrinology* **129**, 1481–1488.

Papadopoulos, V., Nowzari, F.B. and Krueger, K.E. (1991b). Hormone-stimulated steroidogenesis is coupled to mitochondrial benzodiazepine receptors. *J. Biol. Chem.* **266**, 3682–2687.

Papadopoulos, V., Guarneri, P., Krueger, K.E., Guidotti, A. and Costa, E. (1992). Pregnenolone biosynthesis in C6-2B glioma cell mitochondria: Regulation by a mitochondrial diazepam binding inhibitor receptors. *Proc. Natl. Acad. Sci. USA* **89**, 5113–5117.

Parola, A.L. and Laird II, H.E. (1991). The bovine peripheral-type benzodiazepine receptor: A receptor with low affinity for benzodiazepines. *Life Sci.* **48**, 757–764.

Parola, A.L., Stump, D.G., Pepperl, D.J., Krueger, K.E., Regan, J.W. and Laird II, H.E. (1991). Cloning and expression of a pharmacologically unique bovine peripheral-type benzodiazepine receptor receptor isoquinoline binding protein. *J. Biol. Chem.* **266**, 14 082–14 087.

Pedersen, R.C. and Brownie, A.C. (1987). Steroidogenesis activator polypeptide isolated from a rat Leydig cell tumor. *Science* **236**, 188–190.

Privalle, C.T., Crivello, J.F. and Jefcoate, C.R. (1983). Regulation of intramitochondrial cholesterol transfer to side-chain cleavage cytochrome P-450 in rat adrenal gland. *Proc. Natl. Acad. Sci. USA* **80**, 702–706.

Puia, G., Santi, M.R., Vicini, S., Pritchett, D.B., Purdy, R.H., Paul, S.M., Seeburg, P.H. and Costa, E. (1990). Neurosteroids act on recombinant human GABA$_A$ receptors. *Neuron* **4**, 759–765.

Rheaume, E., Tonon, M.C., Smith, F., Simard, J., Desy, L., Vaudry, H. and Pelletier, G. (1990). Localization of the endogenous benzodiazepine ligand octadecaneuro-peptide in the rat testis. *Endocrinology* **127**, 1986–1994.

Riond, J., Mattei, M.G., Kaghad, M., Dumont, X., Guillemot, J.C., Le Fur, G., Caput, D. and Ferrara, P. (1991). Molecular cloning and chromosomal localization of a human peripheral-type benzodiazepine receptor. *Eur. J. Biochem.* **195**, 305–311.

Ritta, M.N. and Calandra, R.S. (1989). Testicular interstitial cells as targets for peripheral benzodiazepines. *Neuroendocrinology* **49**, 262–266.

Ritta, M.N., Campos, M.B. and Calandra, R.S. (1987). Effect of GABA and benzodiazepines on testicular androgen production. *Life Sci.* **40**, 791–798.

Sharpe, R.M. (1986). Paracrine control of the testis. *Clin. Endocrinol. Metab.* **15**, 185–207.

Shiraishi, T., Black, K.L. and Ikezaki, K. (1990). Peripheral benzodiazepine receptor ligands induce morphological changes in mitochondria of cultured glioma cells. *Soc. Neurosci. Abst.* **16**(1), 498 (Abstr.).

Simpson, E.R., McCarthy, J.L. and Peterson, J.A. (1978). Evidence that the cycloheximide-sensitive site of adrenocorticotropic hormone action is in the mitochondrion. *J. Biol. Chem.* **253**, 3135–3139.

Slobodyansky, E., Guidotti, A., Wambebe, C., Berkovich, A. and Costa, E. (1989). Isolation and characterization of a rat brain triakontatetraneuropeptide, a post-translational product of diazepam binding inhibitor: specific action at the Ro5-4864 recognition site. *J. Neurochem.* **53**, 1276–1284.

Sprengel, R., Werner, P., Seeburg, P.H., Mukhin, A.G., Santi, M.R., Grayson, D.R., Guidotti, A. and Krueger, K.E. (1989). Molecular cloning and expression of cDNA encoding a peripheral-type benzodiazepine receptor. *J. Biol. Chem.* **264**, 20415–20421.

Verma, A. and Snyder, S.H. (1988). Characterization of porphyrin interactions with peripheral type benzodiazepine receptors. *Mol. Pharmacol.* **34**, 800–805.

Verma, A., Nye, J.S. and Snyder, S.H. (1987). Porphyrins are endogenous ligands for the mitochondrial (peripheral-type) benzodiazepine receptor. *Proc. Natl. Acad. Sci. USA* **84**, 2256–2260.

Warner, M., Tollet, P., Stromstedt, M., Carlstrom, K. and Gustafsson, J.-A. (1989). Endocrine regulation of cytochrome P-450 in the rat brain and pituitary gland. *J. Endocrinol.* **122**, 341–349.

Wikvall, K. (1984). Hydroxylations in biosynthesis of bile acids. *J. Biol. Chem.* **259**, 3800–3804.

Wong, H.Y.C., Nightingdale, T.E., Patel, D.J., Richardi, J.C., Johnson, F.B., Orimilikwe, S.O. and David, S.N. (1980). Long-term effects of diazepam on plasma lipids and atheroma in roosters fed an atherogenic diet. *Artery* **7**, 496–508.

Yanagibashi, K., Ohno, Y., Kawamura, M. and Hall, P.F. (1988). The regulation of intracellular transport of cholesterol in bovine adrenal cells: purification of a novel protein. *Endocrinology* **123**, 2075–2082.

Yanagibashi, K., Phno, Y., Nakamichi, N., Matsui, T., Hayashida, K., Takamura, M., Yamada, K., Tou, S. and Kawamura, M. (1989a). Diazepam potentiates the corticoidogenic response of bovine adrenal fasciculata cells to dibutyryl cyclic AMP. *Jpn. J. Pharmacol.* **51**, 347–355.

Yanagibashi, K., Ohno, Y., Nakamichi, N., Matsui, T., Hayashida, K., Takamura, M., Yamada, K., Tou, S. and Kawamura, M. (1989b). Peripheral-type benzodiazepine receptors are involved in the regulation of cholesterol side chain cleavage in adrenocortical mitochondria. *J. Biochem.* **106**, 1026–1029.

Zafra, M.F., Fernandez-Becerra, M., Castillo, M., Burgos, C. and Garcia-Peregrin, E. (1991). Hypolipidemic activity of dipyridamole: effects on the main regulatory enzymes of cholesterogenesis. *Life Sci.* **49**, 15–21.

CHAPTER 5

EFFECTS OF PERIPHERALLY ACTING BENZODIAZEPINES ON CELL GROWTH AND DIFFERENTIATION

Veijo Saano

Department of Pharmacology and Toxicology, University of Kuopio, P.O. Box 1627, SF-70211 Kuopio, Finland

Table of Contents

5.1 Introduction

From time to time, the indiscriminate use of benzodiazepines has raised worries as to their possible teratogenic and carcinogenic effects. Epidemiological data have been interpreted to suggest that diazepam may act as a tumour promoter in man (Horrobin, 1981a,b). These claims have been refuted (Jackson and Harris, 1981; D'Arcy, 1982), and diazepam use has even been shown to protect from progression of breast cancer (Kleinerman et al., 1984). However, results from animal experiments concerning diazepam-induced teratogenic defects (Nagele et al., 1981) have maintained interest in the topic. Some recent studies have also emphasized the potential genotoxicity of benzodiazepines (Lafi and Parry, 1988).

The pathogenesis of teratogenic effects arises from disturbances in the fundamental phases of development, such as mitosis and cell differentiation (Regan et al., 1990) and benzodiazepines have indeed been shown to have effects on mitosis and cell differentiation (Ober, 1974; Andersson et al., 1981).

5.2 Benzodiazepine receptors

Therapeutically, the most important actions of benzodiazepines, e.g. anxiolytic and anticonvulsive effects, are mediated by the central-type benzodiazepine receptors (CBR; see e.g. Costa and Guidotti, 1991) which bind these compounds with a very high affinity. Low-affinity neuronal benzodiazepine binding sites, also known as micromolar benzodiazepine receptors, mediate the benzodiazepine-induced inhibition of presynaptic Ca^{2+} influx and Ca^{2+}-dependent neurotransmitter release (Bowling and DeLorenzo, 1982; Taft and DeLorenzo, 1984; Johansen et al., 1985). Considerably higher, micromolar concentrations of benzodiazepines are needed to produce these effects.

Mitochondrial benzodiazepine receptors (MBR), also known as peripheral benzodiazepine receptors and as omega$_3$ receptors (Langer and Arbilla, 1988), are the third binding site for benzodiazepines (see Saano et al., 1989). The existence of three types of benzodiazepine binding sites raises the question as to their roles in the actions of benzodiazepines on cell growth and differentiation.

5.2.1 Mitochondrial benzodiazepine receptors in proliferating cells

The potencies of benzodiazepines to alter cell proliferation do not correlate with their activities in the central nervous system (Miernik et al., 1986). The affinities of benzodiazepines for these binding sites are very high, with K_i values in the nanomolar range. The effects of benzodiazepines on cell proliferation have been observed at micromolar concentrations (e.g. Ober, 1974). This suggests that the neuronal, central-type benzodiazepine receptors of the GABA-regulated chloride channel receptor complex are not involved in the actions of benzodiazepines on cell growth. It is also thought that the relatively little known neuronal micromolar benzodiazepine receptor probably does not mediate the antiproliferative effects of benzodiazepines (Gorman *et al.*, 1989).

The possible role of MBR in the actions of benzodiazepines is more intriguing. Considering the possible significance of the MBR on cell growth and differentiation, blood cells and tissues that are involved in the production of blood cells are among the most interesting cell types. MBR are found in lymphocytes (Moingeon *et al.*, 1983), mast cells (Taniguchi *et al.*, 1980), granulocytes (Bond *et al.*, 1985), macrophages (Zavala *et al.*, 1984) and platelets (Wang *et al.*, 1980; Benavides *et al.*, 1984; Moingeon *et al.*, 1984), as well as in the rapidly regenerating epithelial tissue of the olfactory bulbs, nose and salivary glands (Anholt *et al.*, 1985b).

Very high densities of the MBR are also found in steroidogenic tissues such as the adrenal cortex and the testes (Anholt *et al.*, 1985a; De Souza *et al.*, 1985; Mukhin *et al.*, 1989; Papadopoulos *et al.*, 1990). MBR in these tissues seem to be located in cells which are trophically governed by the hypophysis because hypophysectomy causes depletion of MBR (Anholt *et al.*, 1985a).

Other tissues rich in MBR include several epithelia. These localizations fit with actions of peripherally acting benzodiazepines on hormonal secretion and tissue differentiation (Anholt, 1986; Cook *et al.*, 1979; Marc and Marselli, 1969; Shibata *et al.*, 1983).

5.2.2 Mitochondrial benzodiazepine receptors in malignant cells

Several malignant cell lines (e.g. brain tumours in rat and man) contain MBR (Black *et al.*, 1990). Both rat astrocytoma (Strittmatter *et al.*, 1979) and neuroblastoma (Larcher *et al.*, 1989) cells possess a large number of these binding sites. High-affinity MBR have been found in rat basophilic leukaemia (RBL-1) cells (Miller *et al.*, 1988), rat Nb2 lymphoma cells (Laird *et al.*, 1989; Gerrish *et al.*, 1990), as well as in rat (Syapin and Skolnick, 1979; Gallager *et al.*, 1981; Majewska and Chuang, 1985; Starosta-Rubinstein *et al.*, 1987) and human glial neoplasms (Benavides *et al.*, 1988; Olson *et al.*, 1988). Also thymoma (Wang *et al.*, 1984), mouse neuro-2A neuroblastoma (Gorman *et al.*, 1989), rat PC12 phaeochromocytoma (Morgan *et al.*, 1985), mouse Friend erythroleukaemia cells (Clarke and Ryan, 1980) and melanoma cells (Matthew *et al.*, 1981) contain MBR.

Our studies showed a relatively high density of MBR in cultured mouse T-cell lymphoma cells EL-4 and R1.1, as well as in cultured mouse lymphoma cells L-1210. On the other hand, the cultured human Burkitt lymphoma cell lines Raji, Daudi and Ramos, the human acute lymphoblastic leukaemia cell line MOLT-4, and the human promyelocyte leukaemia cell line HL-60 did not differ in their [^{3}H]Ro5-4863 binding from normal human lymphocytes in peripheral blood, which have relatively low number of MBR (V. Saano and Kajander, O. unpubl. res.).

5.3 Effects of benzodiazepines on cell cycle and differentiation: mediation via MBR?

Already early results from primitive organisms, such as *Scenedesmus obliquus* (Ober, 1974) and *Dunaliella* (Marano *et al.*, 1984) show that diazepam, often considered to be a representative of the large group of 1,4-benzodiazepines, inhibits cell proliferation. There are relatively few data on the effects of selectively peripherally acting benzodiazepines on cell differentiation. Clarke and Ryan (1980) showed that diazepam and temazepam induce cell differentiation in Friend erythroleukaemia cells, but several other benzodiazepines are inactive.

Benzodiazepines affect mainly the proliferation of somatic cells (e.g. cultured thymoma cells; Wang *et al.*, 1984) with the main effects focusing on mitosis. Meiosis of oocytes seems to be inhibited, but at diazepam concentrations that are probably too high to be of clinical significance (Stenchever and Smith, 1981).

Diazepam blocks mitogenesis in 3T3 cells (Clarke and Ryan, 1980), according to Andersson *et al.* (1981), by arresting mitosis in prometaphase. This finding has been confirmed, and the effect has been shown to be reversible (Hsu *et al.*, 1983). Also induction of abnormal mitosis has been observed (Callaini *et al.*, 1989). Diazepam decreases thymidine incorporation in mouse spleen lymphocytes (Pawlikowski *et al.*, 1986). Although diazepam has high affinity for CBR and MBR (Saano, 1986), the effects on mitosis seem to be mediated by MBR since they have also been observed using cells that possess MBR only. In human glioma cells (Pawlikowski *et al.*, 1988a), peripherally acting benzodiazepines were slightly more effective in suppressing [³H]thymidine incorporation than other benzodiazepines.

5.3.1 Inhibition of mitosis by benzodiazepines

The data about the effects of diazepam on mitosis are, however, somewhat conflicting. Pawlikowski *et al.* (1987) showed that diazepam decreases mitotic activity in rat cerebral cortex and pituitary gland, but increases it in thymus. Moreover, the mediator of the observed effects is unclear.

The micromolar concentrations of diazepam needed to inhibit mitosis have been too high with regard to the high affinity of diazepam for MBRs. Additionally, diazepam-induced arrest of mitosis cannot be antagonized with

Ro5-4864, a well-known high-affinity ligand of MBR (Marano *et al.*, 1984) and an antagonist in certain cases. In thymus, however, the effects of diazepam (increase in cell proliferation) and Ro5-4864 (decrease) were antagonistic (Stepien *et al.*, 1988). Flumazenil, a selective antagonist of CBR, was inactive.

In the study of Miernik *et al.* (1986), only diazepam and medazepam caused abnormal mitotic figures. Flunitrazepam, in spite of its high affinity for the MBR (Saano, 1986), did not, although it prevented cell proliferation (Miernik *et al.*, 1986). Pawlikowski *et al.* (1988a,b) showed incorporation of thymidine to be inhibited in glioma cells by diazepam and Ro5-4864. Unexpectedly, it was also inhibited by clonazepam, a benzodiazepine with very low affinity for MBR. These effects were not antagonized by flumazenil.

Although Wang *et al.* (1984) studied 15 benzodiazepines and found that affinity for MBR positively correlated with the inhibition of thymoma cell proliferation, Gorman *et al.* (1989) found that the antiproliferative action of benzodiazepines is not mediated through the MBR. In their study on rat C_6 glioma cells and on mouse neuro-2A neuroblastoma cells, MBR ligands diazepam, Ro5-4864 and PK 11195 inhibited cell proliferation, but micromolar concentrations were required. At these concentrations the non-MBR ligand clonazepam also exhibited inhibitory effects. Moreover, the inhibition was similar in SP2 hybridoma and NCTC epithelial cell cultures which are devoid of MBR.

Inhibition of mitosis by high concentrations of diazepam has another interesting property: in male rats this action is abolished by adrenalectomy (Zieleniewski *et al.*, 1990a,b).

5.3.2 Induction of mitosis by low concentrations of mitochondrial benzodiazepine receptor ligands

These findings give a perplexing picture of the role of the MBR in cell proliferation. Indeed, recent studies have shown that low concentrations (10^{-9} M) of diazepam induce mitoses and cell proliferation (Stepien *et al.*, 1988). The MBR ligand PK 11195 also has a similar effect: in rat C_6 glioma cells it stimulates the growth rate and [^{3}H]thymidine incorporation at 10 nM concentrations. At the same time, MBR are downregulated. Higher concentrations of PK 11195 inhibit DNA synthesis (Ikezaki and Black, 1990).

These results suggest that the MBR are, after all, involved in the actions of peripherally acting benzodiazepines and isoquinolines on mitosis. Recent autoradiographic findings by Rägo *et al.* (1990) might support the theory. Their data show that tritiated MBR ligand Ro5-4864 is either bound to a MBR located inside the nucleus of a lymphocyte, or the receptor–ligand complex is transported rapidly after binding from the outer nuclear membrane to inside the nucleus.

Also the indirect actions of peripherally acting benzodiazepines may be mediated through the MBR: Stepien *et al.* (1986) showed that Ro5-4864 increases oestradiol-induced DNA synthesis, but the CBR ligand flumazenil does not. In a study on NB2 node lymphoma cells, 10^{-6} M Ro5-4864 inhibited and 10^{-9} M enhanced the mitogenic action of prolactin; the isoquinoline PK 11195 had similar effects, but clonazepam was without activity and affinity (Laird *et al.*, 1989). It is interesting to note that in addition to these effects, MBR ligands stimulate the release of prolactin (Calogero *et al.*, 1990). Part of the ambiguity about the role of the MBR in the mediation of benzodiazepine-induced effects on mitosis may be due to the existence of receptor subtypes (Mukhin *et al.*, 1990).

Although benzodiazepines show selectivity between CBR and MBR, it is possible that the variation between subpopulations within the MBR in the binding site structure makes a receptor in a particular peripheral tissue appear as a central-type receptor to some ligands. This could explain why, for example, clonazepam in some tissues exerts effects that one might expect only from compounds with affinity for the MBR, such as diazepam.

5.3.3 Mitochondrial benzodiazepine receptors and malignant cell growth

On cultured cells, benzodiazepines exert a variety of slowly appearing effects which may be of importance with regard to promotion of malignancy and malignant cell proliferation. Melanoma cells possess MBR, and melanogenesis in these cells can be enhanced with Ro5-4864 (Matthew *et al.*, 1981). Ro5-4864 stimulates the phospholipid methylation in C6 astrocytoma cells (Strittmatter *et al.*, 1979). Benzodiazepines induce chemotaxis of human monocytes (Ruff *et al.*, 1985) and stimulate humoral immune response (Zavala *et al.*, 1984). MBR ligands have been reported to enhance the respiratory burst of macrophage-like P388D$_1$ cells stimulated by arachidonic acid; this action can be antagonized with PK 11195 (Zavala and Lenfant, 1987). Since O$_2$-radical production is a biochemical indicator of the defensive activity of macrophages against malignant cell growth, the potential immunopharmacological usefulness of MBR ligands is of great interest.

Diazepam has been shown to suppress the X-ray induced transformation in mouse C3H10T1/2 cells (Radner and Kennedy, 1990). Diazepam and clonazepam have been shown to be antiproliferative, but not cytotoxic (Regan *et al.*, 1990). Peripherally acting benzodiazepines suppress the growth of human melanoma cells (Solowey *et al.*, 1990). Dipyridamole, a non-BZ with low affinity for MBR, has been used as an antimelanoma agent with some success (Rhodes *et al.*, 1985); dipyridamole also enhances the cytotoxic action of methotrexate (Kennedy *et al.*, 1986) and 5-fluorouracil (Grem and Fischer,

1989). These findings may indicate a pharmacological effect related to interaction with MBR. In some cases, however, the potency of effect does not correlate with the affinity for MBR; e.g. inhibition of neurite outgrowth in PC12 cell cultures (Morgan et al., 1985) and benzodiazepine effect on mitotic arrest (Lafi et al., 1987).

On the other hand, diazepam and oxazepam act as promoters of hepatocarcinogenesis in rats (Diwan et al., 1986; Preát, 1987). This action may not be mediated through the MBR, as the affinity of oxazepam for the MBR is, however, markedly lower than that of diazepam (K_i values 36.8 and 0.013 nM, respectively; Saano, 1986).

5.4 Actions of putative endogenous mitochondrial benzodiazepine receptor ligands: direct effects and potentiation of other effectors

The role of the group of proposed endogenous MBR ligands, porphyrins (Verma et al., 1987), is still unclear, Protoporphyrin IX, which shows the highest affinity for the MBR (Verma and Snyder, 1988), does not affect prolactin-induced mitogenesis in NB2 lymphocytes (Gerrish et al., 1990). In mouse spleen lymphocytes, however, deuteroporphyrin IX, mesoporphyrin IX, protoporphyrin IX and haematoporphyrin concentration-dependently inhibit [^{3}H]thymidine incorporation (Stepien et al., 1991). Some porphyrins have been shown to have cytotoxic effects (Pinelli et al., 1987).

Most of the actions of peripherally acting benzodiazepines that seem to be biologically relevant are not due to direct effects, but involve modulation of actions of other substances. For instance, in PC-12 phaeochromocytoma cells, nerve growth factor enhances expression of the c-fos oncogene (Curran and Morgan, 1985). Ro5-4864 in nanomolar concentrations markedly augments this effect. PK 11195 which alone does not influence oncogene expression, blocks in these conditions the actions of Ro5-4864. PK 11195 inhibits the chemotaxis of monocytes; diazepam and other benzodiazepines induce it (Ruff et al., 1985). MBR ligands enhance the action of various recombinant human interferons on human melanoma cells (Solowey et al., 1990). Treatment with benzodiazepines inhibits the production of tumour necrosis factor (Zavala et al., 1990).

Chronic treatment with non-benzodiazepines which do not directly interact with the MBR, e.g. barbiturates and antipsychotics, affects the MBR (Gavish et al., 1986b; Weizman et al., 1989). The possible significance of these actions remains to be studied. Psychiatric disturbances such as anxiety and stress have been suggested to affect malignant cell growth; at least they affect the MBR (Weizman *et al.*, 1987; Rägo *et al.*, 1989). Even more important with respect to malignant cell growth is the effect of hormone therapy, e.g. with oestrogens which modulates the MBR (Gavish *et al.*, 1986a).

Non-benzodiazepine ligands with an affinity for the MBR are also of interest in MBR research. Azapropazone has, as does indomethacin, some affinity for MBR, while acetylsalicyclic acid does not (Saano, 1986). Azapropazone inhibits the migration of neutrophiles (Mousa *et al.*, 1990). Indomethacin enhances the antitumour action of interferons for P-388 leukaemia and MC-26 colon carcinoma in mice; acetylsalicyclic acid is inactive (Sidky *et al.*, 1990).

Connections between the MBR and other receptors may also reveal interesting possibilities in cancer therapy or diagnostics. Benzodiazepines antagonize the effects of cholecystokinin (Meldrum *et al.*, 1986). A cholecystokinin antagonist and benzodiazepine analogue, L364 718, decreases cell proliferation (Thumwood *et al.*, 1991) resembling the cholecystokinin antagonists proglumide and benzotript, which have been shown to inhibit colon carcinoma cell proliferation (Hoosein *et al.*, 1988). Blockade of cholecystokinin receptors with a benzodiazepine analogue inhibits mitogenesis (Ferrara *et al.*, 1990).

5.5 Conclusions

The partly conflicting data about the effects of benzodiazepines on cell growth indicate that all benzodiazepines are not identical in their actions. At high drug concentrations, their antimitotic activity seems to be independent of their selectivity for different receptor types. At low concentrations, MBR ligands exert a mitosis-inducing effect which seems to be mediated by the MBR. The selectivity of various compounds in their binding is also exhibited as differences in effects. At least at the moment it is impossible to classify some compounds as agonists and others as antagonists with regard to their actions on mitosis through the MBR.

Although the possible connections between the MBR and the anticancer or procarcinogenic drug effects are yet not known well enough to warrant clinical applications, the MBR binding promises benefits in imaging of brain tumours (Ikezaki *et al.*, 1990a,b) and other diseases involving the brain (Benavides *et al.*, 1987, 1988).

Acknowledgements

I thank Dr Lembit Rägo from Tartu University, Estonia, and Drs Olavi Kajander and Ewen MacDonald from the University of Kuopio, for valuable comments.

References

Andersson, L.C., Lehto, V.-P., Stenman, S., Badley, R.A. and Virtanen, I. (1981). Diazepam induces mitotic arrest at prometaphase by inhibiting centriolar separation. *Nature* **291**, 247–248.

Anholt, R.R.H. (1986). Mitochondrial benzodiazepine receptors as potential modulators of intermediary metabolism. *Trends Pharmacol. Sci.* **7**, 506–511.

Anholt, R.R.H., De Souza, E.B., Kuhar, M.J. and Snyder, S.H. (1985a). Depletion of peripheral-type benzodiazepine receptors after hypophysectomy in rat adrenal gland and testis. *Eur. J. Pharmacol.* **110**, 41–46.

Anholt, R.R.H., De Souza, E.B., Oster-Granite, M.L. and Snyder, S.H. (1985b). Peripheral-type benzodiazepine receptors: autoradiographic localization in whole-body sections of neonatal rats. *J. Pharmacol. Exp. Ther.* **233**, 517–526.

Benavides, J., Fage, D., Carter, C. and Scatton, B. (1987). Peripheral-type benzodiazepine binding sites are a sensitive indirect index on neuronal damage. *Brain Res.* **421**, 167–172.

Benavides, J., Cornu, P., Dennis, T., Dubois, A., Hauw, J.-J., MacKenzie, E.T., Sazdovitch, V. and Scatton, B. (1988). Imaging of human brain lesions with an ω_3 site radioligand. *Ann. Neurol.* **24**, 708–712.

Benavides, J., Quarteronet, D., Plouin, P.-F., Imbault, F., Phan, T., Uzan, A., Renault, C., Dubroeucq, M.-C., Gueremy, C. and Le Fur, G. (1984). Characterization of peripheral type benzodiazepine binding sites in human and rat platelets by using [^{3}H] PK 11195. Studies in hypertensive patients. *Biochem. Pharmacol.* **33**, 2467–2472.

Black, K.L., Ikezaki, K., Santori, E., Becker, D.P. and Vinters, H.V. (1990). Specific high-affinity binding of peripheral benzodiazepine receptor ligands to brain tumors in rat and man. *Cancer* **65**, 93–97.

Bond, P.A., Cundall, R.L. and Rolfe, B. (1985). [^{3}H]Diazepam binding to human granulocytes. *Life Sci.* **37**, 11–16.

Bowling, A.C. and DeLorenzo, R.J. (1982). Micromolar affinity benzodiazepine receptors: identification and characterization in central nervous system. *Science* **216**, 1247–1250.

Callaini, G., Dallai, R. and Riparbelli, M.G. (1989). Diazepam induces abnormal mitosis in the early *Drosophila* embryo. *Biol. Cell* **67**, 313–320.

Calogero, A.E., Kamilaris, T.C., Johnson, E.O., Gold, P.W. and Chrousos, G.P. (1990). 'Peripheral' benzodiazepine receptor ligands stimulate prolactin release in the rat. *J. Neuroendocrinol.* **2**, 745–750.

Clarke, G.D. and Ryan, P.J. (1980). Tranquillizers can block mitogenesis in 3T3 cells and induce differentiation in Friend cells. *Nature* **287**, 160–161.

Cook, P.S., Notelovitz, M., Kalka, P.S. and Kalka, S.P. (1979). Effect of diazepam on serum testosterone and the ventral prostate gland in male rats. *Arch. Androl.* **3**, 31–35.

Costa, E. and Guidotti, A. (1991). Diazepam binding inhibitor (DBI): a peptide with multiple biological actions. *Life Sci.* **49**, 325–334.

Curran, T. and Morgan, J.I. (1985). Superinduction of c-fos by nerve growth factor in the presence of peripherally active benzodiazepines. *Science* **229**, 1265–1268.

D'Arcy, P.F. (1982). Drug reactions and interactions. Diazepam and breast cancer: no real evidence of incrimination. *Pharm. Int.* Sept., 285–286.

De Souza, E.B., Anholt, R.R.H., Murphy, K.M.M., Snyder, S.H. and Kuhar, M.J. (1985). Peripheral-type benzodiazepine receptors in endocrine organs: autoradiographic localization in rat pituitary adrenal and testis. *Endocrinology* **116**, 567–573.

Diwan, B.A., Rice, J.M. and Ward, J.M. (1986). Tumor-promoting activity of benzodiazepine tranquilizers, diazepam and oxazepam, in mouse liver. *Carcinogenesis* **7**, 789–794.

Ferrara, A., McMiller, M.A., Schaefer, H.C., Zucker, K.A. and Modlin, I.M. (1990). Effect of cholecystokinin receptor blockade on human lymphocyte proliferation. *J. Surg. Res.* **48**, 354–357.

Gallager, D.W., Mallorga, P., Oertel, W., Henneberry, R. and Tallman, J. (1981). ^{3}H-Diazepam binding in mammalian central nervous system: a pharmacological characterization. *J. Neurosci.* **1**, 218–225.

Gavish, M., Okun, F., Weizman, A. and Youdim, M.B.H. (1986a). Modulation of peripheral benzodiazepine binding sites following chronic estradiol treatment. *Eur. J. Pharmacol.* **127**, 147–151.

Gavish, M., Weizman, A., Karp, L., Tyano, S. and Tanne, Z. (1986b). Decreased peripheral benzodiazepine binding sites in platelets of neuroleptic-treated schizophrenics. *Eur. J. Pharmacol.* **121**, 275–279.

Gerrish, K.E., Putnam, C.W. and Laird, H.E. (1990). Prolactin-stimulated mitogenesis in the NB2 rat lymphoma cell: lack of protoporphyrin IX effects. *Life Sci.* **47**, 1647–1653.

Gorman, A.M.C., O'Beirne, G.B., Regan, C.M. and Williams, D.C. (1989). Antiproliferative action of benzodiazepines in cultured brain cells is not mediated through the peripheral-type benzodiazepine acceptor. *J. Neurochem.* **53**, 849–855.

Grem, J.L. and Fischer, P.H. (1989). Enhancement of 5-fluorouracil's anticancer activity by dipyridamole. *Pharmacol. Ther.* **40**, 349–371.

Hoosein, N.M., Kiener, P.A., Curry, R.C., Rovati, L.C., McGilbra, D.K. and Brattain, M.G. (1988). Antiproliferative effects of gastrin receptor antagonists and antibodies to gastrin on human colon carcinoma cell lines. *Cancer Res.* **48**, 7179–7183.

Horrobin, D.F. (1981a). Diazepam as tumour promoter. *Lancet* **i**, 277–278.

Horrobin, D.F. (1981b). The possible effect of diazepam on cancer development and growth. *Med. Hypoth.* **7**, 115–125.

Hsu, T.C., Liang, J.C. and Shirley, L.R. (1983). Aneuploidy induction by mitotic arrestants. Effects of diazepam on diploid Chinese hamster cells. *Mutation Res.* **122**, 201–209.

Ikezaki, K. and Black, K.L. (1990). Stimulation of cell growth and DNA synthesis by peripheral benzodiazepine. *Cancer Lett.* **49**, 115–120.

Ikezaki, K., Black, K.L., Santori, E.M., Smith, M.L., Becker, D.P., Payne, B.A. and Toga, A.W. (1990a). Three-dimensional comparison of peripheral benzodiazepine binding and histological findings in rat brain tumor. *Neurosurgery* **27**, 78–82.

Ikezaki, K., Black, K.L., Toga, A.W., Santori, E.M., Becker, D.P. and Smith, M.L. (1990b). Imaging peripheral benzodiazepine receptors in brain tumors in rats: *in vitro* binding characteristics. *J. Cereb. Blood Flow Metab.* **10**, 580–587.

Jackson, M.R. and Harris, P.A. (1981). Diazepam and tumour promotion. *Lancet* **i**, 445.

Johansen, J., Taft, W.C., Yang, J., Kleinhaus, A.L. and DeLorenzo, R.J. (1985). Inhibition of Ca^{2+} conductance in identified leech neurons by benzodiazepines. *Proc. Natl. Acad. Sci. USA* **82**, 3935–3939.

Kennedy, D.G., Van den Berg, H.W., Clarke, R. and Murphy, R.F. (1986). Enhancement of methotrexate cytotoxicity towards the MDA.MB.436 human breast cancer cell line by dipyridamole. The role of methotrexate polyglutamates. *Biochem. Pharmacol.* **35**, 3053–3056.

Kleinerman, R.A., Brinton, L.A., Hoover, R. and Fraumeni, J.F. Jr. (1984). Diazepam use and progression of breast cancer. *Cancer Res.* **44**, 1223–1225.

Lafi, A. and Parry, J.M. (1988). A study of the induction of aneuploidy and chromosome aberrations after diazepam, medazepam, midazolam and bromazepam treatment. *Mutagenesis* **3**, 23–27.

Lafi, A., Parry, E.M. and Parry, J.M. (1987). The effects of benzodiazepines upon the fidelity of mitotic cell division in cultured Chinese hamster cells. *Mutation Res.* **189**, 319–332.

Laird II, H.E., Gerrish, K.E., Duerson, K.C., Putnam, C.W. and Russell, D.H. (1989). Peripheral benzodiazepine binding sites in Nb 2 node lymphoma cells: effects on prolactin-stimulated proliferation and ornithine decarboxylase activity. *Eur. J. Pharmacol.* **171**, 25–35.

Langer, S.Z. and Arbilla, S. (1988). Limitations of the benzodiazepine receptor nomenclature: a proposal for a pharmacological classification as omega receptor subtypes. *Fundam. Clin. Pharmacol.* **2**, 159–170.

Larcher, J.-C., Vayssier, J.-L., Le Marquer, F.J., Cordeau, L.R., Keane, P.E., Bachy, A., Gros, F. and Croizat, B.P. (1989). Effects of peripheral benzodiazepines upon the O_2 consumption of neuroblastoma cells. *Eur. J. Pharmacol.* **161**, 197–202.

Majewska, M.D. and Chuang, D.M. (1985). Benzodiazepines enhance the muscimol-dependent activation of phospholipase A_2 in glioma C_6 cells. *J. Pharmacol. Exp. Ther.* **232**, 650–655.

Marano, R., Santa-Maria, A. and Fries, W. (1984). Effects of diazepam on mitosis and basal body duplication of synchronously dividing flagellate cells. *Biol. Cell* **50**, 163–172.

Marc, V. and Marselli, P.L. (1969). Effect of diazepam on plasma corticosterone levels in the rat. *J. Pharm. Pharmacol.* **21**, 784–786.

Matthew, E., Laskin, J.D., Zimmerman, E.A., Weinstein, I.B., Hsu, K.D. and Englehardt, D.L. (1981). Benzodiazepines have high-affinity binding sites and induce melanogenesis in B16/C3 melanoma cells. *Proc. Natl. Acad. Sci. USA* **78**, 3935–3939.

Meldrum, L.A., Bojarski, J.C. and Calam, J. (1986). Effects of benzodiazepines on responses of guinea-pig ileum and gallbladder and rat pancretic acini to cholecystokinin. *Eur. J. Pharmacol.* **123**, 427–432.

Miernik, A., Santa-Maria, A. and Marano, F. (1986). The antimitotic activities of some benzodiazepines. *Experientia* **42**, 956–958.

Miller, L.G., Lee-Paritz, A., Greenblatt, D.J., Theoharides, T.C. (1988). High-affinity benzodiazepine binding sites on rat peritoneal mast cells and RBL-1 cells: binding characteristics and effects of granule secretion. *Pharmacology* **36**, 52–60.

Moingeon, Ph., Bidart, J.M., Alberici, G.F. and Bohuon, C. (1983). Characterization of a peripheral-type benzodiazepine binding site on human circulating lymphocytes. *Eur. J. Pharmacol.* **92**, 147–149.

Moingeon, Ph., Dessaux, J.J., Fellous, R., Alberici, G.F., Bidart, J.M., Motté, Ph. and Bohoun, C. (1984). Benzodiazepine receptors on human blood platelets. *Life Sci.* **35**, 2003–2009.

Morgan, J.I., Johnson, M.D., Wang, J.K.T., Sonnenfeld, K.H. and Spector, S. (1985). Peripheral-type benzodiazepines influence ornithine decarboxylase levels and neurite outgrowth in PC12 cells. *Proc. Natl. Acad. Sci. USA* **82**, 5223–5226.

Mousa, S.A., Brown, R., Thoolen, M.J.M. and Smith, R.D. (1990). Evaluation of the effect of azapropazone on neutrophil migration in regional myocardial ischemia/reperfusion injury in rabbits. *Br. J. Pharmacol.* **100**, 379–382.

Mukhin, A.G., Papadopoulos, V., Costa, E. and Krueger, K.E. (1989). Mitochondrial benzodiazepine receptors regulate steroid biosynthesis. *Proc. Natl. Acad. Sci. USA* **86**, 9813–9816.

Mukhin, A.G., Zhong, P. and Krueger, K.E. (1990). Cofractionation of the 17-kD

PK14105 binding site protein with solubilized peripheral-type benzodiazepine binding sites. *Biochem. Pharmacol.* **40**, 983–989.

Nagele, R.G., Pietrolungo, J.F., Lee, H. and Roisen, F. (1981). Diazepam-induced neural tube closure defects in explanted early chick embryos. *Teratology* **23**, 343–349.

Ober, K. (1974). Effects of diazepam on cell division rates and productivity of *Scenedesmus obliquus* in synchronous cultures. *Arch. Microbiol.* **99**, 369–378.

Olson, J.M., Junk, L., Young, A.B., Penny, J.B. and Mancini, W.R. (1988). Isoquinoline and peripheral-type benzodiazepine binding in gliomas: implications for diagnostic imaging. *Cancer Res.* **48**, 5837–5841.

Papadopoulos, V., Mukhin, A.G., Costa, E. and Krueger, K.E. (1990). The peripheral-type benzodiazepine receptor is functionally linked to Leydig cell steroidogenesis. *J. Biol. Chem.* **265**, 3772–3779.

Pawlikowski, M., Stepien, H. and Kunert-Radek, J. (1986). Diazepam inhibits proliferation of the mouse spleen lymphocytes *in vitro*. *Pol. J. Pharmacol. Pharm.* **38**, 167–170.

Pawlikowski, M., Stepien, H., Mroz-Wasilewska, Z. and Pawlikowska, A. (1987). Effects of diazepam on cell proliferation in cerebral cortex, anterior pituitary and thymus of developing rats. *Life Sci.* **40**, 1131–1135.

Pawlikowski, M., Kunert-Radek, J., Radek, A. and Stepien, H. (1988a). Inhibition of cell proliferation of human gliomas by benzodiazepines *in vitro*. *Acta Neurol. Scand.* **77**, 231–233.

Pawlikowski, M., Lyson, K., Kunert-Radek, J. and Stepien, H. (1988b). Effect of benzodiazepines on the proliferation of mouse spleen lymphocytes *in vitro*. *J. Neural Transm.* **73**, 161–166.

Pinelli, A., Trivulzio, S., Von Hoff, D.D. and Warfel, L. (1987). Comparison of two methods to evaluate drug-cytotoxicity on tumor cell lines cultured *in vitro*. *Pharmacol. Res. Commun.* **19**, 913–923.

Preát, V., de Gerlache, J., Lans, M. and Roberfroid, M. (1987). Promoting effect of oxazepam in rat hepatocarcinogenesis. *Carcinogenesis* **8**, 97–100.

Radner, B.S. and Kennedy, A.R. (1990). Suppression of X-ray induced transformation by Valium and Aspirin in mouse C3H10T1/2 cells. *Cancer Lett.* **51**, 49–57.

Rägo, L., Kiivet, R.-A., Harro, J. and Pöld, M. (1989). Central- and peripheral-type benzodiazepine receptors: similar regulation by stress and GABA agonists. *Pharmacol. Biochem. Behav.* **32**, 879–883.

Rägo, L., Adojaan, A. and Masso, R. (1990). Location in nucleus of ^{3}H-Ro5-4864 binding sites in rat lymphocytes. *Eur. J. Pharmacol.* **187**, 561–562.

Regan, C.M., Gorman, A.M.C., Larsson, O.M., Maguire, C., Martin, M.L., Schousboe, A. and Williams, D.C. (1990). *In vitro* screening for anticonvulsant-induced teratogenesis in neural primary cultures and cell lines. *Int. J. Dev. Neurosci.* **8**, 143–150.

Rhodes, E.L., Misch, K.J., Edwards, J.M. and Jarrett, P.E.M. (1985). Dipyridamole for treatment of melanoma. *Lancet* **i**, 693.

Ruff, M.R., Pert, C.B., Weber, R.J., Wahl, L.M., Wahl, S.M. and Paul, S.M. (1985). Benzodiazepine receptor-mediated chemotaxis of human monocytes. *Science* **229**, 1281–1283.

Saano, V. (1986). Affinity of various compounds for benzodiazepine binding sites in rat brain, heart and kidneys *in vitro*. *Acta Pharmacol. Toxicol.* **58**, 333–338.

Saano, V., Rägo, L. and Räty, M. (1989). Peripheral benzodiazepine binding sites. *Pharmacol. Ther.* **41**, 503–514.

Shibata, H., Kojima, I. and Ogata, E. (1983). Diazepam inhibits potassium-induced aldosterone secretion in adrenal glomerulosa cells. *Biochem. Biophys. Res. Commun.* **116**, 555–562.

Sidky, Y.A., Borden, E.C., Reilly, S. and Bryan, G.T. (1990). Influence of aspirin and indomethacin on the antitumor effects of interferons (IFNs) for P-388 leukemia and MC-26 colon carcinoma. *Proc. Am. Soc. Clin. Oncol.* **9**, 195.

Solowey, W.E., Pestka, S., Spector, S., Fryer, R.I. and Fisher, P.B. (1990). Peripheral-acting benzodiazepines inhibit the growth of human melanoma cells and potentiate the antiproliferative activity of recombinant human interferons. *J. Interferon Res.* **10**, 269–280.

Starosta-Rubinstein, S., Ciliax, B.J., Penney, J.B., McKeever, P. and Young, A.B. (1987). Imaging of a glioma using peripheral benzodiazepine receptor ligands. *Proc. Natl. Acad. Sci. USA* **84**, 891–895.

Stenchever, M.A. and Smith, W.D. (1981). The effect of diazepam on meiosis in the CF-1 mouse. *Teratology* **23**, 279–281.

Stepien, H., Kunert-Radek, J. and Pawlikowski, M. (1986). Enhancement of estradiol-induced DNA synthesis in the anterior pituitary gland by the peripheral-type benzodiazepine receptor ligand Ro 5-4864. *J. Neural. Transm.* **66**, 303–307.

Stepien, H., Pawlikowski, A. and Pawlikowski, M. (1988). Effects of benzodiazepines on thymus cell proliferation. *Thymus* **12**, 117–121.

Stepien, H., Kunert-Radek, J., Stanisz, A., Zerek-Melen, G. and Pawlikowski, M. (1991). Inhibitory effect of porphyrins on the proliferation of mouse spleen lymphocytes *in vitro*. *Biochem. Biophys. Res. Commun.* **174**, 313–322.

Strittmatter, W.J., Hirata, F., Axelrod, J., Mallorga, P., Tallman, J.F. and Henneberry, R.C. (1979). Benzodiazepine and beta-adrenergic receptor ligands independently stimulate phospholipid methylation. *Nature* **282**, 851–859.

Syapin, P.S. and Skolnick, P. (1979). Characterization of benzodiazepine binding sites in cultured cells of neural origin. *J. Neurochem.* **32**, 1047–1051.

Taft, W.C. and DeLorenzo, R.J. (1984). Micro-molar affinity benzodiazepine receptors regulate voltage-sensitive calcium channels in nerve terminal preparations. *Proc. Natl. Acad. Sci. USA* **81**, 3118–3122.

Taniguchi, T., Wang, J.K.T. and Spector, S. (1980). Properties of [^{3}H]diazepam binding to rat peritoneal mast cells. *Life Sci.* **27**, 171–178.

Thumwood, C.M., Hong, J. and Baldwin, G.S. (1991). Inhibition of cell proliferation by the cholecystokinin antagonist L-364,718. *Exp. Cell Res.* **192**, 189–192.

Verma, A. and Snyder, S.H. (1988). Characterization of porphyrin interactions with peripheral type benzodiazepine receptors. *Mol. Pharmacol.* **34**, 800–805.

Verma, A., Nye, J.S. and Snyder, S.H. (1987). Porphyrins are endogenous ligands for the mitochondrial (peripheral-type) benzodiazepine receptor. *Proc. Natl. Acad. Sci. USA* **84**, 2256–2260.

Wang, J.K.T., Morgan, J.I. and Spector, S. (1984). Benzodiazepines that bind at peripheral sites inhibit cell proliferation. *Proc. Natl. Acad. Sci. USA* **81**, 753–756.

Wang, J.K.T., Taniguchi, T. and Spector, S. (1980). Properties of [^{3}H]diazepam binding sites on rat blood platelets. *Life Sci.* **27**, 1881–1888.

Weizman, A., Fares, F., Pick, C.G., Yanai, J. and Gavish, M. (1989). Chronic phenobarbital administration affects GABA and benzodiazepine receptors in the brain and periphery. *Eur. J. Pharmacol.* **169**, 235–240.

Weizman, A., Tanne, Z., Granek, M., Karp, L., Golomb, M., Tyano, S. and Gavish, M. (1987). Peripheral benzodiazepine binding sites on platelet membranes are increased during diazepam treatment of anxious patients. *Eur. J. Pharmacol.* **1138**, 289–292.

Zavala, F. and Lenfant, M. (1987). Peripheral benzodiazepines enhance the respiratory burst of macrophage-like P388D$_1$ cells stimulated by arachidonic acid. *Int. J. Immunopharmacol.* **9**, 269–274.

Zavala, F., Haumont, J. and Lenfant, M. (1984). Interaction of benzodiazepines with mouse macrophages. *Eur. J. Pharmacol.* **106**, 561–566.

Zavala, F., Taupin, V. and Descamps-Latscha, B. (1990). *In vivo* treatment with benzodiazepines inhibits murine phagocyte oxidative metabolism and production of interleukin 1, tumor necrosis factor and interleukin-6. *J. Pharmacol. Exp. Ther.* **255**, 442–450.

Zieleniewski, J., Nowakowska-Jankiewitz, B. and Stepien, H. (1990a). Influence of diazepam on regenerating adrenal cortex in Wistar rats. *Cytobios* **61**, 85–88.

Zieleniewski, J., Nowakowska-Jankiewitz, B., Stepien, H. and Zieleniewski, W. (1990b). Influence of diazepam on the mitotic activity of adrenocortical cells in rats after unilateral adrenalectomy. *Cytobios* **64**, 81–85.

―――――――――――――― CHAPTER 6 ――――――――――――――
IMMUNOMODULATING EFFECTS OF PERIPHERALLY ACTING BENZODIAZEPINES

Marek Pawlikowski

Institute of Endocrinology, Medical Academy of Lodz, Poland

Table of Contents

6.1 Introduction

Mitochondrial benzodiazepine receptors (MBRs) are present in the central nervous system (mainly on glial cells) as well as on numerous tissues and organs outside the central nervous system (for review, see Verma and Snyder, 1989) and even on circulating blood cells. MBRs have been identified on the lymphoid cells and macrophages (Wang *et al.*, 1981; Moingeon *et al.*, 1983; Lenfant *et al.*, 1985; Benavides *et al.*, 1990) which suggests a possible involvement in the control of the immune system. Table 1 gives a summary of the available binding data. The distribution of MBRs in the immune organs has been investigated by autoradiography and demonstrates their prevalence on T lymphocytes and monocytes (Benavides *et al.*, 1990). There are several lines of evidence that benzodiazepines affect immunocyte growth and function. However, the data available at present are still scarce and sometimes controversial.

PERIPHERAL BENZODIAZEPINE RECEPTORS
ISBN 0-12-282630-2

Table 1   Presence of MBR on immunocompetent cells.

Type of cell	Reference
Lymphocyte (rat)	Rägo et al. (1990)
Lymphocyte (human)	Moingeon et al. (1983)
Nb2 node lymphoma cells (rat)	Laird et al. (1989)
L57/B16 thymocyte (mouse)	Wang et al. (1984)
Splenocyte (mouse)	Wang et al. (1984)
Lymphocyte (human)	Wang et al. (1984)
Peritoneal mast cell (rat)	Miller et al. (1988)
RBL-I mast cell line	Miller et al. (1988)
Peritoneal mast cell (rat)	Taniguchi et al. (1980)
Peritoneal macrophage (mouse)	Zavala et al. (1984)
Granulocyte (human)	Bond et al. (1985)
Thymic cortex (rat spleen, lymph nodes)	Benavides et al. (1990)
Increase of MBR after LPS, IL_1, TNF injection into brain	Bourdiol et al. (1991)
Increase of MBR in rat thymus of arthritic rats	Scatton et al. (1990)

Nanomolar affinities for different lineages of white blood cells and resident lymphocytes have been reported in the literature.

6.2   Effects of MBR ligands on proliferation of lymphoid cells *in vitro*

Stimulatory as well as inhibitory effects of MBR ligands on the proliferative index of immune cells have been reported. In general, immune-stimulating effects were observed with nanomolar concentrations of Ro5-4864, whereas immunosuppressive effects were measured at micromolar concentrations of diazepam or Ro5-4864.

Wang *et al.* (1984) have found that benzodiazepines inhibit the proliferation of murine thymoma cells *in vitro* and their antiproliferative potency is strongly positively correlated with their affinity ranking for MBR; however micromolar concentration had to be used. In 1986, we reported that diazepam, an agent which interacts with both CBR and MBR, inhibits [^{3}H]thymidine incorporation into mouse splenocytes in a dose-dependent manner (Pawlikowski *et al.*, 1986).

Further studies, using the same experimental model but different BZ receptor ligands, showed that the agents with higher affinity to MBR, like Ro5-4864 or PK 11195, exert a stronger antiproliferative effect than substances which selectively bind to CBR (for instance, Ro15-1788 or clonazepam—see Table 1). A carboline derivative, ethyl-β-carboline-3-carboxylate (β-CCE), which acts as an 'inverse agonist' on CBR, failed to affect splenocyte proliferation.

We studied also the effect of a putative endogenous ligand of BZ receptors, octadecaneuropeptide (ODN) (see also Chapter 10). ODN is a peptide generated *in vivo* by the cleavage of larger precursor molecule called diazepam binding inhibitor—DBI (Ferrero *et al.*, 1986). ODN (DBI 33–50) acts as anxiogenic agent and is recognized as an inverse agonist of CBR. ODN failed to influence the spontaneous proliferation of murine splenocytes. Moreover, ODN, when added together with diazepam, did not block the antiproliferative effect of the latter (Lyson *et al.*, 1989), which is in keeping with its CBR selectivity. The effects on lymphoid cell proliferation of another molecule generated from DBI, trikontatetraneuropeptide (TTN, DBI 17–50), which has a higher affinity for MBR than for CBR, have not yet been studied.

Verma *et al.* (1987) have shown that porphyrins bind with submicro-molar affinities to the mitochondrial receptors and proposed that these compounds may be the endogenous ligands for MBR. Recently, we have found that deuteroporphyrin IX, mesoporphyrin IX, protoporphyrin IX and haematoporphyrin exert antiproliferative effects on the mouse splenocytes with EC_{50} values ranging from 9.2×10^{-6} M (deuteroporphyrin IX) to 9.4×10^{-5} M (mesoporphyrin IX) (Stepien *et al.*, 1992).

Low concentrations (10^{-9} M) of Ro5-4864 and PK 11195 potentiate the effect of prolactin on mitogenesis of Nb2 node lymphoma cells (Laird *et al.*, 1989). At higher concentration (10^{-6} M) Ro5-4864 inhibited prolactin-induced proliferation of Nb2 cells, whereas PK 11195 was ineffective at this concentration. Clonazepam was totally ineffective in this system. Interestingly, protoporphyrin IX, a putative endogenous ligand for the MBR, had no effect on prolactin-stimulated mitogenesis of Nb2 cells in spite of the fact that it binds to Nb2 cell MBR (Gerrish *et al.*, 1990). The quoted authors suggest that protoporphyrin IX could be rather an antagonist than agonist of MBR.

Summing up, MBR ligands are found to exert both potentiating and inhibitory effects on the lymphoid cell proliferation (Table 2). The former requires nanomolar, while the latter requires micromolar concentrations of the ligand. The bell-shaped dose–response curve raises the question whether both effects are mediated via the same receptor or via two independent sites. In the first paradigm, the low dose immunostimulatory effect would represent the 'direct', e.g. receptor mediated action which, through prolonged contact or higher concentrations activates a control, feed-back, mechanism and thus exerts at micromolar concentrations immunodepressant effects.

In the second setting, the stimulatory effect would occur via 'classical' MBR, e.g. with nanomolar affinity for its ligands, while the inhibitory effect

Table 2 Antiproliferative potencies of MBR ligands on mouse splenocytes.

Ligand	Type	Range of concentration studied (M)	EC_{50} (M)	Reference
Diazepam	MX	10^{-4}–10^{-10}	5×10^{-6}	Lyson et al. (1989)
Ro5-4864	P	10^{-4}–10^{-10}	9×10^{-6}	Lyson et al. (1989)
PK 11195	P	10^{-4}–10^{-9}	5×10^{-6}	Lyson et al. (1989)
Dipyridamole	P	10^{-5}–10^{-10}	10^{-6}	Unpublished
Deuteroporphyrin IX	P	10^{-4}–10^{-11}	9.2×10^{-6}	Stepien et al. (1991)
Mesoporphyrin IX	P	10^{-4}–10^{-11}	9.4×10^{-5}	Stepien et al. (1991)
Protoporphyrin IX	P	10^{-4}–10^{-10}	8.7×10^{-5}	Stepien et al. (1991)
Haematoporphyrin	P	10^{-4}–10^{-10}	6.6×10^{-5}	Stepien et al. (1991)
Clonazepam	C	10^{-4}–10^{-10}	$>10^{-4}$	Lyson et al. (1989)
Ro15-1788	C	10^{-4}–10^{-8}	$>10^{-4}$	Lyson et al. (1989)
βCCE	C	10^{-4}–10^{-8}	$>10^{-4}$	Lyson et al. (1989)
ODN(DBI 33–50)	C	10^{-5}–10^{-11}	$>10^{-4}$	Lyson et al. (1989)

The types of BZD receptors ligand are denominated as 'mixed' (MX), 'peripheral' (P) and 'central' (C). EC_{50}, concentration which produced 50% inhibition.

would occur via a different binding site. This binding site may be involved in the control of calcium channels (Rampe and Triggle, 1986). It is indeed known that peripherally acting benzodiazepines inhibit plasma membrane calcium influx in micromolar concentrations (Cantor et al., 1984; Taft and De Lorenzo, 1984). Taft and De Lorenzo (1984) suggested the existence of a distinct 'micromolar' binding site for peripherally acting benzodiazepines, connected with plasma membrane and regulating voltage-sensitive calcium channels. On the other hand, the 'peripheral' benzodiazepines may interact with dihydropyridine binding sites. Cantor et al. (1984) have shown that nifedipine, a dihydropyridine calcium channel blocker, displaced tritiated Ro5-4864 from the membranes of the heart, kidney and brain. Also, the existence of an endogenous protein, which modifies both MBR and dihydropyridine calcium channel antagonist binding sites, has been postulated (Mantione *et al.*, 1988). Moreover, dihydropyridines were shown to inhibit mouse spleen lymphocyte proliferation (Kunert-Radek *et al.*, 1990). It is indeed well documented that cell proliferation is a calcium-dependent process (for review, see Lichtman *et al.*, 1983; Gupta and Dudani, 1989).

The antiproliferative effects of 'peripheral' benzodiazepines are not restricted to the lymphoid cells. It has been shown that these compounds inhibit the proliferation of other cell lines, e.g. 3T3 cells (Clarke and Ryan, 1980), human gliomas (Pawlikowski *et al.*, 1988), rat gliomas and neuroblastomas (Gorman *et al.*, 1989), rat pituitary tumoral cells (Kunert-Radek, 1989). There again, micromolar concentration of compound are required whether the 'peripheral' benzodiazepines exert their antiproliferative action via specific 'micromolar' receptors or via unspecific interaction with dihydropyridine binding site is still under discussion. Also, the molecular intermediates of their lympho-mitogenic action, most likely high-affinity receptor-mediated (cytokines, lymphokines?), are unknown.

6.3 *In vitro* effects of benzodiazepines on immune-cell functions

The MBR has been identified on murine macrophages (Zavala *et al.*, 1984; Lenfant, 1989) and the murine cell line P 388D1, which exhibits many of the properties of macrophages (Zavala and Lenfant, 1987). It has been found that Ro5-4864 at concentrations varying from 10^{-9} to 10^{-7} M enhances the oxidative burst of P 388D1 cells evoked by arachidonic acid. The mixed-type ligand, diazepam, was also active, whereas the central-type ligand, clonazepam, was ineffective (Lenfant, 1989). Diazepam and Ro5-4864 in concentrations as low as 10^{-13} M were active in promoting human monocyte chemotaxis *in vitro* (Ruff *et al.*, 1985). Again a bell-shaped concentration curve was observed. Micromolar concentrating of diazepam and Ro5-4864 (but not clonazepam) inhibited *in vitro* superoxide anion generation and chemiluminescence from human neutrophils stimulated by the formylated oligopeptide or by the calcium ionophore (Laghi Pasini *et al.*, 1987). Diazepam was also found to inhibit the phagocytosis of human polynuclear leukocytes and monocytes (Covelli *et al.*, 1989) and human natural killer cell activity (Stepien *et al.*, 1991). The above-mentioned inhibitory effects were obtained with micromolar concentrations of the drug.

6.4 In vivo immunomodulatory actions of benzodiazepines and other MBR ligands

It has been known for several years that high doses of diazepam or long-term treatment depresses the immune response in animals and humans. Descotes *et al.* (1982) found that chronic treatment with high doses of diazepam (8 mg/kg body weight) impaired both humoral and cellular immune responses in mice. In our laboratory we investigated the effects of high doses of diazepam, Ro5-4864 and Ro15-1788 (5 mg/kg body weight) on the mitotic activity of the rat thymus. We found that diazepam increased, whereas Ro5-4864 decreased, the mitotic rate in the rat thymus; Ro15-1788 was without effect (Stepien *et al.*, 1988).

The group of Lichtensteiger has accumulated evidence that prenatal diazepam leads to immunodepression in offspring (Schlumpf *et al.*, 1989). These workers found that the treatment of pregnant rats with diazepam (1.25 mg/kg per day) from gestational days 14 to 20 resulted in a long-lasting depression of cellular immune responses in offspring. Since cellular immune responses in offspring were similarly affected by clonazepam, it seemed that prenatal effects of benzodiazepines on immunity can be mediated via MBR and CBR. However, the recent data reported by the same group indicate a preferential involvement of MBR. This type of receptor is found to develop early in fetal immune organs. The prenatal exposure of rats to diazepam resulted in a change in binding characteristics. Moreover, the prenatal treatment with Ro5-4864 resulted in a long-lasting immunosuppression (Schlumpf *et al.*, 1990).

In contrast, the administration of a lower dose of the drug (1 mg/kg body weight) enhanced the humoral response. Similar stimulation was obtained using Ro5-4864 or PK 11195, more selective MBR ligands. The central-type ligands clonazepam and Ro15-1788 were ineffective (Lenfant *et al.*, 1986). Diazepam and flunitrazepam at low doses (0.5 and 1 mg/kg body weight) stimulated, and in a dose of 8 mg/kg suppressed the rosette-forming reaction to immunization with red sheep blood cells in mice (Devoino and Beletskaya, 1988).

Recently, Zavala *et al.* (1990) have found that the intraperitoneal administration of 'peripheral' and 'mixed' benzodiazepines in mice resulted in inhibition of the oxidative response of macrophages. Moreover, the injection of Ro5-4864 inhibited the capacity of macrophages to produce interleukins 1 and 6 and a tumour necrosis factor. Again, clonazepam was ineffective in this experiment (Table 3).

All these data taken together indicate that the effects of benzodiazepines on the immune responses *in vivo* are exerted mainly by 'peripheral' or 'mixed' benzodiazepines. Moreover, they are frequently biphasic: stimulatory at low doses and inhibitory at high doses.

Table 3 Effects of MBR ligands on immunocompetent cells.

Effect	Type of cell	Reference
Stimulation of prolactin dependent mitogenesis (nM) or inhibition (μM)	Nb2 node lymphoma (rat)	Laird et al. (1989)
Stimulation of mitogenesis (nM)	Splenocyte (mouse)	Giesen-Crouse et al. (unpublished)
Inhibition of mitogenesis (μM)	Splenocyte (mouse)	Stepien et al. (1988)
$O_2^{\bullet}$ production (10 nM Ro5-4864, inhibition by PK 11195)	P388D$_1$ macrophage (mouse)	Zavala and Lenfant (1987)
NADPH-oxidase activation ($O_2^{\bullet}$ burst)	PMN (human)	Zavala et al. (1990)
Inhibition of $O_2^{\bullet}$ production (10 μM Ro5-4864)	PMN (human)	Laghi Pasini et al. (1987)
IgM production (1 mg/kg)	Peritoneal macrophage (mouse)	Lenfant et al. (1986)
Chemotaxis (10^{-12} M)	Monocytes (human)	Ruff et al. (1985)
Inhibition of IL$_1$, IL$_6$, TNF release (1 mg/kg)	Macrophage (mouse)	Zavala et al. (1990)

6.5 Concluding remarks

The data discussed above indicate clearly that benzodiazepines affect the immune system. These immunomodulatory effects are exerted mainly by 'peripheral' and 'mixed' type compounds, whereas 'central' type benzodiazepines are rather ineffective, especially *in vitro*. (This situation however does not preclude the possibility that central sites are additional intermediates in the observed effects.) The actions of benzodiazepines on immunocytes are biphasic; stimulatory as well as inhibitory effects of the same compound have been reported, depending upon the concentration used. Generally, to obtain inhibitory effects, higher doses or concentrations of the drug are needed. Although the ability of the 'peripheral' and 'mixed' type of benzodiazepines to produce immunomodulation and the relative ineffectiveness of the 'central' type of benzodiazepines suggest strongly the involvement of MBR, the fact that benzodiazepines generate bell-shaped dose–response curves raises the question whether one or two types of receptor mediate these actions. Given the correlation between acceptor affinity and effective dose, it can be assumed that the immunostimulatory action is exerted via high-affinity MBR. On the other hand, the immunosuppressive effects may depend on the interaction of benzodiazepines with a low-affinity (calcium transport?) site. A question whether the above-mentioned phenomenon is mediated by another subtype of the MBR or results from the unspecific interaction (for instance, with dihydropyridine binding sites) remains to be answered. The immunomodulatory action of benzodiazepines may be of pharmacological importance since during therapeutic applications their plasma concentrations reach micromolar levels. Recently, Taupin *et al.* (1991) reported on the effect of benzodiazepine anaesthesia on cytokine production in human subjects. These authors observed that a single intravenous injection of midazolam in patients undergoing endoscopy of the urinary tract significantly inhibited the monocyte production of cytokines interleukin-1, tumour necrosis factor and interleukin 6 in response to lipopolysaccharide. Ro5-4864 (1 mg/kg i.p.) inhibited significantly TNF, IL_1 and IL_6 production of rat macrophages. The same cytokines in turn sharply increase MBR density in rat cortex and striatum after local injection (Bourdiol *et al.*, 1991). In thymus of rats with adjuvant-induced arthritis, MBR density is significantly increased (Scatton *et al.*, 1990). A similar mechanism may be the source of the increased MBR levels which are observed during the early inflammatory reaction of the brain to injury (Benavides and Toulmond, 1992). (For more detail on MBR in brain injury, refer to Chapter 13.) Are the immunomodulatory benzodiazepine actions physiologically relevant? Low dose effects, MBR response to pathological situations and the existence of endogenous ligands of MBR like porphyrins and TTN speaks in favour of such a hypothesis.

References

Benavides, J., Dubois, A., Dennis, T., Hamel, E. and Scatton, B. (1990). Omega-3 (peripheral type) benzodiazepine binding site distribution in rat immune system: an autoradiographic study with the photoaffinity ligand ^{3}H-PK 14105. *J. Pharmacol. Exp. Ther.* **249**, 333–339.

Benavides, J. and Toulmond, S. (1992). The role of cytokines in the glial reaction. *Neuroimmunomodulation in Pharmacology.* Congress Abstract, Paris.

Bond, P.A., Cundall, R.L. and Rolfe, B. (1985). [^{3}H]Diazepam binding to human granulocytes. *Life Sci.* **37**, 11–16.

Bourdiol, S., Toulmond, S., Serrano, A., Benavides, J. and Scatton, B. (1991). Increase in ω_3 (peripheral type benzodiazepine) binding sites in the rat cortex and striatum after local injection of interleukin-1, tumour necrosis factor-α and lipopolysaccharide. *Brain Res.* **543**, 194–200.

Cantor, E.H., Kenessay, A., Senemuk, G. and Spector, S. (1984). Interaction of calcium channel blockers with non-neuronal benzodiazepine binding sites. *Proc. Natl. Acad. Sci. USA* **81**, 1549–1552.

Clarke, G.D. and Ryan, P.J. (1980). Tranquillizers can block mitogenesis in 3T3 cells and induce differentiation in Friend cells. *Nature* **287**, 160–161.

Covelli, V., Decandia, P., Altamura, M. and Jirillo, E. (1989). Diazepam inhibits phagocytosis and killing exerted by polymorphonuclear cells and monocytes from healthy donors. *Immunopharmacol. Immunotoxicol.* **11**, 701–714.

Descotes, J.R., Tedone, R. and Evreux, J.C. (1982). Suppression of humoral and cellular immunity in normal mice by diazepam. *Immunol. Lett.* **5**, 41–42.

Devoino, L.V. and Beletskaya, I.O. (1988). Action of benzodiazepines on the immune response. *Bull. Exp. Biol. Med.* **105**, 440–442.

Ferrero, P., Santi, M.R., Conti-Troconi, B., Costa, E. and Guidotti, A. (1986). Study of an octadecaneuropeptide derived from diazepam binding inhibitor (DBI). *Proc. Natl. Acad. Sci. USA* **83**, 827–831.

Gerrish, K.E., Putnam, C.W. and Laird II, H.E. (1990). Prolactin-stimulated mitogenesis in the Nb2 rat lymphoma cell: lack of protoporphyrin IX effects. *Life Sci.* **47**, 1647–1653.

Gorman, A.M.C., O'Beirne, G.B., Regan, C.M. and Williams, D.C. (1989). Anti-proliferative action of benzodiazepines in cultured brain cells in not mediated through the peripheral-type benzodiazepine receptor. *J. Neurochem.* **53**, 849–855.

Gupta, R.S. and Dudani, A.K. (1989). Mechanism of action of antimitotic drugs: a new hypothesis based on the role of cellular calcium. *Med. Hypotheses* **28**, 57–69.

Kunert-Radek, J. (1989). Effects of benzodiazepines and calcium channel modulators on proliferation of rat prolactin-secreting tumor. Thesis, Medical Academy of Lodz, 1989 (in Polish).

Kunert-Radek, J., Stepien, H., Lyson, K. and Pawlikowski, M. (1990). Effects of calcium channel modulators on the profile-ration of mouse spleen lymphocytes *in vitro. Agents and Actions* **29**, 254–258.

Laghi Pasini, F., Ceccatelli, L., Capecci, P.L., Orrico, A., Pasqui, A.L. and Di Pierri, T. (1987). Benzodiazepines inhibit *in vitro* free radical formation from human neurophiles induced by FLMP and A23187. *Immunopharmacol. Immunotoxicol.* **9**, 104–114.

Laird II, H.E., Gerrish, K.E., Duerson, K.C., Putnam, C.W. and Russell, D.H. (1989). Peripheral benzodiazepine binding sites in Nb2 node lymphoma cells: effects on prolactin-stimulated proliferation and ornithine decarboxylase activity. *Eur. J. Pharmacol.* **171**, 25–35.

Lenfant, M. (1989). Neuroimmunomodulation: a search for the targets of PK 11195 a structurally unrelated ligand for 'peripheral benzodiazepine' binding sites. In *Interactions Among CNS Neuroendocrine and Immune System* (eds Hadden, J.W., Marek, K. and Nistico, G.), pp. 275–281. Pythagora Press, Rome-Milan.

Lenfant, M., Zavala, F., Haumont, J. and Potier, P. (1985). Presence d'un site de fixation de type peripherique des benzodiazepines sur les macrophages. *C.R. Acad. Sci. Paris* **300**, 309–314.

Lenfant, M., Haumont, J. and Zavala, F. (1986). *In vivo* immunomodulating activity of PK 11195, a structurally unrelated ligand for 'peripheral' benzodiazepine binding sites. *Int. J. Immunopharmacol.* **8**, 825–829.

Lichtman, A.H., Segel, G.B. and Lichtman, M.A. (1983). The role of calcium in lymphocyte proliferation. *Blood* **61**, 413–422.

Lyson, K., Kunert-Radek, J., Stepien, H., Pawlikowski, M. and Stanisz, A. (1989). The effects of central, peripheral and mixed-type benzodiazepine receptor ligands on the proliferation of mouse spleen lymphocytes *in vitro*. In *Interactions Among CNS Neuroendocrine and Immune System* (eds Hadden, J.W., Marek, K. and Nistico, G.), pp. 275–281. Pythagora Press, Rome-Milan.

Mantione, C.R., Goldman, M.E., Martin, B., Bolger, G.T., Leuddens, H.W.M., Paul, S. and Skolnick, P. (1988). Purification and characterization of an endogenous protein modulator of radioligand binding to 'peripheral-type' benzodiazepine receptor and dihydropiridine Ca^{2+}-channel antagonist binding sites. *Biochem. Pharmacol.* **37**, 339–347.

Miller, L.G., Lee-Paritz, A., Greenblatt, D.J. and Theoharides, T.C. (1988). High-affinity benzodiazepine binding sites on rat peritoneal mast cells and RBL-1 cells: binding characteristics and effects on granule secretion. *Pharmacology* **36**, 52–60.

Moingeon, P., Bidart, J.M., Alberici, G.F. and Bohuon, C. (1983). Characterization of peripheral-type benzodiazepine binding site on human circulating lymphocytes. *Eur. J. Pharmacol.* **92**, 147–152.

Pawlikowski, M., Stepien, H. and Kunert-Radek, J. (1986). Diazepam inhibits proliferation of the mouse spleen lymphocytes *in vitro*. *Pol. J. Pharmacol.* **38**, 167–170.

Pawlikowski, M., Kunert-Radek, J., Radek, A. and Stepien, H. (1988). Inhibition of cell proliferation of human gliomas by benzodiazepines *in vitro*. *Acta Neurol. Scand.* **77**, 231–236.

Rägo, L., Adojaan, A. and Masso, R. (1990). [^{3}H]Ro5-4864 binding sites in the nucleus of rat lymphocytes. *Eur. J. Pharmacol.* **187**, 561–562.

Rampe, D. and Triggle, D.J. (1986). Benzodiazepine and calcium channel function. *Trends Pharmacol. Sci.* **7**, 461–463.

Ruff, M.R., Pert, C.B., Weber, R.J., Wahl, L.M., Wahl, S.M. and Paul, S.M. (1985). Benzodiazepine receptor-mediated chemotaxis of human monocytes. *Science* **229**, 1281–1283.

Scatton, B., Benavides, J., Dubois, A. and Bourdiol, F. (1990). Proliferation of ω_3 binding sites in the immune organs and leg infiltrate of rats with adjuvant induced arthritis. *Int. J. Tiss. Reac.* **XII**, 15–20.

Schlumpf, M., Ramseier, H. and Lichtensteiger, W. (1989). Prenatal diazepam induced persisting depression of cellular immune responses. *Life Sci.* **44**, 493–501.

Schlumpf, M., Parmar, R., Ramseier, H.R. and Lichtensteiger, W. (1990). Prenatal benzodiazepine immunosuppression: possible involvement of peripheral benzo-diazepine site. *Dev. Pharmacol. Ther.* **15**, 178–185.

Stepien, H., Pawlikowska, A. and Pawlikowski, M. (1988). Effects of benzodiazepines on thymus cell proliferation. *Thymus* **12**, 117–121.

Stepien, H., Kunert-Radek, J., Stanisz, A., Zerek-Melen, G. and Pawlikowski, M. (1991). Inhibitory effect of porphyrins on the proliferation of mouse spleen lymphocytes *in vitro*. *Biochem. Biophys. Res. Commun.* **174**, 313–322.

Stepien, H., Agro, A., Padol, I. and Stanisz, A. (1992). Inhibitory effects of diazepam on human natural killer activity *in vitro*. *Clin. Exp. Immunol.* (submitted).

Taft, W.C. and De Lorenzo, R.J. (1984). Micromolar-affinity benzodiazepine receptors regulate voltage-sensitive calcium channels in nerve terminal preparations. *Proc. Natl. Acad. Sci. USA* **81**, 3118–3122.

Taniguchi, T., Wang, J.K.T. and Spector, S. (1980). Properties of [^{3}H]diazepam binding to rat peritoneal mast cells. *Life Sci.* **27**, 171–178.

Taupin, V., Jayais, P., Descamps-Latscha, B., Cazalaa, J.B., Barrier, G., Bach, J.F. and Zavala, F. (1991). Benzodiazepine anesthesia in humans modulates the interleukin-1-beta, tumor necrosis factor alpha and interleukin-6 responses of blood monocytes. *J. Neuroimmunol.* **35**, 13–19.

Verma, A. and Snyder, S.H. (1989). Peripheral type benzodiazepine receptors. *Ann. Rev. Pharmacol. Toxicol.* **29**, 307–322.

Verma, A., Nye, J.S. and Snyder, S.W. (1987). Porphyrins are endogenous ligands for the mitochondrial (peripheral type) benzodiazepine receptor. *Proc. Natl. Acad. Sci. USA* **84**, 2256–2260.

Wang, J.K.T., Taniguchi, T., Sagikura, M. and Spector, S. (1981). Presence of benzodiazepine binding sites in mouse thymocytes. *The Pharmacologist* **23**, 160–161.

Wang, J.K.T., Morgan, J.I. and Spector, S. (1984). Benzodiazapines that bind at peripheral sites inhibit cell proliferation. *Proc. Natl. Acad. Sci. USA* **81**, 753–756.

Zavala, F. and Lenfant, M. (1987). Peripheral benzodiazepines enhance the respiratory burst of macrophages-like P 388 D_1 cells stimulated by arachidonic acid. *Int. J. Immunopharmacol.* **9**, 269–274.

Zavala, F., Haumont, J. and Lenfant, M. (1984). Interaction of benzodiazepines with mouse macrophages. *Eur. J. Pharmacol.* **106**, 561–566.

Zavala, F., Taupin, V. and Descamps-Latscha, B. (1990). *In vivo* treatment with benzodiazepines inhibits murine phagocyte oxidative metabolism and production of interleukin 1, tumor necrosis factor and interleukin 6. *J. Pharmacol. Exp. Ther.* **255**, 442–450.

CHAPTER 7

MODULATION OF MEMORY CONSOLIDATION BY CENTRAL AND PERIPHERAL BENZODIAZEPINE RECEPTOR LIGANDS

Jorge H. Medina[1] and Ivan Izquierdo[2]

[1] Laboratorio de Neuroreceptores, Instituto de Biologia Celular, Facultad de Medicina, Universidad de Buenos Aires, Paraguay 2155, (1121) Buenos Aires, Argentina
[2] Centro de Memoria, Departamento de Bioquimica, Instituto de Biociencias, Universidade Federal do Rio Grande do Sul, 90049 Porto Alegre RS, Brasil

Table of Contents

7.1 Introduction

Benzodiazepines (BZs) have usually been thought to exert their pharmacological actions in the central nervous system mainly through the central BZ receptors (CBR) (Haefely et al., 1985). However, peripheral BZ receptors (MBR), characterized both in the brain and the periphery (Verma and Snyder, 1989), may mediate some of the central effects elicited by BZ derivatives (Benavides et al., 1984; Weissman et al., 1985; Basile et al., 1989; Hirsch et al., 1989; Krueger and Papadopoulos, 1990).

PERIPHERAL BENZODIAZEPINE RECEPTORS
ISBN 0-12-282630-2

MBR are mainly located in mitochondria (Verma and Snyder, 1989), however a significant proportion of binding sites are also found in synaptosomal fraction (Basile and Skolnick, 1986). In addition, the archetypic MBR ligand Ro5-4864 allosterically modulates the Cl^- channel-coupled t-butylbicyclophosphorotionate binding site (Gee, 1987; Gee et al., 1988; Basile et al., 1989). The binding of Ro5-4864 to its Cl^- channel site results in a blockade of Cl^- conductance, which in turn is antagonized by the isoquinoline, PK 11195 (Gee, 1987; Basile et al., 1989; Slobodyansky et al., 1989). Thus, this binding site could represent an additional locus for the regulation of GABA-gated Cl^- fluxes (Gee et al., 1988; Puia et al., 1989). The Ro5-4864 and PK 11195 binding site in the Cl^- channel is different from the picrotoxin binding site; however, this binding results in a blockade of the channel both in the case of Ro5-4864 and picrotoxin. As a consequence, Ro5-4864, at appropriate doses, may induce generalized convulsions similar to those brought about by picrotoxin (see Basile et al., 1989 for references).

7.2 Benzodiazepines and memory modulation

Benzodiazepines have been known for 30 years to induce anterograde amnesia without affecting acquisition performance (Heise and McConnell, 1961; Randall et al., 1961). Over the past two decades, the amnesic effect of BZ has been confirmed by many authors in different learning tasks in humans and animals (Thiebot, 1985; Lister, 1985; Cahill et al., 1986; Izquierdo and Ferreira, 1989; Izquierdo et al., 1990a,c). Moreover, low doses of several CBR inverse agonists, such as β-carbolines, provoke anterograde memory facilitation (Venault et al., 1986; Jensen et al., 1987; File and Pellow, 1988; Pereira et al., 1989). Recently, we have postulated that CBR agonists and inverse agonists affect learning and memory mainly by modulating $GABA_A$ receptor-mediated regulation of memory storage processes (Izquierdo et al., 1990a,c; Izquierdo and Medina, 1991; Izquierdo et al., 1992; Wolfman et al., 1991).

This hypothesis is based on different pieces of evidence:

(1) picrotoxin, a GABA-gated Cl^- channel blocker, elicits retrograde memory facilitation in several learning paradigms (Breen and McGaugh, 1961; McGaugh, 1989; Da Cunha et al., 1991a);
(2) the intracerebral microinjection of the $GABA_A$ receptor agonist muscimol depresses memory consolidation whereas that of the $GABA_A$ receptor antagonist, bicuculline, causes memory facilitation in aversively motivated (Brioni et al., 1989) and spatial learning (Brioni et al., 1990) tasks;

(3) endogenous BZ-like molecules are present in brain (Sangameswaran *et al.*, 1986; De Robertis *et al.*, 1988; Medina *et al.*, 1988; Olasmaa *et al.*, 1990; Unseld *et al.*, 1990) and are mainly located in the synaptic vesicle fraction (Medina *et al.*, 1988); the brain levels of these molecules undergo rapid changes suggestive of a release during different training paradigms correlating with the degree of anxiety that accompanies the learning tasks (Izquierdo *et al.*, 1990a,b; 1991; Izquierdo and Medina, 1991; Wolfman *et al.*, 1991);

(4) the microinjection of flumazenil, a CBR antagonist, into the amygdala, hippocampus and septum produces selective enhancements of memory consolidation depending on the tasks and the region (Izquierdo *et al.*, 1990a; Da Cunha *et al.*, 1991b; Wolfman *et al.*, 1991).

All these findings supported the proposal that amygdala, hippocampal and septal $GABA_A$ receptor mechanisms are involved in the modulation of memory consolidation (McGaugh, 1989; Izquierdo *et al.*, 1990b,c; Izquierdo and Medina, 1991). Since the amygdala has long been though to play a pivotal role in memory consolidation of aversively motivated learning tasks (Gray, 1982; McGaugh, 1988, 1989; Cahill and McGaugh, 1990; Izquierdo *et al.*, 1990c), we decided to study the effects of the systemic and intra-amygdala administration of CBR and MBR ligands on the step-down inhibitory avoidance task.

7.2.1 Effect of the systemic administration of CBR and MBR ligands on memory

Pre-training i.p. administration of the CBR agonists diazepam or clonazepam hinders retention of step-down inhibitory avoidance learning (Izquierdo *et al.*, 1990c). This amnesic effect was antagonized by a very low dose of flumazenil (Figure 1). In addition, it blocks the enhancement of memory, induced by very low doses of CBR inverse agonist *n*-butyl-β-carboline carboxylate (CCB) (5–20 times lower than those reported to be anxiogenic; Novas *et al.*, 1988). Pre-training i.p. administration of non-anxiogenic doses of flumazenil also enhances memory consolidation (Figure 1). Therefore, the effect of flumazenil on its own at doses that were antagonistic to CBR ligands strongly suggests that learning of this task is normally downregulated by an endogenous mechanism involving BZ agonists (Pereira *et al.*, 1989; Izquierdo *et al.*, 1990c). Systemic flumazenil also affects retention of habituation to a buzzer (Izquierdo *et al.*, 1990c), an active avoidance task (Lal *et al.*, 1988) and spatial learning in a water tank (Brioni *et al.*, 1991), but does not affect retention for other less stressful learning paradigms, such as habituation to an open field (Pereira *et al.*, 1989). It therefore seems to affect retention only of

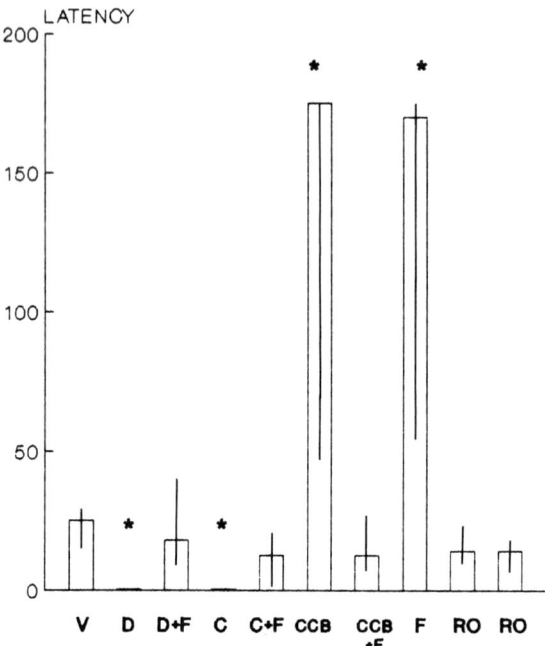

Figure 1 Effect of the vehicle, diazepam (D; 1 mg/kg), clonazepam (C; 1 mg/kg), β carboline butyl ester (CCB; 0.5 mg/kg), flumazenil (F; 5 mg/kg), Ro5-4864 (Ro; 2.5 and 6.25 mg/kg), D + F (2 mg/kg), C + F (2 mg/kg) and CCB + F (2 mg/kg) given i.p. 30 min prior to training, on retention test performance in a step-down inhibitory avoidance task in rats. Data are expressed as median (interquartile range) test minus training step-down latency. N = 14 for the vehicle group, N = 10 for all other groups. *$P < 0.02$ in a two tailed Mann-Whitney U test with respect to vehicle group.

stressful or anxiogenic behaviours. (For more information on endogenous MBR ligands and stress refer to Chapter 10.)

On the other hand, while pre-training i.p. administration of flumazenil enhances retention, pre-training i.p. administration of Ro5-4864 has no effect on inhibitory avoidance learning (Izquierdo *et al.*, 1990b) (Figure 1). When administered post-training, all CBR ligands tested are devoid of influence on retention scores (Figure 2) (Izquierdo *et al.*, 1990c). These results led to the hypothesis that CBR ligands modulate acquisition rather than consolidation mechanisms (Thiebot, 1985; Cahill *et al.*, 1986; Pereira *et al.*, 1989). However, all these studies did not take into account the time at which these drugs reach

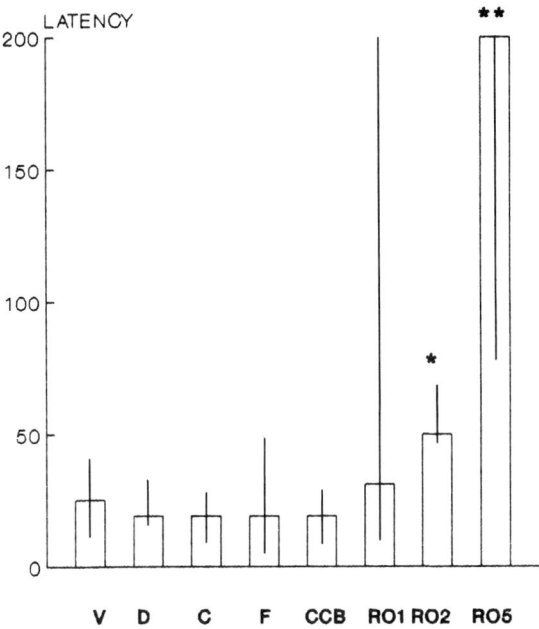

Figure 2 Effect of the immediate post-training i.p. administration of the same drugs as Figure 1 on retention of step-down inhibitory avoidance in rats. Ro5-4864 (Ro; 1, 2 and 5 mg/kg) produces a clear cut enhancement of retention at the highest dose used (5 mg/kg). Data are expressed as median test minus training step-down latency (interquartile range). N = 12 for all other groups. *$P < 0.05$, **$P < 0.002$ (Mann-Whitney U test, two tailed).

peak concentrations in brain: 10–30 min after administration (Izquierdo and Medina, 1991). Thus, it is likely that their post-training systemic injection reaches the brain too late, once the consolidation phase is over; whereas their pre-training administration would instead reach the brain in time to affect consolidation (Izquierdo and Medina, 1991).

As will be seen below, compounds acting on the different components of the $GABA_A/CBR/Cl^-$ channel complex mainly affect the early events in memory storage, e.g. during and/or immediately after training.

The immediate post-training i.p. administration of Ro5-4864 produced a dose-dependent increase of memory consolidation of inhibitory avoidance learning (Figure 2). It is important to stress that these data are reminiscent of those obtained by Breen and McGaugh more than 30 years ago with picrotoxin. This seminal work, conducted much before it was known that

picrotoxin acts on the GABA-gated Cl⁻ channel, clearly showed that
systemically administered picrotoxin facilitates retention of maze learning in
rats. More recent research by the same group extended this finding to various
other forms of learning (see McGaugh, 1989). This post-training effect of
Ro5-4864 is in contrast to that obtained with flumazenil. Influences on
retention of post-training treatments are usually taken to reflect specific effects
on memory consolidation (McGaugh, 1988, 1989; Izquierdo and Medina,
1991). It is thus likely that Ro5-4864 reaches the brain faster than flumazenil
or other BZ upon systemic administration.

7.2.2 Effect of the intra-amygdala administration of CBR and MBR ligands on memory

The amygdala becomes 100 times more sensitive to the deleterious effect of
muscimol on retention of inhibitory avoidance learning when this $GABA_A$
receptor agonist is microinjected immediately post-training than when it is
administered pre-training (Izquierdo et al., 1990b; Figure 3). Also, intra-
amygdala injection of muscimol hinders memory without altering acquisition
of active avoidance (Izquierdo et al., 1990a). These results suggest that in
amygdala $GABA_A$ receptors become sensitized during training. On the other
hand, bicuculline microinjected post-training into the amygdala facilitates
retention of this task (Brioni et al., 1989).

Quite similar enhancements of memory consolidation processes have
recently been shown to occur after the immediate post-training intra-amygdala
administration of flumazenil (see Figure 4) (Izquierdo et al., 1990a; Wolfman
et al., 1991).

The post-training intra-amygdala injection of the MBR agonist Ro5-4864
produced a dose-dependent increase of memory scores (Figure 4). Picrotoxin
caused a similar effect. However, PK 11195 at a dose of 8 ng/amygdala
blocked only the Ro5-4864 action without affecting picrotoxin-induced
enhancement of retention (Figure 4). Taken together, the intra-amygdala
injection experiments strongly support the hypothesis that post-training
storage processes are physiologically downregulated by $GABA_A/BZ/Cl^-$
ionophore receptor complex. Also, they demonstrated that both CBR and
MBR ligands are involved in the modulation of memory consolidation. Because
the amygdala has been shown to be involved in the processing of aversive
behaviours (Cahill and McGaugh, 1990; Izquierdo et al., 1992), our results
endorse the assumption that amygdalar $GABA_A$ mechanisms are especially
important in the post-training memory modulation of aversively-motivated
learning tasks.

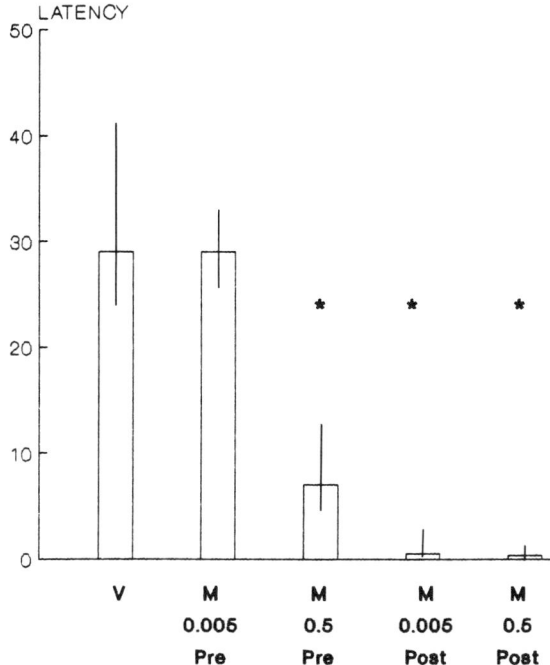

Figure 3 Effect of pretraining (Pre) or post-training (Post) intra-amygdala uscimol (M) administration on retention of a step-down inhibitory avoidance task in rats. Data are expressed as median (interquartile range) test minus training step-down latency. N = 9 for all the groups. *P < 0.02 in a two tailed Mann-Whitney U test.

7.3 Conclusion

In conclusion, both systemic and local administration of CBR and MBR ligands strongly affect post-training memory processes. The present findings suggest that these effects are mediated through their interactions with the different components of the $GABA_A / BZ / Cl^-$ ionophore receptor complex. The precise role of the MBR on these phenomena, if any, remains to be determined.

Concerning the possible clinical significance of these findings, clearly both Ro5-4864 and flumazenil fall into the general class of memory-enhancing drugs (e.g. McGaugh, 1989). These drugs belong to a number of distinct pharmacological families (hormones, neurotransmitters, neurotransmitter antagonists, channel blockers, mood enhancers, etc.); it is therefore difficult

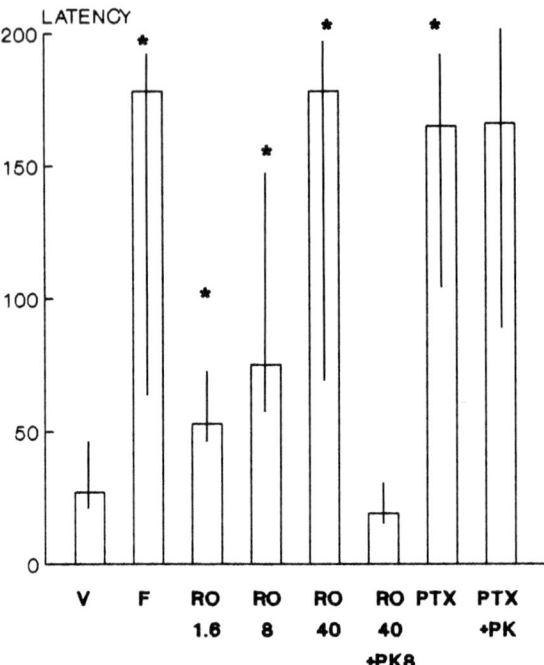

Figure 4 Effect of post-training intra-amygdala administration of flumazenil (F; 10 nM/each side), Ro5-4864 (Ro; 1.6, 8 and 40 nM/each side) and picrotoxin (PTX; 80 nM/each side) on retention of a step-down inhibitory avoidance task. PK 11195 (PK; 8 ng/amygdala) blocked Ro5-4864 enhancement of retention. Data are expressed as median (interquartile range). N = 10 for all the groups. *$P < 0.02$ in a two tailed Mann-Whitney U test.

to establish comparisons among them, or to formulate predictions from animal experiments regarding their therapeutic potential. In particular, it would be important to evaluate the possible anxiogenic effect of the chronic administration of Ro5-4864 in humans before considering its eventual use in memory disorders.

References

Basile, A.S., Bolger, G.T., Lueddens, W.M. and Skolnick, P. (1989). Electrophysiological actions of Ro 5-4864 on cerebellar Purkinje neurons: evidence for 'peripheral' benzodiazepine-mediated depression. *J. Pharmacol. Exp. Ther.* **248**, 463–469.

Basile, A.S. and Skolnick, P. (1986). Subcellular localization of peripheral-type binding sites for BZD in rat brain. *J. Neurochem.* **46**, 305–308.

Benavides, J., Guilloux, F., Allam, D.E., Uzan, A., Mizoule, J., Renault, C., Dubroeucq, M.C., Gueremy, C. and Le Fur, G. (1984). Opposite effects of an agonist Ro 5-4864 and an antagonist PK 11195 of the peripheral type benzodiazepine binding sites on audiogenic seizures in DBA/21 mice. *Life Sci.* **34**, 2613–2620.

Breen, R.A. and McGaugh, J.L. (1961). Facilitation of maze learning with posttrial injection of picrotoxin. *J. Comp. Physiol. Psychol.* **54**, 498–501.

Brioni, J.D., Arolfo, M.P., Medina, J.H. and Izquierdo, I. (1991). The effect of flumazenil on acquisition, retention and retrieval of spatial information. *Behav. Neural. Biol.* **56**, 329–335.

Brioni, J.D., Decker, M.W., Gamboa, L.P., Izquierdo, I. and McGaugh, J.L. (1990). Muscimol injections in the medial septum impair spatial learning. *Brain Res.* **522**, 227–234.

Brioni, J.D. Nagahara, A.H. and McGaugh, J.L. (1989). Involvement of the amygdala GABAergic system in the modulation of memory storage. *Brain Res.* **487**, 105–112.

Cahill, L., Brioni, J.D. and Izquierdo, I. (1986). Retrograde memory enhancement by diazepam: its relation to anterograde amnesia and some clinical implications. *Psychopharmacology.* **90**, 554–556.

Cahill, L. and McGaugh, J.L. (1990). Amygdaloid complex lesions differentially affect retention of tasks using appetitive and aversive reinforcement. *Behav. Neurosci.,* **104**, 532–543.

Da Cunha, C., De Paula, L.P., Medina, J.H. and Izquierdo, I. (1991a). The amnestic effect of pre- and post-training intraamygdala muscimol. *Comun. Biol. (Buenos Aires),* **9**, 219–225.

Da Cunha, C., Huang, C.H., Walz, R., Fias, M., Koya, R., Bianchin, M., Pereira, M.E., Izquierdo, I. and Medina, J.H. (1991b). Memory facilitation by post-training intraperitoneal, intracerebroventricular and intra-amygdala muscimol. *Brain. Res.,* **544**, 133–136.

De Robertis, E., Peña, C., Paladini, A. and Medina, J.H. (1988). New developments on the search for the endogenous ligand(s) of central benzodiazepine receptors. *Neurochem. Int.,* **13**, 1–11.

File, S. and Pellow, S. (1988). Low and high doses of benzodiazepine receptor inverse agonists respectively improve and impair performance in passive avoidance but do not affect habituation. *Behav. Brain Res.,* **30**, 31–36.

Gee, K.W. (1987). Phenylquinolines PK8165 and PK9084 allosterically modulate [^{35}S]-tert-butylbicyclophosphorothionate binding to a chloride ionophore in rat brain via a novel Ro 5-4864 binding site. *J. Pharmacol. Exp. Ther.,* **240**, 743–753.

Gee, K.W., Brinton, R.E. and McEwen, B.S. (1988). Regional distribution of a Ro 5-4864 binding site that is functionally coupled to the γ-aminobutyric acid/benzodiazepine receptor complex in rat brain. *J. Pharmacol. Exp. Ther.,* **244**, 379–383.

Gray, J.A. (1982). *The neuropsychology of anxiety: an enquiry into the functions of the septo-hippocampal system.* Clarendon Press, Oxford.

Haefely, W., Kybura, E., Gerecke, M. and Mohler, H. (1985). Recent advances in the molecular pharmacology of benzodiazepines receptor and in the structure–activity relationships of their agonist and antagonists. *Adv. Drug. Res.,* **14**, 165–322.

Heise, G.A. and McConnell, H. (1961). Differences between chlordiazepoxide-type and chlorpromazine-type action in "trace" avoidance. *Proc. 3rd World Congress of Psych.,* **Vol.2**, p. 917.

Hirsch, J.D., Beyer, C.F., Malkowitz, L., Beer, B. and Blume, A.L. (1989). Mitochondrial benzodiazepine receptors mediate inhibition of mitochondrial respiratory control. *Mol. Pharmacol.,* **35**, 157–163.

Izquierdo, I., Da Cunha, C. and Medina, J.H. (1990b). Endogenous benzodiazepine modulation of memory processes. *Neurosci. Biobehav. Rev.*, **14**, 419–424.

Izquierdo, I., Da Cunha, C., Huang, C., Walz, R., Wolfman, C. and Medina, J. (1990a). Post-training down-regulation of memory consolidation by a GABA-A mechanism in the amygdala modulated by endogenous benzodiazepines. *Behav. Neural. Biol.* **54**, 105–109.

Izquierdo, I. and Ferreira, M.B.C. (1989). Diazepam prevents post-training drug effects related to state dependency, but no post-training memory facilitation by epinephrine. *Behav. Neural Biol.* **51**, 73–78.

Izquierdo, I. and Medina, J.H. (1991). GABA$_A$ regulation memory: The role of endogenous benzodiazepines. *TIPS*, **12**, 260–265.

Izquierdo, I., Medina, J.H., Da Cunha, C. and Wolfman, C. (1992). Endogenous benzodiazepine/GABA-A systems in the brain that modulate memory storage: localization and mode of action. In *Cellular Aspects of Memory Foundation and Long-term Potentiation*. (Eds. Matthies, H.J. and Kemmerer, F.) p. 315, Springer Verlag, Berlin.

Izquierdo, I., Pereira, M.E. and Medina, J.H. (1990c). Benzodiazepine ligand influences on acquisition: suggestion of an endogenous modulatory mechanism mediated by benzodiazepine receptors. *Behav. Neural. Biol.*, **54**, 27–41.

Jensen, L.H., Stephens, D.N., Santer, M. and Petersen, E.N. (1987). Bidirectional effects of β-carbolines and benzodiazepines on cognitive processes. *Brain Res. Bull.*, **19**, 359–364.

Krueger, K.E. and Papadopoulos, V. (1990). Peripheral-type benzodiazepine receptors mediate translocation of cholesterol from outer to inner mitochondrial membranes in adrenocortical cells. *J. Biol. Chem.*, **265**, 15015–15022.

Lal, H., Kumar, B. and Forester, M.J. (1988). Enhancement of learning in mice by a benzodiazepine antagonist. *FASEB J.*, **2**, 2707–2711.

Lister, R.G. (1985). The amnesic action of benzodiazepine in man. *Neurosci. Biobehav. Rev.*, **9**, 87–94.

McGaugh, J.L. (1988). Modulation of memory storage processes. In *Perspective of Memory Research*. (Eds. Solomon, P.R., Goethals, G.R., Kelley, C.M. and Stephens, B.R.). pp. 33–64, Springer, New York.

McGaugh, J.L. (1989). Involvement of hormonal and neuromodulatory system in the regulation of memory storage. *Annu. Rev. Neurosci.*, **12**, 255–287.

Medina, J.H., Peña, C., Piva, M., Paladini, A.C. and De Robertis, E. (1988). Presence of Benzodiazepine-like molecules in mammalian brain and milk. *Biochem. Biophys. Res. Commun.*, **152**, 534–539.

Novas, M.L., Wolfman, C., Medina, J.H. and De Robertis, E. (1988). Proconvulsant and "anxiogenic" effects of n-butyl β-carboline 3-carboxilate, an endogenous benzodiazepine binding inhibitor from brain. *Pharmacol. Biochem. Behav.*, **30**, 331–336.

Olasmaa, M., Rothstein, J.D., Guidotti, A., Weber, R.J., Paul, S.M., Spector, S., Zeneroli, M.L., Baraldi, M. and Costa, E. (1990). Endogenous benzodiazepines in the human and animal epatic encephatopathy. *J. Neurochem.*, **55**, 2015–2023.

Pereira, M.E., Izquierdo, I. and Medina, J.H. (1989). Effect of pre-training flumazenil administration on retention of three different tasks in rats. *Braz. J. Med. Biol. Res.*, **22**, 1501–1505.

Puia, G., Santi, M.R., Vicini, S., Pritchett, D.B., Seeburg, P.H. and Costa, E. (1989). Differences in the negative allosteric modulation of γ-aminobutyric acid receptors elicited by 4'-chlorodiazepam and by a β-carboline-3-carboxilate ester: a study with natural and reconstituted receptors. *Proc. Natl. Acad. Sci. USA*, **86**, 7275–7279.

Randall, L.O., Heirse, G.A., Schallek, W., Bagdon, R.E., Banziger, R.F., Boris, A.,

Moe, R.A. and Abrams, W.G. (1961). Pharmacological and clinical studies on Valium (TM), a new psychotherapeutic agent of the benzodiazepine class. *Current Ther. Res.*, **3**, 405–425.

Sangameswaran, L., Fales, H.M., Friedrich, P. and De Blas. (1986). Purification of a benzodiazepine from bovine brain and detection of benzodiazepine-like immunoreactivity in human brain. *Proc. Natl. Acad. Sci.* USA, **83**, 9236–9240.

Slobodyansky, E., Guidotti, A., Wambebe, C., Berkovich, A. and Losta, E. (1989). Isolation of a rat brain triakontatetraneuropeptide, a posttranslational product of diazepam binding inhibitor: specific action at the Ro 5-4864 recognition site. *J. Neurochem*, **53**, 1276–1284.

Theibot, M.H. (1985). Some evidences for amnesic-like effects of benzodiazepines in animals. *Neurosci. Biobehav. Rev.*, **9**, 95–100.

Unseld, E., Fischer, C., Rothemund, E. and Klotz, U. (1990). Occurrence of "natural" diazepam in human brain. *Biochem. Pharmacol.*, **39**, 210–212.

Venault, P., Chapoutier, G., Prado de Carvalho, L., Simiand, J., Morre, M., Dodd, R.H. and Rossier, J. (1986). Benzodiazepine impairs and β-carboline enhances performance in learning and memory tasks. *Nature*, **321**, 864–866.

Verma, A. and Snyder, S. (1989). Peripheral type benzodiazepine receptors. *Ann. Rev. Pharmacol. Toxicol.*, **29**, 307–322.

Weissman, B.A., Cott, J., Jackson, J.A., Bolger, R.T., Weber, K.H., Horst, H.D., Paul, S.M. and Skonick, P. (1985). Peripheral-type binding sites for benzodiazepines in brain: relationship to the convulsant action of Ro 50-4864. *J. Neurochem.*, **44**, 1494–1499.

Wolfman, C., Da Cunha, C., Jerusalinsky, D., Levi De Stein, M., Viola, H., Izquierdo, I. and Medina, J.H. (1991). Habituation and inhibitory avoidance training alter brain regional levels of benzodiazepine-like molecules and are affected by intracerebral flumazenil microinjection. *Brain Res.*, **548**, 74–80.

147

CHAPTER 8

MITOCHONDRIAL BENZODIAZEPINE RECEPTORS AND SMOOTH MUSCLE REACTIVITY

Giuseppe Marano

Laboratorio di Farmacologia, Istituto Superiore di Sanità, V.le Regina Elena 299, 00161 Roma, Italy

Table of Contents

8.1 Introduction

In a variety of peripheral tissues, which include diaphragm (Wilkinson *et al.*, 1982), ileal longitudinal muscle (Hullihan *et al.*, 1983) and pulmonary smooth muscle (Mak and Barnes, 1990), high densities of [^{3}H]diazepam binding sites have been found with similar nanomolar affinity for diazepam as central receptors. These binding sites are the peripheral, or 'mitochondrial', benzodiazepine receptors (MBR).

The sparse data on the functional role of MBR indicate, among other effects, their involvement in the relaxation of vascular and non-vascular smooth muscle and in the depression of cardiac muscle contractility.

8.2 Functional role of MBR

Holck and Osterrieder (1985) reported that, at micromolar concentrations, Ro5-4864 caused a negative inotropic effect in isolated guinea-pig papillary muscles, reduced K^+ depolarization-induced contractures of isolated rat aorta, and inhibited [^{3}H]nitrendipine binding to guinea-pig cardiac membranes.

Grupp et al. (1987) reported that Ro5-4864 and PK 11195 increased coronary flow in Langendorff rat heart preparation.

Raeburn et al. (1988) showed that Ro5-4864 relaxed guinea-pig tracheal strips under basal tone; the effect was augmented significantly when the epithelium was removed. Similar results were obtained in tissues pre-contracted with methacholine or KCl. In the same study, midazolam, a central agonist, produced relaxation without showing any evidence of binding to MBR. Both benzodiazepines exhibited Ca^{2+} antagonist activities as assessed by their action on Ca^{2+} movements in depolarized tissues. One major difference from the commonly used Ca^{2+} antagonists was the ability to relax guinea-pig tracheal strips under basal tone. Indeed, it is accepted that basal tone in animal or human airways in vitro or in vivo is not affected by the action of Ca^{2+} antagonists. These data suggest that Ro5-4864 is not only exerting its action by blocking the entry of Ca^{2+} through these channels, but also may be acting in some other way, possibly by preventing the passive accumulation of Ca^{2+} via the so-called leak pathway, which is unaffected by the classical Ca^{2+} antagonists. Another possibility is that benzodiazepines are acting to prevent an increase or to reduce the intracellular concentration of Ca^{2+} after cell stimulation.

It has been also reported that benzodiazepines produced a dose-dependent decrease in the electrically induced contractions of ileal longitudinal muscle strips in guinea-pig, but their potencies in this effect did not correlate with their binding affinities. The inhibition by diazepam (central and peripheral agonist) of Ca^{2+}-induced contractions was partially reversed by increasing concentrations of Ca^{2+} in the bath (Hullihan et al., 1983).

In another study, Ro5-4864, diazepam, clonazepam and PK 11195 inhibited the responses of rat vas deferens following electrical stimulation; the effect of adenosine in rat vas deferens was potentiated by Ro5-4864 and diazepam, but not by clonazepam or PK 11195. Potentiation by Ro5-4864 and diazepam is thought to occur through an inhibition of the adenosine uptake system (Escubedo et al., 1991).

In vivo, Ro5-4864 produced enhancement of pharmacologically induced bronchoconstriction. PK 11195 did not antagonize the enhancing effects of Ro5-4864, but itself enhanced the bronchoconstriction (Marano et al., 1990). Neither compound influenced the effector organ, but both affected the formation or propagation of the stimulus. In other words, Ro5-4864 and PK 11195 behaved as metactoid sensitizers according to Van den Brink (1977). These data may be of practical importance: it is possible that in asthmatic patients, peripherally acting benzodiazepines, by lowering the threshold to bronchoconstriction, may increase the sensitivity to airway effects of various stimuli.

8.3 Conclusion

In conclusion, peripherally acting benzodiazepines exhibit Ca^{2+} antagonist activities in different tissues as assessed by their effects on Ca^{2+} entry in depolarized tissues and possibly, via the [^{3}H]nitrendipine binding site of the Ca^{2+} channel (Holck and Osterrieder, 1985). However, this last finding was not confirmed by Doble et al. (1985). The underlying mechanism interfering with the Ca^{2+} appears to involve antagonism of Ca^{2+} movements through the receptor site for dihydropyridine Ca^{2+} channel blockers. Indeed, it is unlikely that Ro5-4864 interacts with the binding sites through which non-dihydropyridine Ca^{2+} channel blockers, such as diltiazem or tiapamil, exert their actions. This was seen by the lack of effect of Ro5-4864 on tiapamil-induced inhibition of [^{3}H]nitrendipine binding.

Because the effects on vascular and non-vascular smooth muscle are usually seen at benzodiazepine concentrations $> 1\ \mu M$, it is not possible to exclude that these effects are related to binding with a postulated micromolar affinity benzodiazepine receptor associated structurally with the Ca^{2+} channel (Holck and Osterrieder, 1985; Doble et al., 1985).

References

Doble, A., Benavides, J., Ferris, O., Bertrand, P., Menager, J., Vaucher, N., Burgevin, M.C., Uzan, A., Gueremy, C. and Le Fur, G. (1985). Dihidropyridine and peripheral type benzodiazepine binding sites: subcellular distribution and molecular size determination. Eur. J. Pharmacol. 119, 153–167.

Escubedo, E., Camarasa, J., Pallas, M. and Adzet, T. (1991). Peripheral benzodiazepines potentiate the effect of adenosine in rat vas deferens. J. Pharm. Pharmacol. 43, 49–50.

Grupp, I.L., French, J.F. and Matlib, M.A. (1987). Benzodiazepine Ro5-4864 increases coronary flow. Eur. J. Pharmacol. 143, 143–147.

Holck, M. and Osterrieder, W. (1985). The peripheral high affinity benzodiazepine binding site is not coupled to the cardiac Ca^{2+} channel. Eur. J. Pharmacol. 118, 293–301.

Hullihan, J.P., Spector, S., Taniguchi, T. and Wang, J.K.T. (1983). The binding of [^{3}H]diazepam to guinea pig ileal longitudinal muscle and the in vitro inhibition of contraction by benzodiazepines. Br. J. Pharmacol. 78, 321–327.

Mak, J.C.W. and Barnes, P.J. (1990). Peripheral type benzodiazepine receptors in human and guinea-pig lung: characterization and autoradiographic mapping. J. Pharmacol. Exp. Ther. 252, 880–885.

Marano, G., Massotti, M., Spagnolo, A. and Carpi, A. (1990). Enhancement of pharmacologically induced bronchoconstriction by Ro5-4864. Eur. J. Pharmacol. 179, 237–240.

Raeburn, D., Miller, L.G. and Summer, W.R. (1988). Peripheral type benzodiazepine receptor and airway smooth muscle relaxation. J. Pharmacol. Exp. Ther. 245, 557–562.

Wilkinson, M., Grovestine, D. and Hamilton, J.T. (1982). Flunitrazepam binding sites in rat diaphragm. Receptors for direct neuromuscular effects of benzodiazepines? Can. J. Physiol. Pharmacol. 60, 1003–1005.

CHAPTER 9
CARDIOVASCULAR ACTIONS OF PERIPHERAL BENZODIAZEPINES

Gordon T. Bolger

Department of Pharmacology, Bio-Méga Inc., 2100 rue Cunard, Laval, Quebec, H7S 2G5 Canada

Table of Contents

9.1 Introduction

The benzodiazepines have been classified into two large groups, those drugs having a well characterized function in the central nervous system (CNS) (for review see Skolnick and Paul, 1982) and those drugs acting in non-CNS or peripheral tissues whose physiological function is less well characterized (for review, see Rampe and Triggle, 1986a).

Clonazepam (Figure 1) is a prototypic ligand for central benzodiazepine receptors (CBR) and Ro5-4864 and the non-benzodiazepine isoquinoline derivative PK 11195 (Figure 1) are prototypic ligands for mitochondrial benzodiazepine receptors (MBR). In addition, it has been determined that Ro5-4864 and diazepam are agonists and PK 11195 an antagonist of MBR (Lefur *et al.*, 1983a). It is clear that the benzodiazepines and other ligands active at their receptors possess far-reaching cardiovascular actions which may lead to the development of new clinical candidates in the treatment of cardiac and vascular disorders. This chapter will review the evidence supporting a functional activity for peripheral benzodiazepine ligands in the cardiovasculature and provide insight into their possible mechanism(s) of action.

PERIPHERAL BENZODIAZEPINE RECEPTORS
ISBN 0-12-282630-2

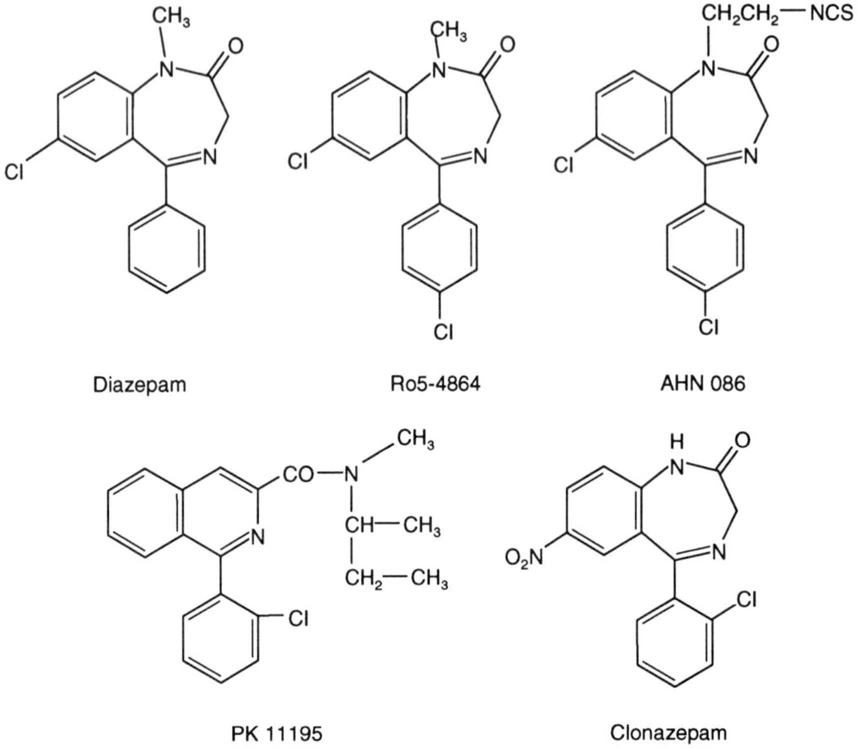

Figure 1 Structures of benzodiazepines.

9.2 Cardiovascular mitochondrial benzodiazepine receptors

Before detailing the cardiovascular effects of peripheral benzodiazepine ligands, it would seem most appropriate to discuss the sites at which these ligands interact. A high affinity and density characterizes the wide tissue distribution of MBR. While early studies used the mixed CBR and MBR ligand [³H]diazepam to characterize MBR (Hullihan et al., 1983; Davies and Huston, 1981), the development of more potent and selective radioligands such as the agonist [³H]Ro5-4864 and antagonist [³H]PK 11195 has led to the characterization of 'high-affinity' MBR in the kidney, adrenal, salivary gland, heart (atrium and ventricle), aorta, lung, intestine, testis, lymphocytes, platelets, mast cells, pineal gland, and brain (LeFur et al., 1983a,b; Bénevidès et al., 1984a,b; Wang et al., 1984; Weissman et al., 1984; Gehlert et al., 1985; Gavish et al., 1986a,b; Basile et al., 1986; Bolger et al., 1989; Doble et al.,

1987a,b; Mihara and Fujimoto, 1989; O'Beirne *et al.*, 1990). The affinity of 'high-affinity' MBR identified with either [^{3}H]Ro5-4864 or [^{3}H]PK 11195 ranges from 1 to 20 nM. Detailed subcellular distribution studies for many of the tissues listed above favours the largest density of MBR on the outer mitochondrial membrane (Hirsch *et al.*, 1988b; Antkiewicz-Michaluk *et al.*, 1988); hence the terminology MBR. This is not to say that MBR are only localized to the outer mitochondrial membrane. In the heart and the liver the subcellular distribution of MBR revealed the existence of small, but significant densities of MBR on the plasma, endoplasmic reticulum, Golgi and lysosomal membranes (Doble *et al.*, 1985; O'Beirne *et al.*, 1990).

A rather interesting finding in cardiovascular tissues is that the distribution of agonist and antagonist receptors for MBR differ in an apparently species-dependent manner. In porcine aortic smooth muscle membranes [^{3}H]PK 11195 bound with a high affinity (9.0 nM), while Ro5-4864 bound with a low (1200 nM) affinity (Mihara and Fujimoto, 1989). In contrast, [^{3}H]Ro5-4864 and [^{3}H]PK 11195 bound to rat aortic smooth muscle membranes with a similar high affinity (≈ 3.0 nM) (French and Matlib, 1988). Other complexities associated with the binding of MBR ligands to cardiovascular and other tissues are those consistent with their agonist and antagonist properties: [^{3}H]Ro5-4864 bound to high- and low-affinity MBR, and PK 11195 only to high-affinity MBR (Basile *et al.*, 1986; Awad and Gavish, 1987). Detailed studies have revealed that both Ro5-4864 and PK 11195 bind to differentially regulated domains of the MBR (LeFur *et al.*, 1983a,b; Bénevidès *et al.*, 1984a; Doble *et al.*, 1987a,b). Thus, the tissue and species specificity of [^{3}H]Ro5-4864 and [^{3}H]PK 11195 binding to MBR may reflect a differential exposure of their binding domains within the membrane (Basile *et al.*, 1986; Awad and Gavish, 1987).

As with other tissues, subcellular distribution studies in porcine and rat aorta revealed that both high- and low-affinity MBR are localized in their highest density on the mitochondrial membrane (French and Matlib, 1988; Mihara and Fujimoto, 1989). In the myocardium, autoradiographic analysis of [^{3}H]Ro5-4864 binding revealed a diffuse distribution of MBR throughout the heart, with the highest densities in the left ventricular wall (Gehlert *et al.*, 1985). MBR in the atrium were of a comparably lower density (Davies and Huston, 1981; Bolger *et al.*, 1989). Detailed molecular studies have revealed that the critical binding site domain of MBR has a molecular weight of ≈ 18 kDa (Doble *et al.*, 1987b).

The physiological regulation of cardiac MBR has been well investigated. Unfortunately, this is not the case for vascular MBR. Stress appears to play a key role in altering the number of cardiac MBR. In behaviourally 'anxious' mice the density of cardiac MBR was reduced compared to that in less anxious controls (Rägo *et al.*, 1989). Stress-inducing parameters in the rat such as inescapable tail shock and forced restraint have also been found to reduce the density of cardiac MBR (Armando *et al.*, 1988; Drugan *et al.*, 1988). The Maudsley reactive rat, a strain selectively bred for a high degree of fearfulness, also demonstrated a reduction in cardiac MBR compared to behaviourally normal controls (Drugan *et al.*, 1987). Administration of clonazepam reversed the stress-induced reduction in cardiac MBR (Drugan *et al.*, 1988). Stress also increased the levels of a putative endogenous ligand in the heart called 'tribulin' which inhibited both monoamine oxidase (MAO) activity and ligand binding to MBR (Armando *et al.*, 1988). These studies suggest that stress has a powerful modulatory influence over MBR in the myocardium. In contrast to stress, changes in blood pressure which often accompany stress do not appear to affect cardiac MBR, as their density was similar in the spontaneously hypertensive rat and its normotensive control (Thyagarajan *et al.*, 1981). However the density of renal MBR was lower. Various drug treatments have also been shown to modulate cardiac MBR. A 21-day treatment with 0.5 mg/kg diazepam in rats produced a significant elevation of cardiac MBR (Weizman and Gavish, 1989). In contrast, a 21-day treatment with 5 mg/kg chlorpromazine in rats produced a significant reduction in cardiac MBR (Gavish and Weizman, 1989). In the case of diazepam, the predominantly mitochondrial localization of MBR has led to speculation that the diazepam-dependent increases in cardiac MBR are associated with changes in the activity of the mitochondrial Na^+/Ca^{2+} exchanger (Weizman and Gavish, 1989). In the case of chlorpromazine, its propensity to produce negative chronotropic and potent arrythmogenic effects may be associated with the reduction of cardiac MBR (Gavish and Weizman, 1989). In contrast to the regulatory effects of diazepam and chlorpromazine on cardiac MBR, chronic oestradiol administration did not affect MBR in the heart (Gavish *et al.*, 1986b). Experimental hyperthyroidism induced by administration of D-thyroxin (T_4) for 10 days resulted in an increase in the density of cardiac MBR (Gavish *et al.*, 1986a), chronic drug treatments having also been shown to modulate non-cardiac MBR (i.e. kidney, testis and brain) in both the presence and absence of changes in the density of cardiac MBR (Gavish *et al.*, 1986a,b; Gavish and Weizman, 1989; Weizman and Gavish, 1989).

Several acute drug interactions worthy of note have also been observed with cardiac MBR, including those with the coronary vasodilator dipyridamole, nitrovasodilators and porphyrins. Dipyridamole displaced [^{3}H]diazepam from cardiac membranes (Davies and Huston, 1981; Doble *et al.*, 1985), suggesting that potentiation of the actions of diazepam and dipyridamole on adenosine release in cardiac and vascular smooth muscles was mediated via an interaction with MBR. In the guinea-pig heart, nitrovasodilators inhibited both the binding of MBR ligands and the functional effects of Ro5-4864 in the heart (Weissman *et al.*, 1990). The naturally occurring porphyrins, by-products of the biosynthesis of haeme (i.e. protoporphyrin IV, mesoporphyrin IX, deuteropophyrin IX and haemin), exert a high affinity for MBR (Snyder *et al.*, 1987; Verma *et al.*, 1987). However, the relevance of this interaction remains unknown, as does the physiological significance of the porphyrins.

A particularly interesting role for MBR in the ontogenetic development of cardiac muscle was proposed by Fares *et al.* (1987). Marked increases in the density of MBR over 31 days following birth were suggested to play a role in the cellular proliferation of the heart. However, caution must be applied to this claim, as Wang *et al.* (1984) observed that MBR ligands inhibited the cellular proliferation of mouse thymoma cells. (Effects of benzodiazepines on cell proliferation are reviewed in Chapter 5.) Thus, in summary, MBR exist in the cardiovasculature, are largely localized to mitochondria and can be regulated by chronic and acute drug treatments. Since the consensus holds that MBR play a regulatory role in the heart at the metabolic level, acute drug interactions with MBR might be postulated to alter cardiac function. A similar scenario might also hold true for vascular MBR. As a discussion of the physiological effects of MBR ligands unfolds it will become apparent that other sites of interaction for MBR ligands exist.

9.3 Cardiovascular actions of ligands acting at MBR

Early studies investigated the cardiovascular effects of diazepam. However, such studies were complicated by its effects on CBR. In general the intravenous administration of diazepam in man was observed to cause a moderate fall in blood pressure accompanied by an increase in pulse rate (Allen *et al.*, 1976; Coté *et al.*, 1976). Flunitrazepam, a benzodiazepine structurally related to diazepam which bound to both MBR and CBR (Basile *et al.*, 1986), had a similar cardiovascular profile to diazepam (Seitz *et al.*, 1977). The finding that combined sympathetic and parasympathetic blockade did not alter the responses to diazepam suggested a direct peripheral action of the drug on arteries and/or veins (Coté *et al.*, 1976). The vasodilatory effects, while not marked, also were not universally observed. Following aorto-coronary vein bypass graft, diazepam did not increase graft blood flow (Stoelting and King, 1977). Subsequent to bilateral carotid occlusion in rabbits, diazepam increased total peripheral resistance and decreased heart rate, but did not alter the response to arterial carotid occlusion (Sakamoto *et al.*, 1990); in this instance the dissociated effects of diazepam on the reflex control of circulation are due to its influence on the central control of sympathetic and vagal mediated pathways. The effects of diazepam at relatively high concentrations

(50–500 μM) were studied *in vitro* on both homotopic and atrial ectopic (injury-induced) ventricular automaticity (Ruiz *et al.*, 1989a,b). While in both cases diazepam abolished the spontaneous contractions, the inhibition by PK 11195 of diazepam's effect in the ventricle and not in the atrium suggested that MBR mediated the actions of diazepam in the ventricle. Several other features of the cardiovascular actions of diazepam include increases in coronary blood flow through potentiation of adenosine responses, decreased myocardial oxygen consumption and a narrowing of the arterial–coronary sinus oxygen differences (Clanchan and Marshall, 1979, 1980; Moritoki *et al.*, 1985).

Focusing on the cardiovascular physiology of the more specific MBR ligands, several effects both *in vitro* and *in vivo* have been documented. LeFur and colleagues have conducted a number of studies in this regard. At the electrophysiological level, Ro5-4864 (3×10^{-9} M–3×10^{-6} M) decreased the duration of the intracellular action potential and the contractility of guinea-pig papillary muscle (Mestre *et al.*, 1984; LeFur *et al.*, 1985). PK 11195 and increases in extracellular calcium blocked the effects of Ro5-4864 and CBR ligands were without effect. The results of these studies strongly support the physiological relevance of MBR in the heart. Saano *et al.* (1989) have also demonstrated that Ro5-4864 produced a negative inotropic effect in both the isolated rat atrium and papillary muscle. In the noradrenaline-stimulated rat atrium Ro5-4864 produced negative chronotropy, an effect which surprisingly was not antagonized by PK 11195 (Elgoyhen and Adler-Graschinsky, 1989). Furthermore, Ro5-4864 induced coronary vasodilation and increased coronary flow in the retrograde-perfused Langendorf rat heart model, an effect which also was not blocked by PK 11195 (Grupp *et al.*, 1987). The reasons why PK 11195 may act as an antagonist of Ro5-4864 in some cases, but not in others may reflect a differential interaction between MBR agonist and antagonist domains dependent on the stimulus being monitored. The chronotropic responses on exposure of the rat atrium either to calcium chloride or the dihydropyridine (DHP) calcium activator BAY K 8644 were not affected by Ro5-4864, but Ro5-4864 did reduce the positive chronotropic response to adenylate cyclase activation and phosphodiesterase inhibition (Elgoyhen and Adler-Graschinsky, 1989). Thus, it was speculated that MBR ligands reduced positive chronotropy in the atrium by interacting with the c-AMP-linked chain of events that follows the activation of β-adrenoceptors.

A modulatory role for MBR during cardiac ischaemia has been proposed. PK 11195 reduced early and delayed cardiac arrhythmias induced by myocardial ischaemia and reperfusion in the dog, prompting speculation that MBR might represent a novel target for the treatment of angina and cardiac ischaemia (Mestre *et al.*, 1985). In support of this concept, Charbonneau *et al.* (1986), using positron emission tomography in dogs and humans, found a direct correlation between increasing regional myocardial perfusion and increasing amounts of PK 11195 bound to the myocardium. In contrast, in human subjects in whom myocardial ischaemia was induced by atrial tachycardia and elongation of the ST-segment of the QRST-complex, PK 11195 did not have an anti-ischaemic effect (Drobinski *et al.*, 1989), casting doubt on the usefulness of PK 11195 as a possible anti-ischaemic agent.

More recent studies with modified MBR ligands containing acylating groups (i.e. AHN086—Figure 1—an isothiocyanate derivative of Ro5-4864; Lueddens *et al.*, 1986; Newman *et al.*, 1987) have revealed some rather interesting observations. Treatment of the isolated spontaneously beating guinea-pig atrium with AHN086 to acylate specifically 'high-affinity' MBR, followed by removal of unbound drug, did not affect atrial inotropy or chronotropy (Bolger *et al.*, 1989). However, the inotropic but not the chronotropic responses to BAY K 8644, the DHP calcium channel antagonist nifedipine and isoproterenol were inhibited. The use of AHN086 may help to sort out the often complex and divergent actions of MBR ligands in the cardiovasculature. In this and many of the *in vitro* studies discussed previously, rather high (1 μM–100 μM) concentrations of MBR ligands were required to modify cardiac responses. There may be several reasons for this, such as the interaction of ligands with MBR in different subcellular localizations (i.e. intracellular versus sarcolemmal) and an interaction with low-affinity micromolar MBR. Through the use of AHN086, it was clearly demonstrated that removal of high MBR ligand concentrations, while maintaining MBR occupancy, alters the response pattern of the atrium to Ro5-4864 and other MBR ligands, perhaps by removing certain sites of interaction such as low-affinity MBR and non-specific sites (Bolger *et al.*, 1989). Despite the complexity of the responses that MBR ligands modulate, and their equivocal activity, it is clear that MBR and the ligands they interact with can modify cardiovascular responses both *in vitro* and *in vivo*.

9.4 Functional sites of interaction for MBR ligands

A number of sites of interaction have been proposed for MBR ligands to account for their cardiovascular actions. As has already been mentioned, high densities of MBR are associated with the mitochondria, raising speculation that MBR may be involved in intermediary metabolism. Hirsch *et al.* (1988) demonstrated that the MBR affinities of Ro5-4864, PK 11195, diazepam, flunitrazepam, dipyridamole, deuteroporphyrin IX and mesoporphyrin IX corresponded well with their affinity to decrease the mitochondrial respiratory control ratio, clonazepam being inactive in this regard. Both Ro5-4864 and

PK 11195 were reported to reduce oxygen consumption in neuroblastoma cells in a dose-dependent manner, a further indication of MBR modulation of mitochondrial function (Larcher *et al.*, 1989). Pyruvate dehydrogenase, a mitochondrial membrane-bound enzyme and key in the oxidation of carbohydrates for energy production, inhibited the binding of [^{3}H]Ro5-4864 to its binding site (Daval *et al.*, 1989). Taken together, these observations strongly imply a functional mitochondrial site of interaction for MBR ligands. Thus, MBR ligands may gain at least a portion of their cardiovascular actions (i.e. reduced cardiac contractility) via MBR-mediated deficits of mitochondrial function. Since mitochondria can serve as important intracellular calcium storage sites, MBR may also modulate the mitochondrial handling of calcium and thus contractility of the myocardium and vascular smooth muscle.

Another postulated site of interaction for MBR ligands, perhaps via a subpopulation of MBR, is the voltage-dependent calcium channel (VDCC). (For a review of MBR as channel proteins refer to Chapter 2.) Extensive studies have been conducted both *in vitro* and *in vivo* strongly favouring an interaction between VDCC and MBR ligands. The subtype of VDCC most likely to interact with MBR ligands is the dihydropyridine-modulated 'L-type' channel (for a review of calcium channel subtypes, see Janis *et al.*, 1987). Mestre *et al.* (1986a,b) have clearly demonstrated that direct activation of cardiac 'L-type' VDCC by BAY K 8644, but not activation of cardiac VDCC by histamine or isoproterenol, was blocked by PK 11195. Bolger *et al.* (1989, 1990) have noted MBR agonist (Ro5-4864, AHN086) potentiation of inotropy and antagonist (PK 11195) inhibition of inotropy, specifically involving 'L-type' VDCC activated by BAY K 8644 in guinea-pig atrium. Cardioselectivity may exist since 'L-type' VDCC in the gut were not affected by MBR ligands in the same way as those in the atrium (Bolger *et al.*, 1989). Receptor-operated calcium channels (ROCC) do not appear to be affected by MBR ligands. Mestre *et al.* (1986a,b) demonstrated that rabbit aortic responses to BAY K 8644, but not to phenylphrine, were blocked by PK 11195.

Despite the wealth of evidence associating MBR with calcium channels, phylogenetic studies suggest that MBR and 'L-type' VDCC-linked DHP receptors are discrete entities (Bolger *et al.*, 1986). Furthermore, while it has been suggested that DHP receptors and MBR are packaged within the same membrane compartment in cardiac tissue (Doble *et al.*, 1985), whether a subpopulation of either high- or low-affinity MBR are coupled to VDCC, and in particular are associated with DHP receptors, remains equivocal (Holck and Osterrieder, 1985; Rampe and Triggle, 1986b; Bolger *et al.*, 1989). However, in addition to *in vitro* studies, *in vivo* studies clearly support an interaction between DHP and MBR ligands. In human subjects both the DHP VDCC blocker nicardipine and PK 11195 reduced the vascular resistance of brachial and carotid arteries. In contrast, a combination of the two ligands blocked the peripheral vasodilating effects of both drugs (Thuillez *et al.*, 1989). In a novel model of cardiac ischaemia in rats described by Abraham *et al.* (1987), Ro5-4864 potentiated, while PK 11195 inhibited the myocardial ischaemic actions of BAY K 8644 (Bolger *et al.*, 1990). These observations coupled with strong evidence for an interaction between 'L-type' VDCC and MBR ligands *in vitro* support the notion that a sarcolemmal subpopulation of MBR of either low- or high-affinity modulate 'L-type' VDCC indirectly through an interaction with DHP receptors.

Transport sites may also be targets for the action of MBR ligands, since tissues involved in endocrine and exocrine functions have substantially higher levels of MBR than those which do not. In the kidney, MBR were shown to interact strongly with anion transport inhibitors (Basile *et al.*, 1988) and to be regulated by anions (Lueddens and Skolnick, 1987). Gut smooth muscle was shown to possess chloride-regulated calcium pools involved in mediating contraction (Rangachari and Triggle, 1986). Thus, it is plausible that MBR ligands interacting with anion transport/anion–cation exchange sites might regulate the contractility of vascular smooth muscle.

9.5 Summary

A summary of the sites of interaction for MBR ligands and the localization of MBR is illustrated in a schematic view of a cell (Figure 2). Both high-affinity (HMBR) and low-affinity (LMBR) MBR may modulate the activity of 'L-type' VDCC as previously discussed. The presence of both plasma membrane and intracellular localizations of MBR will allow for selective ligand interactions depending on the ligand permeability of the cell. The DHP receptor may play a key role in the interaction of MBR ligands with 'L-type' VDCC. In contrast, ROCC would be relatively uninvolved. Intracellular stores of calcium, such as those in the mitochondria and sarcoplasmic reticulum, might be affected by MBR ligands since MBR have been characterized on these intracellular organelles. Although not discussed in detail, HMBR and LMBR may modulate sodium, potassium and chloride transport sites, exchangers or channels as previously mentioned. Mitochondrial function may be altered at the levels of pyruvate dehydrogenase, calcium mobilization monoamine oxidase (MAO) and ATP production. Finally, MBR in the CNS may also have a regulatory activity on cardiovascular function through reflex pathways. Given the multiple sites of interaction for MBR ligands, it is understandable that they possess pleotropic actions in the cardiovasculature. Clearly the efficacy of MBR ligands, both agonist

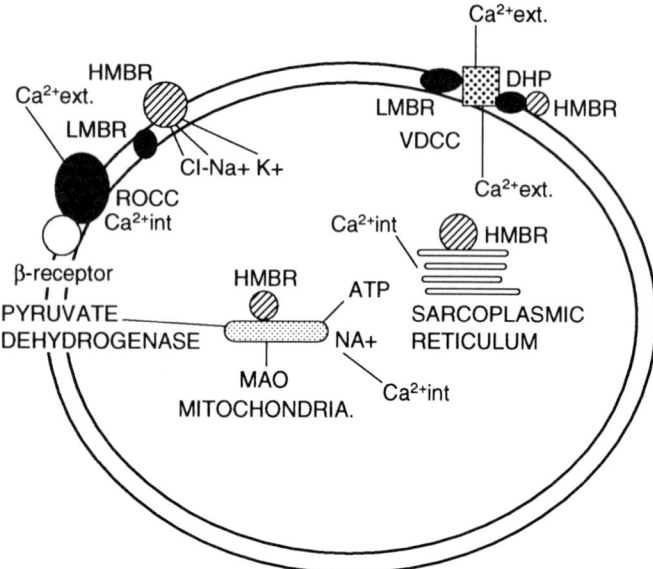

Figure 2 Schematic view of a cell, illustrating the major postulated cellular arrangement and functional aspects of MBR.

and antagonist, to modify cardiovascular responses will depend on their ability to access both extra- and intracellular loci. The isolation of endogenous ligands active at MBR *in vitro* and *in vivo* (Katz *et al.*, 1990) underscores the need for further research into their control of such important events as mitochondrial function, ion transport and the cellular proliferation in the cardiovasculature.

References

Abraham, S., Amitai, G., Oz, N. and Weissman B.A. (1987). BAY K 8644-induced changes in the ECG pattern of the rat and their inhibition by antianginal drugs. *Br. J. Pharmacol.* **92**, 603–608.

Allen, G.D., Everett, G.B. and Butler, L.A. (1976). Human cardiorespiratory and analgesic effects of intravenous diazepam and local anesthesia. *J. Am. Dental Assoc.* **92**, 744–747.

Antikiewicz-Michaluk, L., Guidotti, A. and Krueger, K.E. (1988) Molecular characterization and mitochondrial density of a recognition site for peripheral-type benzodiazepine ligands. *Mol. Pharmacol.* **34**, 272–278.

Armando, I., Levin, G. and Barontini, M. (1988). Stress increases endogenous benzodiazepine receptor ligand-monoamine oxidase inhibitory activity (tribulin) in rat tissues. *J. Neurotransmission* **71**, 29–37.

Awad, M. and Gavish, M. (1987). Binding of [³H]Ro5-4864 and [³H]PK 11195 to cerebral cortex and peripheral tissues of various species: species differences and heterogeneity in peripheral benzodiazepine binding sites. *J. Neurochem.* **49**, 1407–1414.

Basile, A.S., Klien, D.C. and Skolnick, P. (1986). Characterization of benzodiazepine receptors in the bovine pineal gland: evidence for the presence of an atypical binding site. *Mol. Brain Res.* **1**, 127–135.

Basile, A.S., Lueddens, H.W.M. and Skolnick, P. (1988). Regulation of renal peripheral benzodiazepine receptors by anion transport inhibitors. *Life Sci.* **42**, 715–726.

Bénevidès, J., Begassat, F., Phan, T., Tur, C., Uzan, A., Renault, C., Dubroeucq, M.C., Guérémy, C. and LeFur, G. (1984a). Histidine modification with diethylpyrocarbonate induces a decrease in the binding of an antagonist PK 11195, but not of an agonist, Ro5-4864, of the peripheral benzodiazepine receptors. *Life Sci.* **35**, 1249–1256.

Bénevidès, J., Guilloux, F., Rufat, P., Uzan, A., Renault, C., Dubroeucq, M., Guérémy, C. and LeFur, G. (1984b). *In vivo* labelling in several rat tissues of 'peripheral type' benzodiazepine binding sites. *Eur. J. Pharmacol.* **99**, 1–7.

Bolger, G.T., Weissman, B.A., Lueddens, H., Barrett, J.E., Witkin, J., Paul, S.M. and Skolnick, P. (1986). Dihydropyridine calcium channel antagonist binding in non-mammalian vertebrates: characterization and relationship to 'peripheral-type' binding sites for benzodiazepines. *Brain Res.* **368**, 351–356.

Bolger, G.T., Newman, A.H., Rice, K.C., Lueddens, H.W.M., Basile, A.S. and Skolnick, P. (1989). Characterization of the effects of AHN086, an irreversible ligand of 'peripheral' benzodiazepine receptors, on contraction in guinea-pig atria and ileal longitudinal smooth muscle. *Can. J. Physiol. Pharmacol.* **67**, 126–134.

Bolger, G.T., Abraham, S., Oz, N. and Weissman, B.A. (1990). Interactions between peripheral-type benzodiazepine receptor ligands and an activator of voltage-operated calcium channels. *Can. J. Physiol. Pharmacol.* **68**, 40–45.

Charbonneau, P., Syrota, A., Crouzel, C., Valoi, J.-M., Prenant, C. and Crouzel, M. (1986). Peripheral-type benzodiazepine receptors in the living heart characterized by positron emission tomography. *Circulation* **3**, 476–483.

Clanchan, A.S. and Marshall, R.J. (1979). Therapeutic concentrations of diazepam potentiate the effects of adenosine on isolated cardiac and smooth muscle (Abstract). Proceedings of the Biannual Meeting of the British Pharmacological Society April 1980, p. 66.

Coté, P., Noble, J. and Bourassa, M.G. (1976). Systemic vasodilation following diazepam after combined sympathetic and parasympathetic blockade in patients with coronary heart disease. *Catherization Cardiovasc. Diagn.* **2**, 369–380.

Daval, J.-L., Post, R.M. and Marangos, P.J. (1989). Pyruvate dehydrogenase interactions with peripheral-type benzodiazepine receptors. *J. Neurochem.* **52**, 110–116.

Davies, L.P. and Huston, V. (1981). Peripheral benzodiazepine binding sites in heart and their interaction with dipyridamole. *Eur. J. Pharmacol.* **73**, 209–211.

Doble, A., Bénevidès, J., Ferris, O., Bertrand, P., Ménager, J., Vaucher, N., Burgevin, M.-C., Uzan, A., Guérémy, C. and Lefur, C. (1985). Dihydropyridine and peripheral type benzodiazepine binding sites: subcellular distribution and molecular size determination. *Eur. J. Pharmacol.* **119**, 153–167.

G.T. Bolger

Doble, A., Ferris, O., Burgevin, M.-C., Ménager, J., Uzan, A., Dubroeucq, M.C., Renault, C., Guérémy, C., LeFur, G. (1987a). Photoaffinity labeling of peripheral-type benzodiazepine-binding sites. *Mol. Pharmacol.* **31**, 42–49.

Doble, A., Burgevin, M.-C., Ménager, J., Ferris, O., Bégassat, F., Renault, C., Dubroeucq, M.C., Guérémy, C., Uzan, A. and LeFur, G. (1987b). Partial purification and pharmacology of peripheral-type benzodiazepine receptors. *J. Receptor. Res.* **7**, 55–70.

Drobinski, G., Lechat, Ph., Eugène, L. and Grosgogeat, Y. (1989). Absence d'effet anti-ischémique d'un antagoniste de récepteurs périphériques aux benzodiazépines PK 11195. *Therapie* **44**, 263–267.

Drugan, R.C., Basile, A.S., Crawley, J.N., Paul, S.M. and Skolnick, P. (1987). 'Peripheral' benzodiazepine binding sites in the Maudsley reactive rat: selective decrease confined to peripheral tissues. *Brain Res. Bul.* **18**, 143–145.

Drugan, R.C., Basile, A.S., Crawley, J.N., Paul, S.M. and Skolnick, P. (1988). Characterization of stress-induced alterations in [^{3}H]Ro5-4864 binding to peripheral benzodiazepine receptors in rat heart and kidney. *Pharmacol. Biochem. Behav.* **30**, 1015–1020.

Elgoyhen, B. and Adler-Graschinsky, E. (1989). Dimunition by benzodiazepines of the chronotropic responses to noradrenaline in rat isolated atria. *Eur. J. Pharmacol.* **164**, 467–478.

Fares, F., Weizman, A., Zlotogorski, D. and Gavish, M. (1987). Ontogenetic development of peripheral benzodiazepine binding sites in rat brain, heart and lung. *Brain Res.* **408**, 381–384.

French, J.F. and Matlib, M.A. (1988). Identification of a high-affinity peripheral-type benzodiazepine binding site in rat aortic smooth muscle membranes. *J. Pharmacol. Exp. Therap.* **247**, 23–28.

Gavish, M., Weizman, A., Okun, F. and Youdim, M.B.H. (1986a). Modulatory effects of thyroxine treatment on central and peripheral benzodiazepine receptors in the rat. *J. Neurochem.* **47**, 1106–1110.

Gavish, M., Okun, F., Weizman, A. and Youdim, M.B.H. (1986b). Modulation of peripheral benzodiazepine binding sites following chronic estradiol treatment. *Eur. J. Pharmacol.* **127**, 147–151.

Gavish, M., Weizman, A., Youdim, M.B.H. and Okun, F. (1987). Regulation of central and peripheral benzodiazepine receptors in progesterone-treated rats. *Brain Res.* **409**, 386–390.

Gavish, M. and Weizman, R. (1989). Effects of chronic chlorpromazine treatment on peripheral benzodiazepine binding sites in heart, kidney, and cerebral cortex of rats. *J. Neurochem.* **52**, 1553–1558.

Gehlert, D.R., Yamamura, H.I. and Wamsely, J.K. (1985). Autoradiographic localization of 'peripheral-type' benzodiazepine binding sites in the rat brain, heart and kidney. *Naunyn-Schmiedeberg's Arch. Pharmacol.* **328**, 454–460.

Grupp, I.L., Frenchm, J.F. and Matlib, M.A. (1987). Benzodiazepine Ro5-4864 increases coronary flow. *Eur. J. Pharmacol.* **143**, 143–147.

Hirsch, J.D., Beyer, C.F., Malkowtiz, L., Beer, B. and Blume, A.J. (1988a). Mitochondrial benzodiazepine receptors mediate inhibition of mitochondrial respiratory control. *Mol. Pharmacol.* **34**, 157–163.

Hirsch, J.D., Beyer, C.F., Malkowitz, L., Loullis, C.C. and Blume, A.J. (1988b). Characterization of ligand binding to mitochondrial benzodiazepine receptors. *Mol. Pharmacol.* **34**, 164–172.

Holck, M. and Osterrieder, W. (1985). The peripheral, high affinity benzodiazepine binding site is not coupled to the cardiac Ca^{2+} channel. *Eur. J. Pharmacol.* **118**, 293–301.

Hullihan, J.P., Spector, S., Tanaguchi, T. and Wang, J.K.T. (1983). The binding of [³H]-diazepam to guinea-pig ileal longitudinal muscle and the *in vitro* inhibition of contraction by benzodiazepines. *Br. J. Pharmacol.* **78**, 321–327.

Janis, R.A., Silver, P.J. and Triggle, D.J. (1987). Drug action and cellular calcium regulation. *Adv. Drug Res.* **16**, 309–591.

Katz, Y., Allon, N., Abraham, S., Oz, N. and Gavish, M. (1990). Characterization of partially purified peripheral benzodiazepine binding inhibitors from human cerebrospinal fluid (CSF): its arrhythmogenic effects in the rat. In *Calcium Channel Modulators in Heart and Smooth Muscle* (eds S. Abraham and G. Amitai), pp. 253–268. VCH Publishers, New York.

Larcher, J.-C., Vayssiere, J.-L., LeMarquer, F.J., Cordeau, L.R., Keane, P.E., Bachy, A., Grus, F. and Croizat, B.P. (1989). Effects of peripheral benzodiazepines upon the O_2 consumption of neuroblastoma cells. *Eur. J. Pharmacol.* **161**, 197–202.

LeFur, G., Baucher, M.L., Flamier, P.A., Bénevidès, J., Renault, C., Dubroeucq, M.C., Guérémy, C. and Uzan, A. (1983a). Differentiation between two ligands for peripheral benzodiazepine binding sites, [³H]Ro5-4864 and [³H]PK 11195, by thermodynamic studies. *Life Sci.* **33**, 449–457.

LeFur, G., Perrier, M.L., Vaucher, N., Imbault, F., Flamier, A., Bénevidès, J., Uzan, A., Renault, C., Dubroeucq, M.C. and Guérémy, C. (1983b). Peripheral benzodiazepine binding sites: Effect of PK 11195, 1-(2-chlorophenyl)-N-methyl-N-(1-methylpropyl)-3-isoquinolinecarboxamide. *Life Sci.* **32**, 1839–1847.

Lefur, G., Mestre, M., Carriot, T., Belin, C., Renault, C., Dubroeucq, M.C., Guérémy, C. and Uzan, A. (1985). Pharmacology of peripheral type benzodiazepine receptors in the heart. In *Endocoids* (eds H. Lal, F. Labella and J. Lane), pp. 175–186. Alan R. Liss Inc., New York.

Lueddens, H.W.M., Newman, H., Rice, K.C. and Skolnick, P. (1986). AHN 086: An irreversible ligand of 'peripheral': benzodiazepine receptors. *Mol. Pharmacol.* **29**, 540–454.

Lueddens, H.W.M. and Skolnick, P. (1987). 'Peripheral-type' benzodiazepine receptors in the kidney: regulation of radioligand binding by anions and DIDS. *Eur. J. Pharmacol.* **133**, 205–214.

Mestre, M., Carriot, T., Belin, C., Uzan, A., Renault, C., Dubroeucq, M.C., Guérémy, C. and LeFur, G. (1984). Electrophysiological and pharmacological characterization of peripheral benzodiazepine receptors in a guinea pig heart preparation. *Life Sci.* **35**, 953–962.

Mestre, M., Bouefard, G., Uzan, A., Guérémy, C., Renault, C., Dubroeucq, M.-C. and LeFur, G. (1985). PK 11195, an antagonist of peripheral benzodiazepine receptors, reduces ventricular arrhythmias during myocardial ischemia and reperfusion in the dog. *Eur. J. Pharmacol.* **112**, 257–260.

Mestre, M., Belin, C., Uzan, A., Renault, C., Dubroeucq, M.-C., Guérémy, C. and LeFur, G. (1986a). Modulation of voltage-operated, but not receptor-operated, calcium channels in rabbit aorta by PK 11195, an antagonist of peripheral-type benzodiazepine receptors. *J. Cardiovasc. Pharmacol.* **8**, 729–734.

Mestre, M., Carriot, T., Néliat, G., Uzan, A., Renault, C., Dubroeucq, M.-C., Guérémy, C., Doble, A. and LeFur, G. (1986b). PK 11195, an antagonist of peripheral type benzodiazepine receptors, modulates BAY K 8644 sensitive but not β- or H_2-receptor sensitive voltage-operated calcium channels in the guinea pig heart. *Life Sci.* **39**, 329–339.

Mihara, S.-I. and Fujimoto, M. (1989). High-affinity binding sites for PK 11195, but not for Ro5-4864, in porcine aortic smooth muscle. *Life Sci.* **44**, 1713–1720.

Moritoki, H., Fukuda, H., Kotani, M., Veyama, T., Ishida, Y. and Takei, M. (1985). Possible mechanism of action of diazepam as an adenosine potentiator. *Eur. J. Pharmacol.* **113**, 89–98.

Newman, A.H., Lueddens, H.M.W., Skolnick, P. and Rice, K.C. (1987). Novel irreversible ligands specific for 'peripheral' type benzodiazepine receptors: (±)-, (+)-, and (−)-1-(2-chlorophenyl)-N-(1-methylpropyl)-N-(2-isothocyanatoethyl)-3-isoquinolinecarboxamide and 1-(2-isothiocyanatoethyl)-7-chloro-1,3-dihydro-5-(4-chlorophenyl)-2H-1,4-benzodiazepine-2-one. *J. Med. Chem.* **30**, 1901–1905.

O'Beirne, G.B., Woods, M.J. and Williams, D.C. (1990). Two subcellular locations for peripheral-type benzodiazepine acceptors in rat liver. *Eur. J. Biochem.* **188**, 131–138.

Rägo, L., Kiivet, R.-A., Harro, J. and Pöld, M. (1989). Central- and peripheral-type benzodiazepine receptors: similar regulation by stress and GABA receptor agonists. *Pharmacol. Biochem. Behav.* **32**, 879–883.

Rampe, D. and Triggle, D.J. (1986a). Benzodiazepines and calcium channel function. *TIPS* **11**, 461–464.

Rampe, D. and Triggle, D.J. (1986b). Benzodiazepine interactions at neuronal and smooth muscle Ca^{2+} channels. *Eur. J. Pharmacol.* **134**, 189–197.

Rangachari, P.K. and Triggle, D.J. (1986). Chloride removal and excitation-contraction coupling in guinea-pig ileal smooth muscle. *Can. J. Physiol. Pharmacol.* **64**, 1521–1527.

Ruiz, F., Herrandez, J. and Pérez, D. (1989a). The effect of diazepam on atrial automaticity is not mediated by benzodiazepine receptors. *Arch. Int. Pharmacodyn.* **299**, 77–85.

Ruiz, F., Hernandez, J. and Pérez, D. (1989b). The effect of diazepam on ventricular automaticity induced by a local injury. Evidence of involvement of 'peripheral-type' benzodiazepine receptors. *J. Pharm. Pharmacol.* **41**, 306–310.

Saano, V., Räty, M. and MacDonald, E. (1989). Effect of peripheral benzodiazepine receptor ligands on the contraction of isolated heart atrium and papillary muscle of rats. *Pharmacol. Toxicol.* **64**, 147–149.

Sakamoto, M., Ohsumi, H., Yamazaki, T. and Okumura, R. (1990). Effects of diazepam on the carotid sinus baroreflex control of circulation in rabbits. *Acta. Physiol. Scand.* **139**, 281–287.

Seitz, W., Hempelmann, G. and Pipenbrock, S. (1977). Zur kardiovasscularen wirkung von flunitrazepam (Rohypnol, RO-5-4200). *Anaesthetist* **26**, 249–256.

Skolnick, P. and Paul, S.M. (1982). Benzodiazepine receptors in the central nervous system. *Int. Rev. Neurobiol.* **23**, 103–140.

Snyder, S.H., Verma, A. and Trifiletti, R.R. (1987). The peripheral-type benzodiazepine receptor: a protein of mitochondrial outer membranes utilizing prophyrins as endogenous ligands. *FASEB J.* **1**, 282–288.

Stoelting, R.K. and King, R.D. (1977). Aortocoronary vein graft flow in response to peripheral venous administration of sodium nitroprusside or diazepam. *Ann. Thorac. Surg.* **24**, 59–61.

Thuillelz, C., Louieslati, H., Dulazé, P. and Giudicelli, J.F. (1989). Functional antagonism between PK 11195, a peripheral benzodiazepine receptor antagonist, and nicorandil at the vascular level in healthy subjects: a peripheral hemodynamic study. *J. Cardiovasc. Pharmacol.* **13**, 307–313.

Thyagarajan, R., Brehnan, T. and Ticku, R.K. (1983). GABA and benzodiazepine binding sites in spontaneously hypertensive rat. *Eur. J. Pharmacol.* **93**, 127–136.

Verma, A., Nye, J.S. and Snyder, S.J. (1987). Porphyrins are endogenous ligands for the mitochondrial (peripheral-type) benzodiazepine receptor. *Proc. Natl. Acad. Sci. USA* **84**, 2256–2260.

Wang, J.K.T., Magan, J.I. and Spector, S. (1984). Benzodiazepines that bind at peripheral sites inhibit cell proliferation. *Proc. Natl. Acad. Sci. USA* **81**, 753–756.

Weissman, B.A., Bolger, G.T., Isaac, L., Paul, S.M., Skolnick, P. (1984). Characterization of the binding of [^{3}H]Ro5-4864, a convulsant benzodiazepine, to guinea-pig brain. *J. Neurochem.* **42**, 969–975.

Weissman, B.A., Chiang, P.K. and Bolger, G.T. (1990). The effects of nitrovasodilators on peripheral benzodiazepine receptors in guinea-pig heart. In *Calcium Channel Modulators in Heart and Smooth Muscle: Basic Mechanisms and Pharmacological Aspects* (eds S. Abraham and G. Amitai), pp. 269–281. VCH Publishers, New York.

Weizman, R. and Gavish, M. (1989). Chronic diazepam treatment induces an increase in peripheral benzodiazepine binding sites. *Clin. Neuro. Pharmacol.* **12**, 346–351.

.

MODULATION OF RECEPTOR NUMBERS IN RESPONSE TO PATHOPHYSIOLOGICAL SITUATIONS

EFFECT OF CHANGES IN TISSUE CONCENTRATIONS OF PUTATIVE ENDOGENOUS LIGANDS ON MITOCHONDRIAL BENZODIAZEPINE RECEPTORS *EX VIVO*

Tiziana Mennini

Istituto di Ricerche Farmacologiche 'Mario Negri', Via Eritrea 62, 20157 Milano, Italy

Table of Contents

10.1 Introduction

The so-called mitochondrial benzodiazepine receptors (MBR) can be distinguished from the central benzodiazepine receptors (CBR) on the basis of their pharmacology, tissue distribution, cellular and subcellular localization, interaction with other sites and molecular weight of the receptor protein (see Verma and Snyder, 1989 and Giesen-Crouse, 1990 for reviews). Despite a wealth of radioligand binding data, the existence and physiological relevance of these sites has long been a matter of debate. The discovery of an endogenous

PERIPHERAL BENZODIAZEPINE RECEPTORS
ISBN 0-12-282630-2

ligand, specific for MBR, would justify the fact that the MBR is now considered to be a true receptor, rather than an acceptor site.

The presence of endogenous ligands acting on MBR was suggested by early studies showing that MBR binding could be increased by perfusion of tissue (Schoemaker *et al.*, 1983) or by treatment of rat adrenal membrane preparations with freezing and thawing or digitonin as detergent (Gavish and Fares, 1985). The search for endogenous ligands led to the first identification in rat brain extracts of a 9 kDa propeptide, called diazepam binding inhibitor (DBI) (Guidotti *et al.*, 1983), which inhibited benzodiazepine binding to central and peripheral sites with micromolar affinity. A few years later, Verma *et al.* (1987) convincingly demonstrated that haemin, purified from human red blood cells, and porphyrins extracted from different rat tissues, inhibited MBR binding with nanomolar affinity. Other compounds have been proposed, less convincingly, as endogenous ligands, capable of inhibiting MBR binding with micromolar affinity. Among them an endogenous protein, a phospholipase A2 isoenzyme (Mantione *et al.*, 1988); different phospholipids, including phosphatidylethanolamine, and unsaturated fatty acids (Beaumont *et al.*, 1988). Subsequent studies were aimed at further characterizing the interaction of porphyrins (Verma and Snyder, 1988) and DBI (see Costa and Guidotti, 1991, for review) with MBR, using *in vitro* approaches.

If DBI and porphyrins functionally interact *in vivo* with MBR, presumably changes in their tissue concentrations cause modifications to MBR binding measured *ex vivo*. This would provide further evidence of their role *in vivo* as endogenous ligands.

This chapter deals with this topic, analysing available results and speculating on the pharmacological potential of MBR ligands.

10.2 MBR interaction with porphyrins

10.2.1 Background

Haeme serves as the prosthetic group for mitochondrial cytochromes and cytosolic proteins, such as haemoglobin, myoglobin, catalase, tryptophan pyrrolase and several peroxidases; furthermore haeme-containing enzymes are required for steroidogenesis. Porphyrins are tetrapyrrolic pigments formed in the haeme biosynthesis pathway (see Figure 1). The initial and final steps in protoporphyrin IX biosynthesis occur within the mitochondria, so that coproporphyrinogen, a precursor of protoporphyrin IX, must cross mitochondrial membranes. The final product, haeme, formed inside the

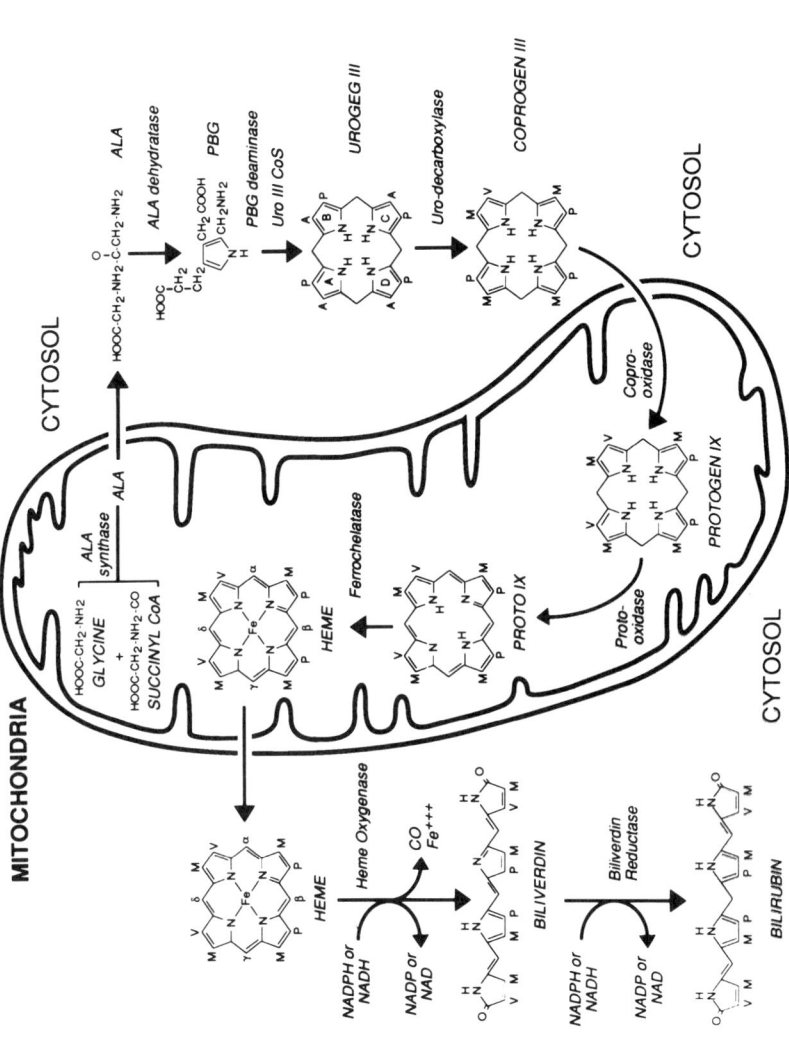

Figure 1 Schematic representation of pathway of haeme biosynthesis. Modified from Ibraham *et al.* (1983).

mitochondria, then crosses the mitochondrial membranes back toward the cytoplasm. These movements correlate with the mitochondrial localization of MBR and the possible action of porphyrins as endogenous ligands. It has been found that the MBR ligand 4-chloro-diazepam (Ro5-4864) induces porphyrin synthesis in animals (Moore et al., 1987). Porphyrins, like other MBR ligands, reduce the rates of respiratory states III and IV in isolated rat kidney mitochondria with a potency strongly correlated to their affinity for MBR (Hirsch et al., 1988).

There are few conditions in which porphyrin concentrations are altered. Inherited or acquired enzymatic defects in their biosynthetic pathways cause the porphyrias, a group of disorders characterized by excessive production of porphyrins or porphyrin precursors, and accompanied by neurological and skin manifestations. In animals 3,5-diethoxycarbonyl-1,4-dihydrocollidine (DDC) has been used as a model for precipitation of the attacks of hereditary human porphyria of the acute varieties and also for variegate porphyria and, in mice, for erythropoietic protoporphyria (Smith and De Matteis, 1980). In laboratory animals this compound causes accumulation of protoporphyrin IX in the liver and erythrocytes (Smith and De Matteis, 1980). The mechanism through which DDC and similarly acting drugs like griseofulvin cause liver protoporphyrin accumulation is complex since it requires that the compounds convert the haeme of cytochrome P-450 into N-methylprotoporphyrin (De Matteis et al., 1987), a powerful inhibitor of the mitochondrial enzyme ferrochetalase (De Matteis et al., 1980). Inhibition of this enzyme causes a block in the metabolism of protoporphyrin and, at the same time, a secondary, compensatory stimulation of ALA-synthase, which results in marked accumulation of protoporphyrin. Porphyria therefore results from a combination of increased biosynthesis and under-utilization of porphyrins.

DDC treatment has been used to raise the liver concentration of protoporphyrin IX. MBR binding was measured in rat liver 5 h after treatment (Cantoni et al., 1992), when the increase in porphyrin concentrations peaks (De Matteis et al., 1973).

10.2.2 Experimental evidence

DDC treatment caused a 10-fold increase in porphyrin accumulation in rat liver (Table 1). Analysis of the hepatic porphyrin pattern by high performance liquid chromatography (HPLC) showed that in control rats the prevalent porphyrin was protoporphyrin, followed by coproporphyrin and uroporphyrin; trace amounts of heptacarboxylic and hexacarboxylic porphyrin were detectable. After treatment, the porphyrin pattern did not change but

Table 1 Effect of DDC treatment on porphyrins content and [^{3}H]PK 11195 binding in rat liver.

	Vehicle	Treated
	[^{3}H]PK 11195 binding	
Bound at 1 nM (pmol/g)	36.80 ± 14.7	**10.40 ± 10.7
B_{max} (pmol/g)	160.30 ± 81.0	*72.10 ± 21.0
K_d (nM)	3.56 ± 0.7	**15.37 ± 2.4
K_i (nM) of Ro5-4864	23.90 ± 5.4	*72.99 ± 37.6
	Porphyrin content (nmol/g)	
Total porphyrin	0.80 ± 0.23	**10.0 ± 3.1
Protoporphyrin IX	0.30 ± 0.09	**7.2 ± 2.6

Data are means ± SD of 6–10 animals per group. *$p < 0.05$; **$p < 0.01$ statistically different from vehicle, Student's t-test.
Male CD-COBS rats, weighing 180–200 g, fasted for 24 h, were given an intraperitoneal injection of DDC (100 mg/kg) dissolved in peanut oil (10 ml/kg) and were killed 5 h later by decapitation. Control animals were given oil alone. Livers were removed, rinsed in ice-cold saline, blotted dry and immediately frozen on dry ice. Tissues were kept at −80 °C until utilized for measuring [^{3}H]PK 11195 binding (Mennini *et al.*, 1989) and porphyrin content (Kennedy *et al.*, 1986). Modified from Cantoni *et al.* (1992).

protoporphyrin increased about 20 times in absolute terms (Table 1). This increase explains the rise in the total porphyrin content.

Table 1 summarizes the results obtained measuring [^{3}H]PK 11195 binding in the liver from rats treated with vehicle or DDC. The amount of [^{3}H]PK 11195 bound at 1 nM was significantly reduced after DDC treatment. This decrease was mainly due to a reduced affinity (K_d) of the ligand (from 3.56 to 15.37 nM, $p < 0.01$) although a reduction in the maximum number of binding sites (B_{max}) was also apparent (55% decrease, $p < 0.05$). The affinity (K_i) of Ro5-4864 for [^{3}H]PK 11195 binding sites was also significantly reduced in the liver from animals treated with DDC (from 23.9 to 72.99 nM, $p < 0.05$).

When protoporphyrin IX and N-methylprotoporphyrin IX were added *in vitro* to membrane preparations from liver of control rats, they reduced the affinity of [^{3}H]PK 11195 binding in a dose-dependent manner, N-methylprotoporphyrin IX being about 20 times more potent than protoporphyrin IX (K_i 0.25 and 4.5 μM respectively for N-methylprotoporphyrin IX and protoporphyrin IX). In the presence of protoporphyrin IX a non-significant lowering of the B_{max} was also observed, suggesting a complex inhibition (mixed type). *In vitro*, competition by porphyrins has already been reported using other tissues (Verma and Snyder, 1988; Hirsch *et al.*, 1988). When liver homogenates from control rats were incubated for 120 min at 4 °C in the absence or presence of two concentrations of N-methylprotoporphyrin IX (1 and 10 μM) and of protoporphyrin IX (5.7 and 16 μM), and membranes prepared by centrifugation and four washes before measuring [^{3}H]PK 11195 binding, binding was still significantly inhibited even after the washing procedure. The inhibition was about 50% when 10 μM N-methylprotoporphyrin IX was added at the beginning and amounted to 15 and 34% after addition of 5.7 or 16 μM protoporphyrin IX respectively (Cantoni *et al.*, 1992).

10.2.3   Discussion

Convincing evidence that porphyrins interact functionally *in vivo* with MBR is provided by the reduction in [^{3}H]PK 11195 binding in the liver of DDC-treated rats as a result of a specific increase of N-methylprotoporphyrin IX and protoporphyrin IX. In fact, other possible modulators of MBR binding, like DBI and corticosterone, were not modified in the liver and plasma, respectively, of these treated rats (Cantoni *et al.*, 1992). Moreover the decrease was not due to a concomitant loss of ATP in DDC-treated rats, since the addition of 1 mM ATP did not affect MBR binding in control animals and did not reverse the decrease in treated animals (Cantoni *et al.*, 1992).

Porphyrins are highly lipophilic compounds, and cannot easily be removed from tissues during membrane preparation, as demonstrated by the significant inhibition of [^{3}H]PK 11195 binding measured in membranes even after four washings. This suggests that the reduction of [^{3}H]PK 11195 binding in the liver of rats treated with DDC might indicate some competition of endogeneous porphyrins, accumulated *in vivo* in the cell and still present in the membrane preparation. This was in fact confirmed by the observation that the affinity of the agonist Ro5-4864 for liver MBR was also reduced in DDC-treated rats. The basal concentration of protoporphyrin IX in rat liver is about 0.3 μM, and increases to 7 μM after DDC. Since its K_i value *in vitro* is 4.5 μM, this makes it unlikely that in basal conditions protoporphyrin IX interacts with MBR but, in conditions where it accumulates, like the porphyrias, this concentration is fully compatible with the possibility that it acts *in vivo* as an endogenous ligand at MBR. The idea that in basal conditions protoporphyrin IX accumulates in selected subcellular fractions where it reaches concentrations high enough to interact with MBR cannot be ruled out. Accumulation of liver porphyrins after *in vivo* treatment with DDC appeared to reduce also the maximum number of [^{3}H]PK 11195 binding sites. This could be related either to the presence of protoporphyrin IX, which shows a similar tendency towards complex inhibition of [^{3}H]PK 11195 binding *in vitro*, or to adaptive changes in receptor number (desensitization) occurring *in vivo* during porphyrins exposure.

Whatever the mechanism resulting in decreased MBR binding, the experiment discussed here provides direct evidence that an *in vivo* increase in liver porphyrins significantly affects MBR binding.

Hepatic protoporphyria is associated with hepatomegaly, cholestasis, alterations in the microsomal cytochromes and tumour formation (Cantoni *et al.*, 1983; Hurst and Paget, 1963). Since the mechanisms underlying these effects have not been clarified, some of them might be the result of alterations by protoporphyrin IX accumulation of physiological hepatic mitochondrial processes modulated by MBR. Antagonists at MBR, like the isoquinoline carboxamide derivative PK 11195, have been proposed in clinical treatment of porphyria (Katz *et al.*, 1989).

Other types of porphyrias are accompanied by neurological symptoms, although there seems to be no relation between increased circulating porphyrins and neurological manifestations (Herrick *et al.*, 1989).

It is important to consider that, even in a condition in which hepatic accumulation of protoporphyrin IX occurs, as in DDC-treated rats, MBR binding in the brain is not affected (Cantoni *et al.*, 1992). This could be because porphyrins do not cross the blood–brain barrier to any significant extent (Golberg *et al.*, 1954). However, they could indirectly affect brain functioning through metabolic changes like, for instance, the reported reduction of hepatic tryptophan pyrrolase activity which results in increased brain tryptophan, serotonin and 5-hydroxy-indole acetic acid (Litman and Correia, 1983).

10.3 MBR interaction with diazepam binding inhibitor

10.3.1 Background

Diazepam binding inhibitor (DBI) (Figure 2) is the precursor of a family of different peptides which interact with either central or peripheral benzodiazepine receptors (Guidotti *et al.*, 1988).

DBI is an effector of ACTH-induced steroidogenesis in the adrenal gland, by binding to MBR (Besman *et al.*, 1989). In glial cells also, DBI stimulates neurosteroidogenesis through the C_{27} side-chain cleavage of cholesterol catalysed by a class of cytochrome P-450 (Baulieu and Robel, 1990). (For the review of MBR in steroidogenesis, see Chapter 5). The rate of activity of this enzyme appears to be proportional to the amount of cholesterol present

```
        1         10        20        30        40        50
        ·         ·         ·         ·         ·         ·
RAT     SQADFDKAAEEVKRLKTQPTDEEMLFIYSHFKQATVGDVNTDRPGLLDLKG
                             └────TTN────
                                          └────ODN────┘

MOUSE   ...E..............................................
BOVINE  ...E......H..K.A.............Y....I..E...M..F..
PORCINE ...E.E.....N..K.A.D..........Y....I..E...I.....
HUMAN   WGDLWLLPPSANPGTGTE.E.....RH...K.S........G.Y......I..E...M..FT.

        60        70        80
        ·         ·         ·
RAT     KAKWDSWNKLKGTSKENAMKTYVEKVEELKKKYGI

MOUSE   ...............S.....D..........
BOVINE  .....A..E......D..A.ID..........
PORCINE .....A..G......D..A.IN..........
HUMAN   .....A..E......D..LA.IN..........
```

Figure 2 Amino acid sequence of DBI and related fragments in different animal species. Solid lines represent the positions of rat DBI 17–50 peptide (TTN) and rat DBI 33–50 peptide (ODN). Modified from Costa and Guidotti (1991).

178

in the inner mitochondrial membrane, where the enzyme is located. Regulation of cholesterol transport to the inner mitochondrial membrane is mediated by DBI through the MBR (Mukhin *et al.*, 1989; Baulieu and Robel, 1990; Papadopoulos *et al.*, 1990). The neurosteroids produced by glial cells through MBR may in turn positively or negatively modulate the $GABA_A$ receptor by acting at its allosteric site (Majewska *et al.*, 1986; Puia *et al.*, 1990).

DBI levels increase parallel to DBI mRNA in cerebellum and hypothalamus of diazepam-tolerant rats (Miyata, 1987). No data are available on MBR binding in these experimental conditions.

To ascertain whether changes in DBI concentrations affect MBR binding in brain areas and adrenals, data obtained in a model of acute noise stress are available (Ferrarese *et al.*, 1991).

10.3.2 Experimental evidence

In the experiments by Ferrarese *et al.* (1991) a model of unstressed animals, the 'handling-habituated' rat (Biggio *et al.*, 1981), was used. The rats were habituated to handling prior to sacrifice twice daily for four consecutive days. On the fifth day, in order to be 'stressed', rats were transferred to new cages and moved into another room. A sharp loud noise (about 100 decibels) was made by quickly running a pencil twice over the cover of the cage. The rats were killed at different times after noise, by the same handling procedure as the one they were habituated to. The 'resting' or non-stressed rats (controls) were left in their home cage and killed in the same way as the 'stressed' animals (Bizzi *et al.*, 1983; Mennini *et al.*, 1989).

The levels of DBI in the different brain areas investigated in basal conditions and at various times after acute noise stress are shown in Table 2. At baseline, DBI levels were highest in hippocampus, cerebellum and hypothalamus. After acute noise stress, in the striatum, cerebral cortex and hypothalamus there was a small, insignificant increase of DBI, whereas in hippocampus the increase was significant. DBI levels began to increase at 15 and 30 min and were significantly higher ($+100\%$) by 90 and 120 min after stress; they returned to normal values by 360 min ($F = 11.4; p < 0.01$). Stress induced no change in DBI levels in the cerebellum.

The levels of DBI in adrenal gland at different times after stress are shown in the upper panel of Figure 3; using the one-way ANOVA test we found that acute noise stress had a significant effect on DBI levels ($F = 7.6; p < 0.01$). The time-course of DBI changes in adrenals was different from that in the hippocampus, with significantly higher levels 30 min after stress, and normal values at 90 min. The DBI peak in adrenals was seen earlier than in hippocampus. The middle panel of Figure 3 shows the time-course of the changes of MBR in adrenal gland. The receptor number parallels the DBI

Table 2 DBI-like immunoreactivity in different areas of rat brain after acute noise stress.

Brain areas	(N)	Baseline	Time after stress (min)				
			15	30	90	120	360
Cortex	(20)	60 ± 5.8	65 ± 6.9	68 ± 7.1	78 ± 8.0	58 ± 6.0	49 ± 5.2
Hippocampus	(30)	180 ± 17	250 ± 26	270 ± 25	380 ± 30*	370 ± 33*	190 ± 20
Striatum	(30)	70 ± 7.3	90 ± 9.2	80 ± 8.8	85 ± 9.2	75 ± 8.1	75 ± 7.7
Hypothalamus	(30)	150 ± 16	180 ± 19	170 ± 20	150 ± 15	160 ± 18	150 ± 14
Cerebellum	(30)	170 ± 18	160 ± 15	180 ± 20	170 ± 16	170 ± 18	170 ± 15

Data are pmol/mg prot. ± standard error of the mean (SEM) of different animals per group (numbers are reported in brackets).
*$p < 0.01$ ANOVA and Tuckey's test.
Peptide extraction, DBI radioimmunoassay and identification by reverse-phase high pressure liquid chromatography were performed according to the methods described by Ferrarese et al. (1987). Modified from Ferrarese et al. (1991).

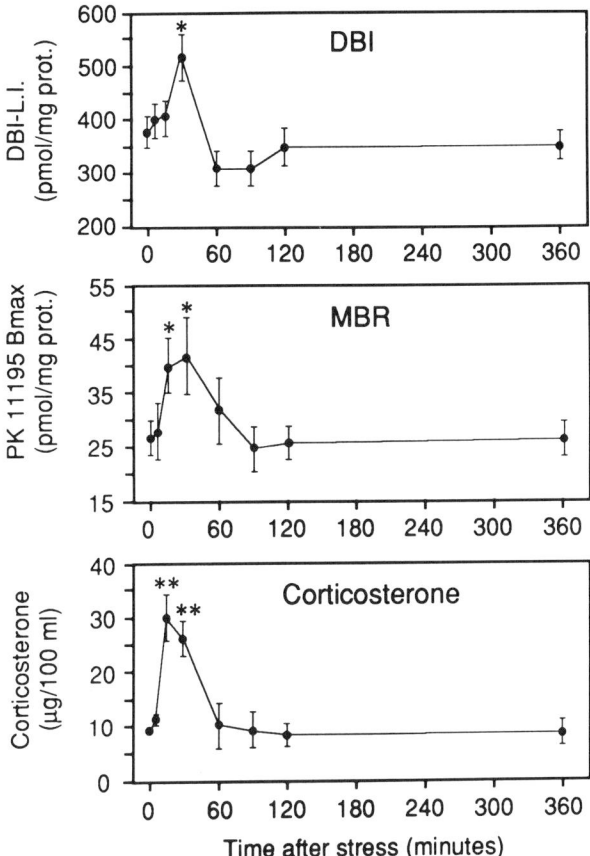

Figure 3 Stress-induced changes in DBI immunoreactivity and MBR binding in rat adrenal gland and stress-induced changes in plasma corticosterone levels. Data are means $\pm$ SEM of 15 animals per group. *$p < 0.05$ and **$p < 0.01$ versus resting animals (ANOVA). MBR were measured by the binding of [³H] PK 11195, as described by Mennini *et al.* (1989). Corticosterone was measured in plasma according to Guillemin *et al.* (1959). Peptide extraction, DBI radioimmunoassays and reverse-phase HPLC were performed according to the methods described by Ferrarese *et al.* (1991) using ¹²⁵I-Bolton–Hunter-labelled DBI. Modified from Ferrarese *et al.* (1991).

content in the same tissue and closely follows corticosterone levels in plasma (Figure 3, bottom panel).

In summary, in these experiments the MBR number increased by 30–40% after stress in both cortex and hippocampus. The increase in cortex was apparent from 15–30 min up till the last time tested (360 min), with maxima at 90 and 120 min; in the hippocampus the MBR number reached its maximum after 30 min, then slowly returned to normal values (Ferrarese *et al.*, 1991).

10.3.3 Discussion

DBI and its processing products increase selectively in the rat hippocampus after stress (Ferrarese *et al.*, 1991). These increases could indicate either reduced utilization or an increased synthesis of the peptide; however, the time-course of the DBI increase, and of its processing products, could indicate an increase of DBI synthesis and processing. Our aim was to correlate the observed DBI changes and the modifications of MBR in the same stress model. MBR increased early in cerebral cortex (Mennini *et al.*, 1989) and hippocampus, but persisted longer in the hippocampus (Ferrarese *et al.*, 1991).

The absence of significant DBI increases in cerebral cortex and the late increase in hippocampus (Ferrarese *et al.*, 1991) seem to exclude any direct relation between DBI and MBR changes in the brain. However, we cannot rule out the possibility of increased release of DBI at earlier times, which could not be detected by measuring the peptide levels in discrete brain regions.

The fact that the time-course of the increase in DBI in adrenals closely parallels the increase of MBR in that tissue (Ferrarese *et al.*, 1991) and the plasma corticosterone increase in the same model of stress (Bizzi *et al.*, 1983; Mennini *et al.*, 1989), further supports the view that DBI mediates the steroidogenic action of ACTH in the adrenal gland (Besman *et al.*, 1989).

In the same animal model of stress, CBR also decrease in the cortex and hippocampus at early times (30 min) and return to normal values within 90–120 min (Mennini *et al.*, 1989; Ferrarese *et al.*, 1991). Since adrenal DBI, MBR and plasma corticosterone undergo rapid modifications at early times after stress (15 min), adrenal steroids or neurosteroids could be responsible for the changes in BZ receptors and DBI in brain areas, particularly in the hippocampus. This is the brain region with the highest density of steroid receptors. It produces 'neurosteroids' (Jung-Testas *et al.*, 1989) which in turn can modulate GABAergic transmission (Majewska *et al.*, 1986). (For more information on the involvement of hippocampal MBR in memory consolidation, refer to Chapter 7.) The increase of DBI in the hippocampus of stressed rats could possibly lead to increased synthesis of 'neurosteroids' through MBR in glial cells, and these neurosteroids might have a negative feedback action, stopping activation of the hypothalamic–pituitary–adrenal axis. Thus, although no clear relationship is evident between the DBI increase in the brain and benzodiazepine receptors, this increase could, directly or indirectly, be related to changes in central and peripheral benzodiazepine receptors, and these concerted modifications might participate in the mechanisms by which the brain regains its 'homeostasis', preventing the nervous system from over-reacting to stressful stimuli (Ferrarese *et al.*, 1991).

10.4 Conclusion

The findings discussed in the present chapter convincingly show that increased tissue concentrations of porphyrins or DBI induce modifications in MBR binding, further supporting their *in vivo* functional interaction with MBR. In both cases a direct relationship could be seen in the periphery (liver and adrenals), although secondary (indirect) effects in the brain may arise through changes in neurosteroids or enzymes involved in brain functioning.

Tissue DBI concentrations in basal conditions are high, approaching its affinity for MBR. Conversely, the basal concentration of protoporphyrin IX in rat liver is about 15 times lower than its affinity for MBR. This makes it unlikely that there is a physiological interaction of protoporphyrin IX with MBR, unless selective accumulation in selected subcellular fractions reaches concentrations high enough to bind to MBR. In conditions where protoporphyrin IX accumulates by a factor of 20 times, like the porphyrias, its concentration is fully compatible with the possibility that it acts *in vivo* as an endogenous ligand at MBR.

The accumulation of porphyrins in the liver is possibly directly related to a decrease in MBR binding, due to increased competition by the endogenous ligands and to possible receptor desensitization. The increase of DBI in the adrenals, on the other hand, is associated with increased MBR binding and increased corticosterone secretion. This suggests that DBI 'unmasks' MBR in order to produce the required physiological response.

MBR is associated with a protein in the outer mitochondrial membrane (Anholt *et al.*, 1986). The conversion of cholesterol into pregnenolone, a common intermediate for all steroid hormone biosynthesis, requires transport of cholesterol from extramitochondrial stores to the inner mitochondrial membrane at the site where the C_{27} side-chain cleavage cytochrome P-450 enzyme is located (Mukhin *et al.*, 1989; Papadopoulos *et al.*, 1990). (Chapter 4 reviews the role of MBR in steroidogenesis.) MBR may participate in a physiologically relevant way in this process by binding to an endogenous ligand, like DBI or its cellular processing products. MBR is linked to the mitochondrial energy metabolism by the close overlapping of the distribution of MBR and cytochrome oxidase activity, shown in whole body section autoradiography (Anholt *et al.*, 1985), and directly demonstrated by Hirsch *et al.* (1988). However, in this case too, subcellular localizations diverge since oxidative phosphorylation is carried out by respiratory assemblies located in the inner membrane of the mitochondrion, while MBR are located on the outer mitochondrion. This makes a direct involvement in oxidative phosphorylation unlikely, although MBR could indirectly interfere with transfer of electrons or other components. If MBR control the exchange of substrates or products between mitochondria and cytoplasm, this might explain their pleiotropic effects on different cells and different tissue, since energy supply influences cellular metabolism. The possible therapeutic use of MBR ligands is not clear. On the basis of the data discussed above, MBR agonists and antagonists might be considered as modulators of cell metabolism, particularly of cholesterol utilization and will fine tune various cellular functions (in response to stress) via steroidogenesis or other intermediates.

References

Anholt, R.R.H., De Souza, E.B., Oster-Granite, M.L. and Snyder, S.H. (1985). Peripheral-type benzodiazepine receptors: autoradiographic localization in whole-body sections of neonatal rats. *J. Pharmacol. Exp. Ther.* **233**, 517–526.

Anholt, P.R.H., Pedersen, P.L., De Souza, E.B. and Snyder, S.H. (1986). The peripheral-type benzodiazepine receptor. Localization to the mitochondrial outer membrane. *J. Biol. Chem.* **261**, 576–583.

Baulieu, E.E. and Robel, P. (1990). Neurosteroids: a new brain function? *J. Steroid Biochem. Mol. Biol.* **37**, 395–403.

Beaumont, K., Skowronski, R., Vaughn, D.A. and Fanestil, D.D. (1988). Interactions of lipids with peripheral type benzodiazepine receptors. *Biochem. Pharmacol.* **37**, 1009–1014.

Besman, M.J., Yanagibashi, K., Lee, T.D., Kawamura, M., Hall, P.F. and Shively, J.E. (1989). Identification of des-(Gly-Ile)-endozepine as an effector of corticotropin-dependent adrenal steroidogenesis: stimulation of cholesterol delivery is mediated by the peripheral benzodiazepine receptor. *Proc. Natl. Acad. Sci. USA* **86**, 4897–4901.

Biggio, G., Corda, M.G., Concas, A., De Montis, G., Rossetti, Z. and Gessa, G.L. (1981). Rapid changes in GABA binding induced by stress in different areas of the rat brain. *Brain Res.* **229**, 363–369.

Bizzi, A., Ricci, M.R., Veneroni, E., Amato, M. and Garattini, S. (1983). Benzodiazepine receptor antagonists reverse the effect of diazepam on plasma corticosterone in stressed rats. *J. Pharm. Pharmacol.* **36**, 134–135.

Cantoni, L., Di Padova, C., Rovagnati, P., Ruggeri, R., Dal Fiume, D. and Tritapepe, R. (1983). Bile secretion and liver microsomal mixed function oxidase system in mice with griseofulvin-induced hepatic protoporphyria. *Toxicology* **27**, 27–39.

Cantoni, L., Rossi, C., Rizzardini, M., Skoruptka, M., Cagnotto, A., Codegoni, A., Pecora, N., Ferrarese, C., and Mennini, T. (1992). Hepatic protoporphyria is associated with decreased ligand binding to the peripheral-type benzodiazepine receptor in the liver. *Biochem. Pharmacol.* (in press).

Costa, E. and Guidotti, A. (1991). Minireview: diazepam binding inhibitor (DBI): a peptide with multiple biological actions. *Life Sci.* **49**, 325–344.

De Matteis, F. and Rimington, C. (1963). Disturbance of porphyrin metabolism caused by griseofulvin in mice. *Br. J. Dermatol.* **75**, 91–104.

De Matteis, F., Abbritti, G. and Gibbs, A.H. (1973). Decreased liver activity of porphyrin–metal chelatase in hepatic porphyria caused by 3,5-diethoxycarbonyl-1,4-dihydrocollidine. Studies in rats and mice. *Biochem. J.* **134**, 717–727.

De Matteis, F., Gibbs, A.H. and Holley, A.E. (1987). Occurrence and biological properties of *N*-methyl propoporhyrin. *Ann. NY Acad. Sci.* **514**, 30–40.

De Matteis, F., Gibbs, A.H. and Tephly, T.R. (1980). Inhibition of protohaem ferro-lyase in experimental porphyria. *Biochem. J.* **188**, 145–152.

Ferrarese, C., Vaccarino, F., Alho, H., Mellstrom, B., Costa, E. and Guidotti, A. (1987). Subcellular location and neuronal release of diazepam binding inhibitor. *J. Neurochem.* **48**, 1093–1102.

Ferrarese, C., Mennini, T., Pecora, N., Pierpaoli, C., Frigo, M., Marzorati, C., Gobbi, M., Bizzi, A., Codegoni, A., Garattini, S. and Frattola, L. (1991). Diazepam binding inhibitor (DBI) increases after acute stress in rats. *Neuropharmacology* **30**, 1445–1452.

Gavish, M. and Fares, F. (1985). Solubilization of peripheral benzodiazepine-binding sites from rat kidney. *J. Neurosci.* **11**, 2889–2893.

Giesen-Crouse, E. (1990). Peripherally acting benzodiazepines: do they hold pharmacological potential? *Rev. Neurosci.* **2**, 215–239.

Golberg, A., Paton, W.D.M. and Thompson, J.W. (1954). Pharmacology of the porphyrins and porphobilinogen. *Br. J. Pharmacol.* **9**, 91–94.

Guidotti, A., Forchetti, C.M., Corda, M.G., Konkel, D., Bennet, C.D. and Costa, E. (1983). Isolation, characterization and purification to homogeneity of an endogenous polypeptide with agonistic action on benzodiazepine receptors. *Proc. Natl. Acad. Sci. USA* **80**, 3531–3533.

Guidotti, A., Berkovich, A., Ferrarese, C., Santi, M.R. and Costa, E. (1988). Neuronal–glial differential processing of DBI to yield ligands to central or peripheral benzodiazepine recognition sites. In *Imidazopyridines in Sleep Disorders* (eds Sauvanet, P., Langer, S.Z. and Morselli, P.L.), pp. 25–38. Raven Press, New York.

Guillemin, R., Clayton, G.W., Lipscomb, H.S. and Smith, I.D. (1959). Fluorimetric measurement of rat plasma and adrenal corticosterone concentration. *J. Lab. Clin. Med.* **53**, 830–832.

Herrick, A.L., McKoll, K.E.L. and Moore, M.R. (1989). Peripheral benzodiazepine receptors and treatment of porphyria. *Lancet* **1**, 1264–1265.

Hirsch, J.D., Beyer, C.F., Malkowitz, L., Loullis, C.S. and Blume, A. (1988). Characterization of ligand binding to mitochondrial benzodiazepine receptors. *Mol. Pharmacol.* **35**, 164–172.

Hurst, E.W. and Paget, G.E. (1963). Protoporphyrins, cirrhosis and hepatomata in the livers of mice given griseofulvin. *Br. J. Dermatol.* **75**, 105–112.

Ibraham, N.G., Friedland, M.L. and Levere, R.D. (1983). Heme metabolism in erythroid and hepatic cells. *Progr. Hematol.* **13**, 75–130.

Jung-Testas, I., Hu, Z.H., Baulieu, E.E. and Rabel, P. (1989). Neurosteroids: biosynthesis of pregnenolone and progesterone in primary cultures of rat glial cells. *Endocrinology* **125**, 2083–2091.

Katz, Y., Weizman, A. and Gavish, M. (1989). Ligands specific to peripheral benzodiazepine receptors for treatment of porphyrias. *Lancet* **1**, 932–933.

Kennedy, S.W., Wigfield, D.C. and Fox, G.A. (1986). Tissue porphyrins pattern determination by high-speed-high-performance liquid chromatography. *Anal. Biochem.* **157**, 1–7.

Litman, D.A. and Correia, M.A. (1983). 1-Tryptophan: a common denominator of biochemical and neurological events of acute hepatic porphyria? *Science* **222**, 1031–1033.

Majewska, M.D., Harrison, N.L., Schwartz, R.D., Barker, J.L. and Paul, S.M. (1986). Steroid hormone metabolites are barbiturate-like modulators of the GABA receptor. *Science* **232**, 1004–1007.

Mantione, C.R., Goldman, M.E., Martin, B., Bolger, G.T., Luddens, H.W., Paul, S.M., Seeburg, P.H. and Costa, E. (1988). Purification and characterization of an endogenous protein modulator of radioligand binding to peripheral type benzodiazepine receptors and dihydropyridine Ca^{2+}-channel antagonist binding sites. *Biochem. Pharmacol.* **37**, 339–347.

Mennini, T., Gobbi, M. and Charuchinda, C. (1989). Noise-induced opposite changes in central and peripheral benzodiazepine receptors in rat brain cortex. *Neurosci. Res. Commun.* **5**, 27–34.

Miyata, M., Mocchetti, I., Ferrarese, C., Guidotti, A. and Costa, E. (1987). Protracted treatment with diazepam increases the turnover of putative endogenous ligands for the benzodiazepine/beta-carboline recognition site. *Proc. Natl. Acad. Sci. USA* **84**, 1444–1448.

Moore, M.R., McColl, K.E.L., Rimington, C. and Golderg, A. (1987). *Disorders of Porphyrins Metabolism*, pp. 139–164. Plenum Press, New York.

Mukhin, A.G., Papadopoulos, V., Costa, E. and Krueger, K.E. (1989). Mitochondrial benzodiazepine receptors regulate steroid biosynthesis. *Proc. Natl. Acad. Sci. USA* **86**, 9813–9816.

Papadopoulos, V., Mukhin, A.G., Costa, E. and Krueger, K.E. (1990). The peripheral-type benzodiazepine receptor is functionally linked to Leydig-cell steroidogenesis. *J. Biol. Chem.* **265**, 3772–3779.

Puia, G., Santi, M.R., Vicini, S., Pritchett, D.B., Purdy, R.H., Paul, S.M., Seeburg, P.H. and Costa, E. (1990). Neurosteroids act on recombinant human $GABA_A$ receptors. *Neuron* **4**, 759–765.

Schoemaker, H., Bolger, R.G., Horst, D. and Yamamura, H.I. (1983). Specific high affinity binding sites for [^{3}H]Ro5-4864 in rat brain and kidney. *J. Pharmacol. Exp. Ther.* **225**, 61–69.

Smith, A.G. and De Matteis, F. (1980). Drugs and the hepatic porphyrias. *Clin. Haematol.* **9**, 399–425.

Verma, A. and Snyder, S.H. (1988). Characterization of phorphyrin interactions with peripheral type benzodiazepine receptors. *Mol. Pharmacol.* **34**, 800–805.

Verma, A. and Snyder, S.H. (1989). Peripheral type benzodiazepine receptors. *Ann. Rev. Toxicol.* **29**, 307–322.

Verma, A., Jeffrey, S.N. and Snyder, S.H. (1987). Porphyrins are endogenous ligands for the mitochondrial (peripheral-type) benzodiazepine receptor. *Proc. Natl. Acad. Sci. USA* **84**, 2256–2260.

_____ CHAPTER 11 _____

MODIFICATIONS OF MITOCHONDRIAL BENZODIAZEPINE RECEPTOR NUMBERS IN STRESSFUL SITUATIONS

Robert H. Dodd and Maryse Lenfant

Institut de Chimie des Substances Naturelles, Centre National de la Recherche Scientifique, 91198 Gif-sur-Yvette Cedex, France

Table of Contents

PERIPHERAL BENZODIAZEPINE RECEPTORS
ISBN 0-12-282630-2

11.1   Introduction

The participation of the central benzodiazepine receptor (CBR) in neurological processes mediating anxiety and stress is widely acknowledged (for a recent review, see Drugan and Holmes, 1991). Ligands binding to this receptor can display either anti-anxiety properties (e.g. benzodiazepines in general, as well as other non-benzodiazepine agonists) or anxiogenic activities (e.g. the inverse agonists, including many derivatives of the β-carboline family). The CBR are coupled to the GABA receptor–chloride ionophore complex, modulating its activity allosterically. Depending on whether an agonist or an inverse agonist is bound to the CBR, GABAergic transmission is either enhanced or diminished, resulting in anxiolytic or anxiogenic activity (Haefely and Polc, 1986; Corda et al., 1986). It has been amply demonstrated that, depending on the characteristics of a stressful situation, CBR can be upregulated or downregulated, a phenomenon which may be important in the physiological management of stress (Braestrup et al., 1979; Soubrié et al., 1980; Medina et al., 1983a,b).

The discovery of apparently selective ligands for the mitochondrial benzodiazepine receptor (MBR) (that is, PK 11195 and Ro5-4864, in particular) (Benavides et al., 1983; Schoemaker et al., 1983) immediately raised the question whether these receptors which, unlike CBR, are not coupled to GABA receptors or chloride channels (Patel and Marangos, 1982; Martini et al., 1986), have a role to play in mediating an organism's response to stress. Indeed, PK 11195 was found to be anxiolytic (Mizoule et al., 1985) in the Vogel conflict model of anxiety. These effects were not antagonized by flumazenil, the prototypic CBR antagonist (Hunkeler et al., 1981), suggesting that the pharmacological actions of these MBR ligands were not due to an interaction with CBR (File and Pellow, 1985a; Mizoule et al., 1985). The modulation of immune response, itself sensitive to stress (Glaser et al., 1986, 1987), by MBR ligands (Lenfant et al., 1986) provides another indication that MBR come into play in the event of stressful stimuli. The recent demonstration that MBR may have a role in steroid synthesis (Mukhin et al., 1989; Yanagibashi et al., 1989) and that some steroid derivatives are known to potentiate GABA responses via a direct interaction with the CBR–GABA receptor complex (for a review, see Lambert et al., 1987), offers a further link between MBR and stress.

In the light of these observations, it is not surprising that, just as possible correlations between stress and CBR variations have been extensively investigated, much research over the last decade has been devoted to studying the manner in which stress affects MBR densities and affinities. This chapter reviews some of the observations, conclusions and speculations which currently pertain to this subject.

11.2 Stressful situations which cause a decrease in MBR densities

11.2.1 The inescapable tailshock model (rat) of acute stress

One of the first clear-cut demonstrations that stress has a measurable effect on MBR was provided by the use of a model in which inescapable tailshocks were used as stressor in rats (Drugan et al., 1988). Thus, after a 2 h session in which 80 five-second inescapable tailshocks were delivered, a reduction in [^{3}H]Ro5-4864 binding was observed in membranes from kidney (31%), heart (19%), pituitary (17%) and cerebral cortex (29%) as compared to naive rats. No changes in the affinity of this ligand for MBR were detected in these tissues. Moreover, there were no significant modifications in [^{3}H]Ro5-4864 binding (B_{max} and K_d) in membranes from hippocampus, lung and adrenal gland. When an anxiolytic dose of clonazepam was administered before the inescapable shock session, the shock-induced decrease in renal MBR binding was attenuated by 50%. Clonazepam itself had no effect on the B_{max} of renal MBR in naive rats. Since clonazepam is considered to be selective for CBR with only very low affinity for MBR (Braestrup and Squires, 1977), this result has been taken as an indication that the effects of stress on MBR are at least partly mediated by the CNS. Further evidence that stress-induced MBR modifications are under central nervous control was obtained when rats were treated with a sedative-hypnotic (20 mg/kg) or hypnotic (60 mg/kg) dose of sodium pentobarbital prior to inescapable shock (Drugan et al., 1990). As with clonazepam, sodium pentobarbital blocked the stress-induced decrease in [^{3}H]Ro5-4864 binding in rat kidney. However, this effect of sodium pentobarbital was much more robust than that of clonazepam since both doses of the former compound restored MBR densities on kidney membranes to control levels. On the other hand, a lower, non-sedative dose of sodium pentobarbital (5 mg/kg) had no effect on the reduction of MBR densities after shock. Although the protective effect of the 60 mg/kg dose of sodium pentobarbital could in part be attributed to analgesia produced at this dose, the 20 mg/kg dose was found to be devoid of analgesic effects as shown by its failure to alter tailflick latencies to radiant heat. The analgesic properties of benzodiazepines have also been demonstrated (Zambotti et al., 1991) and this should, of course, be taken into account in such studies.

The pharmacological actions of both clonazepam and pentobarbital are known to be mediated via their interaction with the benzodiazepine–GABA receptor complex (Study and Barker, 1981; Havoundjian et al., 1987). Since the latter is found only in the central nervous system, it seems probable that stress-evoked modifications of MBR are in some way influenced by activity at the level of the central nervous system. Whether it is exclusively central GABA-coupled benzodiazepine receptors which affect MBR under stressful situations remains to be shown. In this respect, it would be interesting to study the effects of anxiolytics whose mechanism of action involves neuronal systems other than the GABA–benzodiazepine receptor complex (e.g. the serotonergic anxiolytic buspirone) (Traber and Glaser, 1987) on stress-evoked MBR modifications. Non-specific actions of benzodiazepines such as clonazepam cannot be discounted, although this possibility could be obviated if clonazepam's blocking of stress-evoked MBR density reduction were shown to be antagonized, in turn, by flumazenil. It might also be expected that anxiogenic substances, in particular the inverse agonists of the CBR (e.g. the β-carbolines β-CCM or FG7142) (Prado de Carvalho et al., 1983; Corda et al., 1983), should either by themselves perturb MBR densities as a result of acute or chronic administration or have an additive effect with externally applied stressors (e.g. inescapable tailshock).

Inescapable shock has been shown to produce increased levels of circulating ACTH and corticosterone in rats (Maier et al., 1986). Hormones by themselves are known to affect MBR densities. Thus, two weeks after adrenalectomy, the binding of both $[^3H]Ro5$-4864 and $[^3H]PK$ 11195 was decreased in rat kidney by 21% and 23%, respectively (Basile et al., 1987). This effect was reversed after one week of aldosterone administration. Similarly, hypophysectomy diminished the B_{max} of $[^3H]Ro5$-4864 by more than 80% in both the adrenal gland and the testis after 23 days (Anholt et al., 1985). However, in contrast to adrenalectomy, hypophysectomy caused no significant changes in MBR density in the kidney. Based on this evidence, it seems reasonable to conclude that stress-evoked MBR variations may not only be under central nervous control but also under hormonal control. This has been confirmed by studies in which adrenalectomized rats were subjected to the inescapable tailshock paradigm described above (Drugan et al., 1988). These animals demonstrated a much larger reduction in renal $[^3H]Ro5$-4864 binding (43%) compared to sham-operated, shocked animals. Hypophysectomy, on the other hand, had no effect on the inescapable shock-induced MBR density reduction in kidney. However, in view of the observation that hypophysectomy does not by itself affect renal MBR densities, this result may not be surprising. Finally, female rats showed an attenuated reduction in renal MBR densities (23%) compared to males (55%) as a result of inescapable tailshock (Drugan et al., 1991). This sexual dimorphism in stress response may also be hormonal in origin since it has been shown that treatment of male rats with oestradiol over 10 days results in an upregulation of MBR in the kidney but a downregulation in the testis (Gavish et al., 1986a).

11.2.2 The Maudsley reactive rat model of anxiety

The study of Maudsley reactive rats lends credence to the likelihood of a correlation between high levels of chronic stress and a decrease in the densities of MBR in certain tissues. The Maudsley reactive rats have been selectively bred to display a high level of fearfulness or stress (Broadhurst, 1960). Using [^{3}H]Ro5-4864 as specific MBR ligand, it has been shown that these rats display decreased binding levels in heart (17%) and kidney (20%) as compared to the normal, Maudsley non-reactive rat (Drugan et al., 1987). These changes were tissue specific since the B_{max} of MBR in the central nervous system (cortex, hypothalamus, hippocampus) as well as in other peripheral tissues (lung, pituitary, adrenal) were essentially identical in the Maudsley reactive and non-reactive rats. These results parallel those found using inescapable tailshock as stressor as far as heart and kidney are concerned though in the latter paradigm, MBR were also modified in pituitary and cortical membranes. The mechanism by which heart and kidney modulate (or are modulated by) responses to stress is as yet unknown, as is the reason why, in the Maudsley reactive rat, tissues which are known to react under stress do not show MBR modifications. It may simply be that the chronic state of 'fearfulness' in these animals has allowed adaptive changes to occur in these organs.

11.2.3 The baclofen responding mouse model of anxiety

An anxiety model similar to the Maudsley reactive rat is that of the baclofen responding mouse (Rägo, 1986). Mice show different sensitivities to the sedative properties of baclofen, a $GABA_B$ receptor agonist (Hill and Bowery, 1981). Behavioural tests have shown baclofen responders (i.e. those showing a greater decrease in motor activity compared to baclofen non-responders) to be in a higher state of 'anxiousness' (Rägo et al., 1989). As in the Maudsley reactive rat, baclofen responders displayed a significantly lower number of MBR in kidney and heart (Rägo et al., 1989). CBR densities were also lower in the forebrain and cerebellum of the baclofen responders. From these two stress models, it seems that chronic stress can be associated with a decrease in MBR densities in selected organs, paralleled by a similar modification in CBR in brain.

11.2.4 Mediation of stress-related effects by putative endogenous ligands

It has been suggested that the decrease in the number of MBR observed in various tissues after stress is the result of the release of an endogenous ligand which, by occupying some of these sites, leads to an apparent decrease in B_{max} as determined using exogenously applied radioactive ligands (e.g. $[^3H]$Ro5-4864). For example, rats stressed by placing them for 1.5 h in restraint cages at 4 °C (cold-restraint stress) showed, as in the tailshock paradigm, 57% and 43% decreases in the B_{max} of $[^3H]$Ro5-4864 in the heart and kidney, respectively (Armando *et al.*, 1988) with no changes in K_d values in these tissues, as compared to controls. Cold-restraint stress had no effect on $[^3H]$flunitrazepam binding to brain (cerebellum, cerebrum) membranes. Extracts from these stressed tissues were also shown to inhibit both $[^3H]$flunitrazepam binding (though no data are available concerning inhibition of MBR-specific $[^3H]$Ro5-4864) and monoamine oxidase (MAO) activity to a greater extent than the non-stressed tissue extracts. It was thus postulated that an enhanced release, in stressed animals, of tribulin, a putative benzodiazepine receptor endogenous ligand able to displace both $[^3H]$Ro5-4864 (MBR) and $[^3H]$clonazepam (CBR) and which also displays MAO inhibitory activity, could account for these observations (Elsworth *et al.*, 1984; Armando *et al.*, 1986). Tribulin being a non-competitive, reversible inhibitor of Ro5-4864 binding, the B_{max} values for the latter ligand would tend to be lower in peripheral tissues of stressed animals. However, this would not explain why a corresponding decrease in $[^3H]$flunitrazepam's B_{max} was not observed in the brain tissues examined since tribulin also reportedly binds to CBR. This discrepancy was ascribed to an incomplete investigation of B_{max} variations in different brain areas, notably the cortex. This is particularly relevant in view of the fact that stress-induced decreases in CBR B_{max} in cerebral cortex and hippocampus have been described (Medina *et al.*, 1983b). Here again, the role of a stress-provoked release of another possible endogenous ligand of the benzodiazepine receptor, known as diazepam binding inhibitor (DBI) was suggested. Unlike tribulin, DBI interacts only with CBR (Alho *et al.*, 1985). However, a protein structurally related to DBI which enhances steroid synthesis has recently been isolated from bovine adrenal (Besman *et al.*, 1989). MBR having been implicated in steroidogenesis and the products of this metabolism being known to modulate the CBR–GABA receptor complex (Lambert *et al.*, 1987; Mukhin *et al.*, 1989), the possibility of a CBR–MBR–endogenous ligand–steroid network which becomes operational upon an organism's exposure to stress is real. (For more information on MBR and steroidogenesis, see also Chapter 4.)

11.2.5 The starvation-induced chronic stress model (rat)

The effects of starvation-induced stress on MBR densities have also been studied. Thus, after 5 days of food deprivation, a decrease in the B_{max} of MBR of rats was observed in the heart (34%), kidney (33%) and adrenal

(35%) (Weizman *et al.*, 1990). Refeeding for 5 days restored MBR densities to control values in heart and kidney but not in adrenal. No changes were observed in [^{3}H]PK 11195 binding to hypothalamus, cortex and ovary as a result of food deprivation. These results differ from those obtained by stress produced by tailshock wherein no changes were detected in adrenal MBR. Food deprivation is known to produce increases in corticosterone levels, thereby suppressing ACTH release (Young *et al.*, 1987). Since the latter hormone has a positive influence on adrenal MBR densities (Anholt *et al.*, 1985; Fares *et al.*, 1989) the diminished numbers of MBR on this organ in food-deprived animals can in part be explained by a non-stress derived factor. Similarly, MBR modifications in heart and kidney caused by starvation could be related to perturbations in metabolism. Starvation-related reductions in thyroid hormone secretion (Connors *et al.*, 1985) and/or in renal electrolyte transport (Thompson *et al.*, 1987), both systems having known links with MBR (Gavish *et al.*, 1986b; Lueddens and Skolnick, 1987), could account for the observed MBR density variations.

11.3 Stressful situations which cause an increase in MBR densities

11.3.1 The forced swimming model of acute stress (rat)

The use of forced swimming to produce acute stress in rats resulted, after 15 min, in an increase in [^{3}H]Ro5-4864 binding in kidney (50%) and olfactory bulb (37%) with practically no change in the affinity (Novas *et al.*, 1987). No changes in [^{3}H]Ro5-4864 binding characteristics were observed in heart and cortex as a result of swimming stress. On the other hand, the number of CBR in cortex was decreased by 29% (using [^{3}H]flunitrazepam as ligand). It has been hypothesized that the increased release of adrenocortical hormones provoked by stress may be connected with the concomitant enhancements of MBR densities. This is based on the previously cited observation that adrenalectomy reduces the number of renal MBR sites by 21% and that this reduction can be reversed by administration of aldosterone (Basile *et al.*, 1985, 1987). However, this argument was similarly invoked to explain the *decrease* in kidney MBR density following stress (via ACTH, see above). The role of adrenal hormones in these phenomena is thus far from being clear. The recently reported ability of medetomidine, a selective α-2-adrenoceptor agonist, to antagonize increases of MBR binding caused by swimming stress may help to clarify this issue (Rägo *et al.*, 1991).

The effects of forced swimming stress on MBR densities were also studied using a shorter swimming time (5 min instead of 15 min; Rägo et al., 1989). This was done in order to prevent possible complications arising from physical exhaustion in the test animals. In this situation, a significant increase in the density of benzodiazepine receptors both in kidney (42%) and in cortex (29%) was observed using [^{3}H]flunitrazepam as ligand. The latter result may indicate an increase in both CBR and MBR densities since no data were given for [^{3}H]Ro5-4864 binding in cortex. A more robust decrease in binding affinity was observed in kidney in the 5-min stressed animals as compared to the 15-min group, although again, different radioactive ligands were used for these determinations. Interestingly, using [^{3}H]Ro5-4864 as probing ligand, a 110% increase in platelet MBR B_{max} was observed in the 5-min forced swimming test (Rägo et al., 1990b).

Correlations between stress and receptor densities are apparently very sensitive to the test conditions. For instance, when inescapable tailshock (discussed above) was employed as stressor, the number of 5-s shocks delivered had a bearing upon whether MBR densities were decreased or increased in tissues. While 80 shocks decreased the B_{max} of [^{3}H]Ro5-4864 in kidney by 31%, the use of only five shocks led instead to a significant increase (35%) in MBR numbers in this organ (Drugan et al., 1986). An intermediate number of shocks (20) resulted in decreased MBR densities, though less pronounced (22%) than that produced by 80 shocks. The 5 and 20 shock group, however, showed no changes in MBR densities in the cerebral cortex, in contrast to 80 shocks, which effectively decreased cortical MBR numbers (29%) as discussed previously. The immediate conclusion which can be drawn from these studies is that acute stress (e.g. 5 shocks or 5–15 min swimming) increases MBR densities in certain specific tissues (for example, kidney) while situations approaching chronic stress (80 shocks) appear to decrease MBR densities in the same tissues. Adaptive changes, perhaps mediated by hormonal feedback mechanisms, may be implicated in this so-called 'graded' response of MBR to stress duration. MBR may also be responding to stress-evoked modifications in GABA receptor-associated CBR since these also show increases and decreases in density variations depending on the stress situation. However, there is no clear correlation between stress-induced MBR and CBR modifications. Thus, while 5 min swimming stress increases both MBR and CBR densities in kidney and cerebral cortex, respectively, 15 min forced swimming, though still increasing MBR numbers in kidney, decreases CBR numbers in cortex.

11.3.2 Comparison of handling habituated and naive animals

In other cases, changes in CBR occur without there being any changes in MBR. For instance, one stress model in rats involves their repeated exposure to the handling preceding sacrifice. These handling-habituated animals, at the time of actual sacrifice, are considered to be less stressed than naive rats (Biggio, 1983). An investigation of receptor densities in both types of animals showed that while [^{3}H]flunitrazepam binding was 31% higher in the cortex of naive (stressed) animals, the differences in kidney were non-significant (Rägo et al., 1989). This result could be another example of 'graded' stress, the handling stress being just at the threshold of acute (increased MBR B_{max}) and chronic (decreased MBR B_{max}) stress.

When an acute stress (e.g. noise) was inflicted on handling-habituated (non-stressed) rats, an increase of 80% in [^{3}H]PK 11195 binding in cortex was observed (no data being available for kidney tissue) while the B_{max} for [^{3}H]flunitrazepam binding in this organ was decreased by 30–40% (Mennini et al., 1989). The K_d values for both ligands remained unchanged. Noise stress thus represents yet another pattern in the stress-evoked bidirectional regulation of MBR and CBR densities in that this paradigm actually increases MBR densities in cortex. It is also, however, one of the few cases where [^{3}H]PK 11195 was used as probing ligand.

It seems, then, that in addition to the type, intensity and duration of the stress employed in these investigations, the nature of the radioactive ligand used to probe resulting MBR modifications may also affect results. There is considerable evidence that within the same species, distinct binding sites for PK 11195 and for Ro5-4864 exist. Furthermore, across different species, the relative numbers of binding sites for these two ligands are not constant (Awad and Gavish, 1987; Skowronski et al., 1987; Hirsch et al., 1988). Comparisons of variations in binding characteristics of MBR ligands as a result of different types of stress in the same or in different species are clearly subject to caution.

11.3.3 Influence of baclofen on MBR

An intriguing clue as to the possible functional link between CBR and MBR comes from experiments performed in the presence of GABA agonists. It has been well documented that, in vitro, addition of $GABA_A$ agonists to the incubation medium enhances the affinity of benzodiazepine agonists for the CBR and vice versa (Costa and Guidotti, 1979; Costa, 1983). This phenomenon has been interpreted as an indication of the close functional association or coupling which exists between these two types of receptor, both being in fact part of the same macromolecular complex (Costa, 1983; Olsen et al., 1984). On the other hand, no such relationship has yet been found in vitro between GABA receptors and MBR, a not surprising finding since these two receptor types are known to be structurally distinct (Patel and Marangos, 1982; Schoemaker et al., 1983; Martini et al., 1986). (For more information on the structure of the MBR refer to Chapter 1.) However, it has been shown that *in vivo* animals treated with muscimol (a $GABA_A$ agonist) or baclofen (a $GABA_B$ agonist) displayed significantly higher affinities for [^{3}H]flunitrazepam in kidney as well as in cortex compared to untreated animals (Rägo *et al.*, 1989). Baclofen, moreover, produced a decrease in MBR B_{max} in kidney (38%) as well as a decrease in CBR B_{max} in cortex (26%). Baclofen would thus appear to have an effect on MBR opposite to that of acute stress. Since baclofen is also known to inhibit the increase in prolactin levels resulting from stress (D'Eramo *et al.*, 1986), $GABA_B$ agonists via MBR modulation may have an important role in stress protection. This was shown more concretely by use of another $GABA_B$ agonist, phenibut (a non-chlorinated analogue of baclofen) (Rägo *et al.*, 1982). When animals were treated with phenibut before being subjected to 5 min forced swimming, the usual stress-evoked increase in MBR density was almost completely inhibited in platelets as was the increase in CBR densities in cortex and hippocampus (Rägo *et al.*, 1990b). Although phenibut has negligible anxiolytic activity by itself, it can, like baclofen, counteract the anxiogenic effects of DMCM (a β-carboline derivative acting through the CBR) (Petersen, 1983) in an elevated plus-maze. Thus, while phenibut and baclofen interact uniquely with the $GABA_B$ receptor (which, in turn, is independent of the benzodiazepine receptor and the chloride ionophore; Majewska and Chuang, 1984), these compounds are nevertheless able to protect against stress and this protection apparently involves MBR as well as CBR.

11.3.4 Effect of hyperthyroidism on MBR

Hyperthyroidism has been associated with behavioural symptoms resembling anxiety (Singhal and Rastogi, 1981) and, as such, provides another model allowing the study of MBR modifications by stress. Thus, hyperthyroidism induced in rats by chronic administration of D-thyroxin resulted in an increased binding of [^{3}H]PK 11195 to kidney (23%), heart (18%) and testis (15%) (Gavish *et al.*, 1986b). A more modest upregulation of CBR in cortex (13%) was also observed. Paradoxically, though it would seem reasonable to view this model as one of chronic stress, the increased MBR B_{max} produced is more characteristic of acute stress. Barring a direct action of D-thyroxin on benzodiazepine receptors, it can be assumed that the modifications in MBR numbers produced by hyperthyroidism may be due to a compensatory hormonal effect whose relevance to stress is less than certain. It is interesting to note, however, that benzodiazepines are able to antagonize the locomotor effects of hyperthyroidism (Singhal and Rastogi, 1981).

11.3.5 Effect of surgical stress on MBR

Surgery is also a recognized source of stress (Noel *et al.*, 1972; Cohen *et al.*, 1981). Measurements made after abdominal wall surgery in rats showed that MBR densities in cortex and kidney had increased over the first 3 days following surgery (14% on day 1 and 29% on day 3 in kidney; 16% on day 1 and 30% on day 3 in cortex) in parallel to similar changes in cortical CBR (Okun *et al.*, 1988). Both types of receptors returned to control levels 7 days after surgery. This delay, corresponding to that of the end of the healing process, may reflect the presence of MBR on monocytes, directly implicated in this process (Ruff *et al.*, 1985). However, because surgery is a highly traumatic experience, affecting metabolism, hormone levels, respiration, etc., the association of surgery with stress, though no doubt valid, may be secondary to other events affecting MBR.

11.4 Stress, immune functions and MBR

The possibility that stress and MBR are associated in the modulation of the immune response observed in stressful situations is suggested by various studies.

The existence of relationships between the central nervous system and the immune system is now a generally accepted concept. Both central and immune systems share common signal molecules and receptors. It has been demonstrated that the central nervous system regulates the function of the peripheral immune tissues either by direct innervation or by hormonal mechanisms; in this latter case, the release of peptide mediators by the hypothalamus and pituitary can modulate both humoral and cell-mediated immunity (Keller *et al.*, 1980; Cross *et al.*, 1982). Besides a direct effect on the immune organs, the secreted peptides act on glands such as thyroid, adrenals or gonads and modify the concentration of the circulating hormones which are produced by these glands and which interfere with the immune system. These interactions are modulated by complex feedback mechanisms which link, in a reciprocal way, the central nervous to the immune system. In fact, the release of the regulatory peptides at the level of the central nervous system is controlled both by the circulating hormones and by the various cytokines which are secreted by the cells of the immune system during its modulation. Because of their numerous activities, cytokines might function as internal signals between the periphery and the central nervous system. Such complex phenomena have been demonstrated by numerous studies and reported in well documented reviews (Weigent and Blalock, 1987; Rabin *et al.*, 1988; Ader *et al.*, 1990; Spector, 1990).

The effect of stress on the immune system has been observed in clinical studies and animal models and both humoral and cellular immunity appear to be affected. For instance, reduced lymphoproliferative response to mitogens, reduced natural killer cell activity, and lower production of γ-interferon by peripheral leukocytes, have been observed both in college students during examination periods and in depressed patients (Dorian et al., 1982; Glaser et al., 1986, 1987). Similar modifications have been detected in stressful life-events such as bereavement (Bartrop et al., 1977; Irwin et al., 1987) or unemployment (Arnetz et al., 1987). Stress is not always immunosuppressive, and depending upon the stressor, enhanced immune responses have been observed (Monjan and Collector, 1977; Blecha et al., 1982; Croiset et al., 1987). A variety of factors seem to be involved in mediating the association between stress and immunity. The most studied hormones have been those of the hypothalamic–pituitary–adrenocortical axis. Various stressors have been shown to increase the release of corticosteroids which are known for their immunosuppressive effects (for a review, see Stein, 1985). The immune functions can be also modified through corticosteroid-independent mechanisms; moreover, catecholamines, endogenous opioids and pituitary hormones have been shown to modulate immune processes (Dantzer and Kelley, 1989). In particular, the role of growth hormone in the regulation of thymus gland, lymphoid cells, phagocytic cells and stem cells has to be emphasized (Kelley, 1989). These mediators interfere with such cells of the immune system as lymphocytes and macrophages through a direct interaction with specific receptors present on those cells. Different studies have demonstrated that immune responses might be modulated by MBR and several lines of evidence suggest that stress and MBR might be linked in the modulation of the immune response observed in stressful situations. The first important point is related to the immunomodulatory activity of benzodiazepines and the isoquinoline PK 11195 observed in mice (Descotes et al., 1982; Zavala et al., 1984; Lenfant et al., 1986) and to the detection of specific receptors for these molecules on immune tissues such as thymus, spleen, lymph nodes and Peyer patches (Benavides et al., 1989) and isolated cell populations such as thymocytes (Wang et al., 1980), circulating mononuclear cells (Pawlikowski et al., 1986) lymphocytes (Moigeon et al., 1983; Rägo et al., 1990a), macrophages (Zavala, 1984), monocytes (Ruff, 1985), platelets (Wang et al., 1980; Benavides et al., 1984; Moigeon et al., 1984), granulocytes (Bond et al., 1985), and mast cells (Taniguchi et al., 1980; Suzuki-Nishimura et al., 1989). (For a complete review of MBR in the immune system, see Chapter 6.) The influence of stress on the number of these receptors or their affinity for MBR ligands has not been extensively studied; however, interesting data on platelets have been reported. In an acute stress situation, it has been shown that normal human platelet MBR are modified, and the observed effect seems to depend on the nature of the stress applied: the number of MBR was found to be increased in students following examination (Karp *et al.*, 1989) but reduced in soldiers at the beginning of a parachute training course (Dar *et al.*, 1991). In both cases, no correlations between the observed modification and variations in cortisol, growth hormone (GH) or prolactin levels, all of which remained unchanged, could be established. A similar modulation was observed in general anxiety disorders in humans where the number of MBR appeared reduced, the difference being attenuated by treating the patients with diazepam (Weizman *et al.*, 1987). Whether this *in vivo* effect of diazepam is the result of an indirect action mediated by the central nervous system or to a direct action on the platelets, is open to conjecture, as diazepam can interact both with CBR and MBR; a downregulation of MBR through an action of diazepam on platelets cannot be excluded. Such a hypothesis was postulated to explain the effect of long-term diazepam treatment in anxious patients which restored the reduced binding density of benzodiazepine receptors in lymphocytes to normal levels, an effect which persisted after drug withdrawal (Ferrarese *et al.*, 1990). In fact, high concentrations of MBR in the pituitary and adrenals have been observed (De Souza *et al.*, 1985; Tong *et al.*, 1990) and the effect of MBR ligands on the regulation of the hypothalamic–pituitary axis has been studied both *in vivo* and *in vitro* on cultured pituitary corticotrophic cells (Calogero *et al.*, 1990). *In vivo* in rats, Ro5-4864 significantly stimulated ACTH and corticosterone secretion in a dose-dependent fashion. This effect could not be antagonized by PK 11195; however, given alone, this agent was shown to stimulate ACTH and corticosterone secretion. *In vitro*, Ro5-4864 stimulated corticotropin releasing hormone (CRH) but not pituitary ACTH secretion in hypothalamic organ cultures, whereas PK 11195 had no effect on CRH secretion but could antagonize the Ro5-4864 effect on CRH secretion. PK 11195 but not Ro5-4864 stimulated cultured pituitary cortico-trophic cells. These results suggest that Ro5-4864 stimulates the hypothalamic–pituitary axis by acting mainly at the central component of the axis, whereas PK 11195 acts at the pituitary level. These modulations could affect immune response and explain the effect of PK 11195 administered intra-cerebroventricular (i.c.v.) on the humoral response in mice (Lenfant *et al.*, 1988).

The presence of MBR on the key organs which are implicated through hormone secretion in the control of immune functions, and their modulation in stressful situations, suggests another possible mechanism of action for the MBR. For instance, the demonstration that steroid synthesis in adrenocortical cells (Besman *et al.*, 1989; Mukhin *et al.*, 1989; Yanagibashi *et al.*, 1989; Krueger and Papadopoulos, 1990) and testes (Ritta *et al.*, 1987) is controlled by MBR, and that analogues of brain diazepam binding inhibitor (DBI) enhance steroidogenesis (Besman *et al.*, 1989) illustrates this possibility.

11.5 Conclusion

A review of the data, some of it contradictory or at the very least discordant, leads to the suspicion that many of the stress-provoked events described lead to anecdotal modifications in MBR densities, these modifications being in fact the result of other, overriding physiological perturbations engendered during the course of the test (e.g. pain in surgery, analgesia in tailshock, physical exhaustion in forced swimming, hormonal variations in hyperthyroidism, fatigue after intensive preparation for examinations, etc.). Much of the data collected here may, in this light, be considered to be empirical. The only conclusive links between stress and MBR seem to be that firstly, stress modifies only the number of receptors and not the affinities of their ligands, and secondly, the modifications in B_{max} are restricted to certain tissues (though the nature of these tissues may vary depending on the stressor). This latter point would indicate that MBR variations due to stress are not simply a non-specific phenomenon.

There is also the question of whether the so-called selective MBR ligands used in all these studies are as specific as claimed. Besides the separate MBR binding sites for PK 11195 and Ro5-4864 already evoked above, the picture is further complicated by the possibility that the anxiolytic and anxiogenic properties, respectively, of these ligands are the result of their binding to a low-affinity site which is in fact allosterically coupled to the GABA receptor in a manner analogous to CBR (Gee, 1987; File and Pellow, 1985a).

A general pattern nevertheless emerges from these studies and that is, that acute stress tends to increase MBR densities while chronic stress leads to a decrease in MBR densities in selected tissues. A mutual interaction or feedback process between CBR and MBR in response to a stressful situation seems likely. 'Communication' between the peripheral and central benzodiazepine receptors could be assured by stress-promoted release or synthesis of endogenous ligands binding to both types of receptors (DBI and analogues, steroids, tribulin, etc.). Hormonal (particularly adrenal) and immune systems, whose responses to stress have already been well characterized, may also control or be controlled by MBR.

Finally, a functional interpretation of the role of CBR and MBR in stress response has recently been formulated (Drugan and Holmes, 1991). In this model, only CBR can respond to low-level, anticipatory or psychological stress while MBR and CBR together come into play as a result of reaction to more severe or physical stress, allowing the organism to cope. Though attractive, this model cannot yet account for the bi-directionality and tissue selectivity of MBR density variations in response to stress, a phenomenon which remains, for the moment, puzzling.

References

Ader, R., Felton, D. and Cohen, N. (1990). Interactions between the brain and the immune system. *Ann. Rev. Pharmacol. Toxicol.* **30**, 561–602.

Alho, H., Costa, E., Ferrero, P., Fujimoto, M., Cosenza-Murphy, D. and Guidotti, A. (1985). Diazepam-binding inhibitor: a neuropeptide located in selected neuronal populations of rat brain. *Science* **229**, 179–182.

Anholt, R.R.H., De Souza, E.B., Kuhar, M.J. and Snyder, S.H. (1985). Depletion of peripheral-type benzodiazepine receptors after hypophysectomy in rat adrenal gland and testes. *Eur. J. Pharmacol.* **110**, 41–46.

Armando, I., Glover, V. and Sandler, M. (1986). Distribution of endogenous benzodiazepine receptor ligand–monoamine oxidase inhibitory activity (Tribulin) in tissue. *Life Sci.* **38**, 2063–2067.

Armando, I., Levin, G. and Barontini, M. (1988). Stress increases endogenous benzodiazepine receptor ligand–monoamine oxidase inhibitory activity (tribulin) in rat tissues. *J. Neural. Transm.* **71**, 29–37.

Arnetz, B.B., Wasserman, J., Petrini, B., Brenner, S.O., Levi, L., Eneroth, B., Solavaana, H., Helm, R., Solovaana, L., Theorell, T. and Petterson, I.L. (1987). Immune function in unemployed women. *Psychosomatic Med.* **49**, 3–11.

Awad, M. and Gavish, M. (1987). Binding of [3H]Ro5-4864 and [3H]PK 11195 to cerebral cortex and peripheral tissues of various species: species differences and heterogeneity in peripheral benzodiazepine binding sites. *J. Neurochem.* **49**, 1407–1414.

Bartrop, R.W., Lazarus, L., Lukherst, E. and Kiloh, L.G. (1977). Depressed lymphocyte function after bereavement. *Lancet* **1**, 834–836.

Basile, A.S., Paul, S.M. and Skolnick, P. (1985). Adrenalectomy reduces the density of peripheral type binding sites for benzodiazepines in the rat kidney. *Eur. J. Pharmacol.* **110**, 149–150.

Basile, A.S., Ostrowski, N.L. and Skolnick, P. (1987). Aldosterone reversible decrease in the density of renal peripheral benzodiazepine receptors in the rat after adrenalectomy. *J. Pharmacol. Exp. Ther.* **240**, 1006–1013.

Benavides, J., Quarteronet, D., Imbault, F., Malgouris, C., Uzan, A., Renault, C., Dubroeucq, M.C., Gueremy, C. and Le Fur, G. (1983). Labeling of peripheral-type benzodiazepine binding sites in the rat brain by using [3H]PK 11195, an isoquinoline carboxamide derivative: kinetic studies and autoradiographic localization. *J. Neurochem.* **41**, 1744–1750.

Benavides, J., Quateronet, D., Plouin, P.F., Imbault, F., Phan, T., Uzan, A., Renault, C., Dubroeucq, M.C., Gueriny, C. and Le Fur, G. (1984). Characterization of peripheral type benzodiazepine binding sites in human and rat platelets by using [3H]PK 11195. Studies in hypertensive patients. *Biochem. Pharmacol.* **33**, 2467–2472.

Benavides, J., Dubois, A., Dennis, T., Hamel, E. and Scatton, B. (1989). ω3 (Peripheral type benzodiazepine binding) site distribution in the rat immune system: an autoradiographic study with photoaffinity ligand [3H]PK 14105. *J. Pharmacol. Exp. Ther.* **249**, 333–339.

Besman, M.J., Yanagibashi, K., Lee, T.D., Kawamura, M., Hall, P.F. and Shively, J.E. (1989). Identification of des-(Gly-Ile)endozepine as an effector of corticotropin-dependent adrenal steroidogenesis: stimulation of cholesterol delivery is mediated by the peripheral benzodiazepine receptor. *Proc. Natl. Acad. Sci. USA* **86**, 4897–4901.

Biggio, G. (1983). The action of stress, beta-carbolines, diazepam and Ro15-1788 on

GABA receptors in the rat brain. In *Benzodiazepine Recognition Site Ligands: Biochemistry and Pharmacology* (eds Biggio, G. and Costa, E.), pp. 400–403. Raven Press, New York.

Blecha, F., Barry, R.A. and Kelley, K.W. (1982). Stress-induced alterations in delayed-type hypersensitivity to SRBC and contact sensitivity to DNFB in mice. *Proc. Soc. Exp. Biol. Med.* **169**, 239–246.

Bond, P.A., Cundall, R.L. and Rolfe, B. (1985). [^{3}H]Diazepam binding to human granulocytes. *Life Sci.* **37**, 11–16.

Braestrup, C. and Squires, R.F. (1977). Specific benzodiazepine receptors in rat brain characterized by high-affinity [^{3}H]diazepam binding. *Proc. Natl. Acad. Sci. USA* **74**, 3805–3809.

Braestrup, C., Nielsen, M., Nielsen, E.B. and Lyon, M. (1979). Benzodiazepine receptors in the brain are affected by different experimental stresses: the changes are small and not unidirectional. *Psychopharmacology (Berlin)* **65**, 273–277.

Broadhurst, P.L. (1960). Experiments in personality. In *Psychogenetics and Psychopharmacology* (ed. Eysenck, H.J.), Vol. 1. Routledge and Keegan Paul, London.

Calogero, A.E., Kamilaris, T.C., Bernardini, R., Johnson, E.D., Chrousos, B.P. and Gold, P.W. (1990). Effects of peripheral benzodiazepine receptor ligands on hypothalamic–pituitary–adrenal axis function in rats. *J. Pharmacol. Exp. Ther.* **253**, 729–737.

Cohen, M.R., Pickar, D. and Dubois, M. (1981). Surgical stress and endorphins. *Lancet* **1**, 213–214.

Connors, J.M., De Vito, W.J. and Hedge, G.A. (1985). Effects of food deprivation on the feedback regulation of the hypothalamic–pituitary–thyroid axis of the rat. *Endocrinology* **117**, 900–906.

Corda, M.G., Blaker, W.D., Mendelson, W.B., Guidotti, A. and Costa, E. (1983). β-Carbolines enhance shock-induced suppression of drinking in rats. *Proc. Natl. Acad. Sci. USA* **80**, 2072–2076.

Corda, M.G., Giorgi, O. and Biggio, G. (1986). Behavioural and biochemical evidence for a long-lasting decrease in GABAergic function elicited by chronic administration of FG7142. *Brain Res.* **384**, 60–67.

Costa, E. (1983). The supramolecular organization of receptors for gamma aminobutyric acid (GABA). In *Receptors as Supramolecular Entities* (eds Biggio, G., Costa, E., Gessa, G.L. and Spano, P.F.), pp. 213–235. Pergamon Press, New York.

Costa, E. and Guidotti, A. (1979). Molecular mechanism in the receptor action of benzodiazepines. *Ann. Rev. Pharmacol. Toxicol.* **19**, 531–545.

Croiset, G., Heijnen, C.J., Veldhuis, H.D., de Wied, D. and Ballieux, R.E. (1987). Modulation of the immune response by emotional stress. *Life Sci.* **40**, 775–782.

Cross, R.I., Brooks, W.H., Roszman, T.L. and Markesbery, R. (1982). Hypothalamic immune interactions. Effect of hypophysectomy on neuroimmunomodulation. *J. Neurol. Sci.* **53**, 557–586.

Dantzer, R. and Kelley, W. (1989). Stress and immunity, an integrated view of relationships between the brain and the immune system. *Life Sci.* **44**, 1995–2008.

Dar, D.E., Weizman, A., Karp, L., Grinshpoon, A., Bidder, M., Kotler, M., Tyano, S., Bleich, A. and Gavish, M. (1991). Platelet peripheral benzodiazepine receptors in repeated stress. *Life Sci.* **48**, 341–346.

D'Eramo, J., Somaza, G.M., Kertesz, E. and Libertun, C. (1986). Baclofen, a GABA derivative, inhibits stress-induced prolactin release in the rat. *Eur. J. Pharmacol.* **120**, 81–85.

Descotes, J., Tedone, R. and Evreux, J.C. (1982). Suppression of humoural and cellular immunity in normal mice by diazepam. *Immunol. Lett.* **5**, 41–43.

De Souza, E.B., Anholt, R.R.H., Murphy, K.M.M., Snyder, S.H. and Kuhar, M.J. (1985). Peripheral type benzodiazepine receptors in endocrine organs: autoradiographic localisation in rat pituitary, adrenals and testis. *Endocrinology* **116**, 567–573.

Dorian, B.L., Garfinkel, P., Brown, C., Shorea, A., Gladman, D. and Keystone, E. (1982). Aberrations in lymphocyte sub-populations and function during psychological stress. *Clin. Exp. Immunol.* **50**, 132–138.

Drugan, R.C. and Holmes, P.V. (1991). Central and peripheral benzodiazepine receptors: Involvement in an organism's response to physical and psychological stress. *Neurosci. Biobehav. Rev.* **15**, 277–298.

Drugan, R.C., Basile, A.S., Crawley, J.N., Paul, S.M. and Skolnick, P. (1986). Inescapable shock reduces [^{3}H] Ro5-4864 binding to 'peripheral type' benzodiazepine receptors in the rat. *Pharmacol. Biochem. Behav.* **24**, 1673–1677.

Drugan, R.C., Basile, A.S., Crawley, J.N., Paul, S.M. and Skolnick, P. (1987). 'Peripheral' benzodiazepine binding sites in the Maudsley reactive rat: selective decrease confined to peripheral tissues. *Brain Res. Bull.* **18**, 143–145.

Drugan, R.C., Basile, A.S., Crawley, J.N., Paul, S.M. and Skolnick, P. (1988). Characterization of stress-induced alterations in [^{3}H] Ro5-4864 binding to peripheral benzodiazepine receptors in rat heart and kidney. *Pharmacol. Biochem. Behav.* **30**, 1015–1020.

Drugan, R.C., Holmes, P.V. and Stringer, A.P. (1990). Pentobarbital blocks the stress-induced decrease in [^{3}H]Ro5-4864 binding in rat kidney. *Brain Res.* **535**, 151–154.

Drugan, R.C., Holmes, P.V. and Stringer, A.P. (1991). Sexual dimorphism of stress-induced changes in renal peripheral benzodiazepine receptors in rat. *Neuropharmacology* **30**, 413–416.

Elsworth, J.D., Clow, A., Glover, V. and Sandler, M. (1984). Characterization of an endogenous inhibitor of monoamine oxidase and benzodiazepine receptor binding. *J. Pharm. Pharmacol.* **36** (Suppl). 1W:59W.

Fares, F., Bar-Ami, S., Haj-Yehia, Y. and Gavish, M. (1989). Hormonal regulation of peripheral benzodiazepine binding sites in female rat adrenal gland and kidney. *J. Recept. Res.* **9**, 143–157.

Ferrarese, C., Appollonio, I., Frigo, M., Perego, M., Piolti, R., Trabucchi, M. and Frattola, L. (1990). Decreased density of benzodiazepine receptors in lymphocytes of anxious patients: reversal after chronic diazepam treatment. *Acta Psychiatr. Scand.* **82**, 169–173.

File, S.E. and Pellow, S. (1985a). The effects of PK 11195, a ligand for benzodiazepine binding sites, in animal tests of anxiety and stress. *Pharmacol. Biochem. Behav.* **23**, 737–741.

File, S.E. and Pellow, S. (1985b). The anxiogenic action of Ro5-4864 in the social interaction test: effect of chlordiazepoxide, Ro15-1788 and CGS 8216. *Naunyn-Schmiedeberg's Arch. Pharmacol.* **328**, 225–229.

Gavish, M., Okun, F., Weizman, A. and Youdim, M.B.H. (1986a). Modulation of peripheral benzodiazepine binding sites following chronic estradiol treatment. *Eur. J. Pharmacol.* **127**, 147–151.

Gavish, M., Weizman, A., Okun, F. and Youdim, M.B.H. (1986b). Modulatory effects of thyroxine treatment on central and peripheral benzodiazepine receptors in the rat. *J. Neurochem.* **47**, 1106–1110.

Gee, K.W. (1987). Phenylquinolines PK 8165 and PK 9084 allosterically modulate [35S]-*t*-butylbicyclophosphorothionate binding to a chloride ionophore in rat brain

R.H. Dodd and M. Lenfant

via a novel Ro5-4864 binding site. *J. Pharmacol. Exp. Ther.* **240**, 747–753.

Glaser, R., Rice, J., Sheridan, J., Fertel, R., Stout, J.C. and Kiecolt-Glaser, J.K. (1986). Stress depresses interferon production by leukocytes concomitant with a decrease in natural killer cell activity. *Behav. Neurosci.* **100**, 675–678.

Glaser, R., Rice, J., Sheridan, J., Fertel, R., Stout, J., Speicher, C., Pinsky, D., Kotur, M., Post, A. and Beck, M. (1987). Stress-related immune suppression: health implications. *Brain Behav. Immunol.* **1**, 7–20.

Haefely, W. and Polc, P. (1986). Physiology of GABA enhancement by benzodiazepines and barbiturates. In *Benzodiazepine/GABA Receptors and Chloride Channels: Structural and Functional Properties* (eds Olsen, R.W. and Ventner, C.J.), pp. 97–133. Alan R. Liss, New York.

Havoundjian, H., Reed, G.F., Paul, S.M. and Skolnick, P. (1987). Protection against the lethal effects of pentobarbital in mice by a benzodiazepine receptor inverse agonist, 6,7-dimethoxy-4-ethyl-3-carbomethoxy-β-carboline. *J. Clin. Invest.* **79**, 473–477.

Hill, D.R. and Bowery, N.G. (1981). [^{3}H]Baclofen and [^{3}H]GABA bind to bicuculline-insensitive $GABA_B$ sites in rat brain. *Nature* **290**, 149–152.

Hirsch, J.D., Beyer, C.F., Malkowitz, L., Loullis, C.C. and Blume, A.J. (1988). Characterization of ligand binding to mitochondrial benzodiazepine receptors. *Mol. Pharmacol.* **34**, 164–172.

Hunkeler, W., Mohler, H., Pieri, L., Polc, P., Bonetti, E.P., Cumin, R., Schaffner, R. and Haefely, W. (1981). Selective antagonists of benzodiazepines. *Nature* **290**, 514–516.

Irwin, M., Smith, T.I. and Gillin, J.C. (1987). Low natural killer cytotoxicity in major depression. *Life Sci.* **41**, 2127–2133.

Karp, L., Weizman, A., Tyano, S. and Gavish, M. (1989). Examination stress, platelet peripheral benzodiazepine sites, and plasma hormone levels. *Life Sci.* **44**, 1077–1082.

Keller, S.E., Stein, M., Camerino, M.S., Schleifer, S.J. and Scherman, J. (1980). Suppression of lymphocyte stimulation by anterior hypothalamic lesions in the guinea pig. *Cell Immunol.* **52**, 334–340.

Kelley, K.W. (1989). Growth hormone lymphocytes and macrophages. *Biochem. Pharmacol.* **38**, 705–713.

Krueger, K.E. and Papadopoulos, V. (1990). Peripheral type benzodiazepine receptors mediate translocation of cholesterol from outer to inner mitochondrial membranes in adrenocortical cells. *J. Biol. Chem.* **205**, 1505–1523.

Lambert, J.J., Peters, J.A. and Cotrell, G.A. (1987). Actions of synthetic and endogenous steroids on the $GABA_A$ receptor. *Trends Pharmacol. Sci.* **8**, 224–227.

Lenfant, M., Haumont, J. and Zavala, F. (1986). *In vivo* immunomodulatory activity of PK 11195, a structurally unrelated ligand for peripheral benzodiazepine binding site. I. Potentiation in mice of the humoral response to sheep red blood cells. *Int. J. Immunopharmacol.* **8**, 825–829.

Lenfant, M., Haumont, J., Horak, P., Sebestova, L. and Masek, K. (1988). *In vivo* immunomodulatory activity of PK 11195, a structurally unrelated ligand for peripheral benzodiazepine binding sites. II. The possible involvement of central nervous system receptors. *Int. J. Immunother.* **4**, 125–129.

Lueddens, H.W.M. and Skolnick, P. (1987). 'Peripheral-type' benzodiazepine receptors in the kidney: regulation of radioligand binding by anions and DIDS. *Eur. J. Pharmacol.* **133**, 205–214.

Maier, S.F., Ryan, S.M., Barksdale, C.M. and Kalin, N.H. (1986). Stressor controllability and the pituitary–adrenal system. *Behav. Neurosci.* **100**, 669–674.

Majewska, M.D. and Chuang, D.M. (1984). Modulation by calcium of gamma aminobutyric acid (GABA) binding to $GABA_B$ and $GABA_A$ recognition sites in rat brain. *Mol. Pharmacol.* **25**, 352–359.

Martini, C., Lucacchini, A., Hrelia, S. and Rossi, C.A. (1986). Central- and peripheral-type benzodiazepine receptors. In *GABAergic Transmission and Anxiety* (eds Biggio, G. and Costa, E.), pp. 1–11. Raven Press, New York.

Medina, J.H., Novas, M.L. and DeRobertis, E. (1983a). Changes in benzodiazepine receptors by acute stress. *Eur. J. Pharmacol.* **96**, 181–185.

Medina, J.H., Novas, M.L., Wolfman, C., Levi de Stein, M. and DeRobertis, E. (1983b). Benzodiazepine receptors in rat cerebral cortex and hippocampus undergo rapid and reversible changes after acute stress. *Neuroscience* **9**, 331–335.

Mennini, T., Gobbi, M. and Charuchinda, C. (1989). Noise-induced opposite changes in central and peripheral benzodiazepine receptors in rat brain cortex. *Neurosci. Res. Commun.* **5**, 27–34.

Mizoule, J., Gauthier, A., Uzan, A., Renault, C., Dubroeucq, M., Gueremy, C. and Le Fur, G. (1985). Opposite effects of two ligands for peripheral type benzodiazepine binding sites, PK 11195 and Ro5-4864, in a conflict situation in the rat. *Life Sci.* **36**, 1059–1068.

Moigeon, Ph., Bidart, J.M., Alberici, G.F. and Bohuon, C. (1983). Characterization of a peripheral type benzodiazepine binding site on human circulating lymphocytes. *Eur. J. Pharmacol.* **92**, 142–149.

Moigeon, Ph., Dessaux, J.J., Fellous, R., Alberici, G.F., Bidart, J.M., Motte, Ph. and Bohuon, C. (1984). Benzodiazepine receptors on human platelets. *Life Sci.* **35**, 2003–2009.

Monjan, A.A. and Collector, W.I. (1977). Corticosterone concentrations in the mouse. *Science* **196**, 307–308.

Mukhin, A.G., Papadopoulos, V., Costa, E. and Krueger, K.E. (1989). Mitochondrial benzodiazepine receptors regulate steroid biosynthesis. *Proc. Natl. Acad. Sci. USA* **86**, 9813–9816.

Noel, G.L., Suh, H.K., Stone, J.C. and Frantz, A.G. (1972). Human prolactin and growth hormone release during surgery and other conditions of stress. *J. Clin. Endocrinol.* **35**, 840–851.

Novas, M.L., Medina, J.H., Calvo, D. and De Robertis, E. (1987). Increase of peripheral type benzodiazepine binding sites in kidney and olfactory bulb in acutely stressed rats. *Eur. J. Pharmacol.* **135**, 243–246.

Okun, F., Weizman, R., Katz, Y., Bomzon, A., Youdim, M.B.H. and Gavish, M. (1988). Increase in central and peripheral benzodiazepine receptors following surgery. *Brain Res.* **485**, 31–36.

Olsen, R.W., Wong, E., Stauber, G. and King, R. (1984). Biochemical pharmacology of the gamma aminobutyric acid receptor/ionophore complex. *Fed. Proc.* **43**, 2773–2778.

Patel, J. and Marangos, P.J. (1982). Differential effects of GABA on peripheral and central type benzodiazepine binding sites in brain. *Neurosci. Lett.* **30**, 157–160.

Pawlikowski, M., Stepien, H. and Kunert-Radek, J. (1986). Diazepam inhibits proliferation of the mouse spleen lymphocytes *in vitro*. *Pol. J. Pharmacol. Pharm.* **28**, 167–170.

Pellow, S. and File, S. (1984). Behavioural actions of Ro5-4864: a peripheral type benzodiazepine? *Life Sci.* **35**, 229–240.

Petersen, E. (1983). DMCM: A potent convulsive benzodiazepine receptor ligand. *Eur. J. Pharmacol.* **94**, 117–124.

Prado de Carvalho, L., Grecksch, G., Chapouthier, G. and Rossier, J. (1983).

Anxiogenic and non-anxiogenic benzodiazepine antagonists. *Nature* **301**, 64–66.

Rabin, B., Gangult, R., Cunnick, J. and Lysle, D.T. (1988). The central nervous system–immune system relationship. *Clin. Lab. Med.* **8**, 253–267.

Rägo, L.K., Nurk, A.M., Korneyev, A.Y. and Allikmets, L. (1982). Binding of phenibut with bicuculline insensitive GABA receptors in the rat brain. *Bull. Exp. Biol. Med.* **11**, 58–59.

Rägo, L.K., Kiivet, R.A. and Harro, J.E. (1986). Variation in behavioural response to baclofen: correlation with benzodiazepine binding sites in mouse forebrain. *Naunyn Schmiedeberg's Arch. Pharmacol.* **333**, 303–306.

Rägo, L., Kiivet, R.A., Harro, J. and Pold, M. (1989). Central- and peripheral-type benzodiazepine receptors: similar regulation by stress and GABA receptor agonists. *Pharmacol. Biochem. Behav.* **32**, 879–883.

Rägo, L., Adojaan, A. and Masso, R. (1990a). [^{3}H]Ro5-4864 binding sites in the nucleus of rat lymphocytes. *Eur. J. Pharmacol.* **187**, 561–562.

Rägo, L., Kiivet, R.A., Adojaan, A., Harro, J. and Allikmets, L. (1990b). Stress-protective action of β-phenyl (GABA): involvement of central and peripheral type benzodiazepine binding sites. *Pharmacol. Toxicol.* **66**, 41–44.

Rägo, L., McDonald, E., Saano, V. and Airaksinen, M.M. (1991). The effect of metetomidine on GABA and benzodiazepine receptors *in vivo*: lack of anxiolytic but some evidence of possible stress-protective activity. *Pharmacol. Toxicol.* **69**, 81–86.

Ritta, M.N., Campos, M.B. and Calondra, R.S. (1987). Effect of GABA and benzodiazepines on testicular androgen production. *Life Sci.* **40**, 791–798.

Ruff, M.R., Pert, C.B., Weber, R.J., Wahl, L.M., Wahl, S.M. and Paul, S.M. (1985). Benzodiazepine receptor mediated chemotaxis of human monocytes. *Science* **229**, 1281–1286.

Schoemaker, H., Boles, R.G., Horst, D. and Yamamura, H.I. (1983). Specific high-affinity binding sites for [^{3}H]Ro5-4864 in rat brain and kidney. *J. Pharmacol. Exp. Ther.* **225**, 61–66.

Singhal, R.L. and Rastogi, R.B. (1981). Thyroid hormone in the regulation of neurotransmitter function and behaviour. In *Neuroendocrine Regulation and Altered Behaviour* (eds Herdina, P.D. and Singhal, R.L), pp. 206–221. Plenum Press, New York.

Skowronski, R., Beaumont, K. and Fanestil, D. (1987). Modification of the peripheral-type benzodiazepine receptor by arachidonate, diethylpyrocarbonate and thiol reagents. *Eur. J. Pharmacol.* **143**, 305–314.

Soubrié, P., Thiebot, M.J., Jobert, A., Montastruc, J.L., Hery, F. and Hamon, M. (1980). Decreased convulsant potency of picrotoxin and pentetrazol and enhanced [^{3}H]flunitrazepam cortical binding following stressful manipulations in rats. *Brain Res.* **189**, 505–517.

Spector, N.H. (1990). Basic mechanisms and pathways of neuroimmunomodulation (NIM): triggers (and problems for future research). *Int. J. Neurosci.* **51**, 335–337.

Stein, M. (1985). Bereavement, depression, stress and immunity. In *Neural Modulation of Immunity* (eds Guillemin, R., Cohn, M. and Melnechuk, T.), pp. 23–44. Raven Press, New York.

Study, R.E. and Barker, J.L. (1981). Diazepam and (−)-pentobarbital: fluctuation analysis reveals different mechanisms for potentiation of GABA responses in cultured central neurons. *Brain Res.* **268**, 171–176.

Suzuki-Nishimura, T., Sano, T. and Uchida, M.K. (1989). Effects of benzodiazepines on serotonin release from rat mast cells. *Eur. J. Pharmacol.* **167**, 75–85.

Taniguchi, T., Wang, J.K.T. and Spector, S. (1980). Properties of [^{3}H]diazepam binding to rat peritoneal mast cells. *Life Sci.* **27**, 171–178.

Thompson, C.S., Mikailidis, D.P., Jeremy, J.Y., Bell, J.L. and Dandona, P. (1987). Effect of starvation on biochemical indices of renal function in the rat. *Br. J. Exp. Pathol.* **68**, 767–775.

Tong, Y., Tonon, M.C., Desy, L., Vaudry, H. and Pelletier, G. (1990). Localization of the endogenous benzodiazepine receptor ligand octadecaneuropeptide and peripheral type benzodiazepine receptors in the rat pituitary. *J. Neuroendocrinol.* **2**, 189–192.

Traber, J. and Glaser, T. (1987). 5-HT$_{1A}$ receptor-related anxiolytics. *Trends Pharmacol. Sci.* **8**, 432–437.

Wang, J.K.T., Taniguchi, T. and Spector, S. (1980). Properties of [^{3}H]diazepam binding to rat blood platelets. *Life Sci.* **27**, 1881–1884.

Weigent, D.A. and Blalock, J.E. (1987). Interactions between the neuroendocrine and immune systems: Common hormones and receptors. *Immunol. Rev.* **100**, 79–108.

Weizman, R., Tanne, Z., Granek, M., Karp, L., Colomb, M., Tyano, S. and Gavish, M. (1987). Peripheral benzodiazepine binding sites in platelet membrane are increased during diazepam treatment in anxious patients. *Eur. J. Pharmacol.* **138**, 282–292.

Weizman, A., Bidder, M., Fares, F. and Gavish, M. (1990). Food deprivation modulates gamma-aminobutyric acid receptors and peripheral benzodiazepine binding sites in rats. *Brain Res.* **535**, 96–100.

Yanagibashi, K., Ohno, Y., Nakamichi, N., Matsui, T., Hayashida, K., Takamura, M., Yamada, K., Ton, S. and Kawamura, M. (1989). Peripheral type benzodiazepine receptors are involved in the regulation of cholesterol side chain cleavage in adrenocortical mitochondria. *J. Biochem.* **106**, 1026–1029.

Young, I., Malozowski, S., Winterer, H., Nicoletti, M.C., Kibarian, M. and Cassorla, F. (1987). Acute starvation affects rat adrenal steroidogenesis. *Horm. Metab. Res.* **19**, 21–23.

Zambotti, F., Zonta, N., Tammiso, R., Conci, F., Hafner, B., Zecca, L., Ferrario, P. and Mantegazza, P. (1991). Effects of diazepam on nociception in rats. *Naunyn-Schmiedeberg's Arch. Pharmacol.* **344**, 84–89.

Zavala, F., Haumont, J. and Lenfant, M. (1984). Interaction of benzodiazepines with mouse macrophages. *Eur. J. Pharmacol.* **106**, 561–566.

CHAPTER 12

PATHOPHYSIOLOGICAL AND ENDOCRINOLOGICAL ASPECTS OF PERIPHERAL-TYPE BENZODIAZEPINE RECEPTORS

Moshe Gavish,[1,2] Shalom Bar-Ami[1,3] and Ronit Weizman[4]

[1] Rappaport Family Institute for Research in the Medical Sciences and [2] Department of Pharmacology, Bruce Rappaport Faculty of Medicine, Technion-Israel Institute of Technology, Haifa, Israel
[3] Department of Obstetrics and Gynecology, Rambam Medical Center, Haifa, Israel
[4] Tel Aviv Community Mental Health Center and Sackler Faculty of Medicine, Tel Aviv University, Tel Aviv, Israel

Table of Contents

PERIPHERAL BENZODIAZEPINE RECEPTORS
ISBN 0-12-282630-2

12.1 Introduction

Benzodiazepines (BZ) are widely used clinically as muscle relaxants, anticonvulsants, sedative-hypnotics, and anxiolytics. These effects are mediated via binding to specific receptors located only in the central nervous system (CNS)—the central BZ receptors (CBR). CBR are coupled to γ-aminobutyric acid (GABA) receptors and to the chloride ion channel (Tallman *et al.*, 1980). Surprisingly, [^{3}H]diazepam has been found to bind not only to CBR, but also to specific sites in the kidney (Braestrup and Squires, 1977) and to various other peripheral tissues (Verma and Snyder, 1989). These sites originally were called 'peripheral BZ binding sites', but since various studies showed that these sites have a physiological role, they are now termed 'mitochondrial BZ receptors' (MBR) (Verma and Snyder, 1989). MBR bind with high affinity Ro5-4864 (4'-chlorodiazepam) and PK 11195 (an isoquinoline carboxamide derivative), but exhibit very low affinity for clonazepam; the reverse is true with regard to CBR. MBR are not coupled to GABA or the chloride ion channel, and structurally are dissimilar to CBR. Their localization is also different: CBR are located on the plasma membrane in the CNS, whereas MBR are located mainly on the outer membrane of the mitochondria (Anholt *et al.*, 1986) in various peripheral tissues and in non-neuronal cells in the brain (Bender and Hertz, 1985). This chapter will deal with pathophysiological and endocrinological aspects of MBR.

12.2 MBR and cancer

MBR ligands at micromolar concentration inhibit the proliferation of thymoma cells *in vitro* (Wang *et al.*, 1984a), enhance melanogenesis in B16/C3 melanoma cells (Matthew *et al.*, 1981), and increase the synthesis of haemoglobin in Friend erythroleukaemia cells (Wang *et al.*, 1984b). MBR ligands stimulate [^{3}H]thymidine incorporation into DNA and increase cell proliferation in C6 glioma at nanomolar concentration. At micromolar concentration, they inhibit DNA synthesis (Ikezaki and Black, 1990). (For more details, see Chapter 5.)

MBR binding is increased significantly in experimental brain tumours (Starosta-Rubinstein *et al.*, 1987; Black *et al.*, 1989) and in brain tumours in humans (Black *et al.*, 1990). In human brains at post-mortem examination, [^{3}H]PK 11195 binds mainly to glioma or astrocytoma (Starosta-Rubinstein *et al.*, 1987).

We have measured MBR density and affinity in renal cell carcinoma, colonic adenocarcinoma, and ovarian carcinoma as compared to normal tissues. There was no difference in the affinity for PK 11195 in the latter two cancer tissues as compared to the respective normal tissue; on the other hand, we found dramatic alterations in MBR density. In renal cell carcinoma we could not determine any binding of [^{3}H]PK 11195, but in benign renal tumours the binding was similar to normal tissue (Katz *et al.*, 1989). MBR density in ovarian carcinoma (Katz *et al.*, 1990b) and in colonic adenocarcinoma (Katz *et al.*, 1990c) was increased by 2- to 3-fold as compared to normal control tissues. The robust increase in MBR densities in ovarian carcinoma and colonic adenocarcinoma may be related to higher metabolic rates in neoplasm (Gold, 1974) or, alternatively, may reflect alteration in the structure of mitochondria in the tumour, termed 'mitochondrial pyknosis', which is revealed by microscopic ultrastructural examination (Ghadially, 1975).

12.3   The involvement of MBR in neuropsychiatric disorders

12.3.1   MBR and neuroleptic treatment

Neuroleptics are dopamine-blocking agents which are used in the treatment of schizophrenia. Besides their activity at the dopamine receptor, they possess antagonistic activity at the muscarinic, cholinergic, α_1-adrenergic, histaminergic, and serotonin-S_2 receptors. The antischizophrenic activity of neuroleptics correlates with their affinity to the dopamine-D_2 receptors. Some of the side-effects of neuroleptics, such as convulsions and dystonia, may be alleviated at least temporarily by diazepam treatment. The beneficial actions of BZ on the above-mentioned side-effects of neuroleptics on the CNS are probably mediated via central BZ/GABA receptors. In addition to the centrally mediated side-effects of neuroleptics, there are peripheral side-effects such as cardiac arrhythmias, gastrointestinal disturbances, and menstrual irregularities.

Since MBR have been localized in the heart, ileum, and endocrine organs (Verma and Snyder, 1989), we tested the possibility that neuroleptic treatment modulates MBR. In the first stage we examined platelet MBR density in nine schizophrenic patients maintained on long-term (>2 years) neuroleptic treatment and in six untreated schizophrenics as compared to age- and sex-matched normal control subjects (Gavish *et al.*, 1986b). A 30% decrease in platelet MBR density was observed in the chronically treated patients in comparison to the untreated patients and the healthy controls. The alteration in the maximal binding capacity of [^{3}H]PK 11195 was not accompanied by any change in the equilibrium dissociation constant.

211

In vitro competition experiments showed that neuroleptics (chlorpromazine and haloperidol) are ineffective displacers of [³H]PK 11195 from MBR (Benavides *et al.*, 1984). Thus, it was concluded that the neuroleptic-induced downregulation of platelet MBR is not related to the presence of the drug in the assay. Furthermore, since the reduction in MBR density was confined to chronically medicated schizophrenics, and was not found in untreated patients, it would appear that the decrease in platelet MBR is not attributable to the disorder, but rather to the chronic neuroleptic treatment.

In a subsequent study, we investigated the effect of short-term (28 days) neuroleptic treatment on platelet MBR. The characteristics of platelet MBR were assessed in 11 schizophrenic inpatients before and after 4 weeks of neuroleptic treatment (Tanne *et al.*, 1987). Short-term neuroleptic administration to schizophrenic patients was not sufficient to induce any detectable alteration in platelet MBR. The beneficial antipsychotic effects of neuroleptics were not accompanied by any alteration in [³H]PK 11195 binding parameters. Thus, it seems that these sites, at least those present on platelets, are not involved in the antipsychotic activity of antidopaminergic agents.

Apparently a longer duration (>4 weeks) of exposure to neuroleptics is required to achieve a decrease in PBR density. Alternatively, it is possible that neuroleptic-induced denervation supersensitivity of dopamine D_2 receptors is needed to downregulate PBR density. In order to test this possibility, we evaluated platelet MBR in chronically treated schizophrenic patients with and without tardive dyskinesia (TD) (Weizman *et al.*, 1986). TD is a serious side-effect of long-term neuroleptic treatment that is characterized by involuntary repetitive movements. This complication is related to denervation hypersensitivity of striatal D_2 receptors. Decreased platelet MBR density was observed in treated schizophrenics with TD (-30%) as well as in patients without TD (-20%), when compared to untreated patients and normal controls. These results indicate that the modulatory effect of long-term neuroleptic treatment on MBR is not associated with the development of TD.

Human studies could not clarify whether the changes observed following neuroleptic treatment are confined to platelets or occur in other tissues too. To this end, we evaluated the effect of daily intraperitoneal administration to rats of 5 mg/kg of chlorpromazine (CPZ) for 21 days on MBR in the cerebral cortex, heart, and kidney (Gavish and Weizman, 1989). An increase of 51% in the maximal binding capacity was observed in the cerebral cortex, while decreases of 25% and 14% were obtained in cardiac and renal MBR, respectively. [³H]PK 11195 binding to cerebral cortex returned to control values following 5 days of CPZ withdrawal, while MBR density in the heart and kidney remained reduced. CBR in the cerebral cortex were not affected by the chronic neuroleptic treatment.

The next step was to investigate the effect of a non-phenothiazine neuroleptic, the butyrophenone haloperidol, on MBR in rat cerebral cortex and heart (Gavish et al., 1988). A significant increase of 38% in maximal binding capacity of [^{3}H]PK 11195 was observed in the cerebral cortex, but not in the heart, following 21 days of haloperidol administration (0.5 mg/kg daily). Similarly to CPZ, treatment with haloperidol did not affect cerebral cortical CBR. This line of evidence indicates that MBR, in contrast to CBR, are sensitive to long-term neuroleptic administration.

It is possible that the alterations in [^{3}H]PK 11195 binding sites are associated with the marked hyperprolactinaemia induced by antidopaminergic agents. MBR in the brain are localized mainly on non-neuronal glial cells (Schoemaker et al., 1983a; Anholt et al., 1984). Thus, the changes observed in brain tissue may be pertinent to glial cell function rather than to neuronal cell function. It may be that the alterations in cerebral cortical MBR are associated with the effect of chronic neuroleptic treatment on brain mitochondria (Byczkowski and Borysewicz, 1979) or on the voltage-regulated calcium channel complex (Cantor et al., 1984; Bender and Hertz, 1985).

The CPZ-induced downregulation of MBR in the heart is in contrast to the lack of effect observed during chronic haloperidol treatment (Gavish et al., 1988). Ro5-4864 decreases in a dose-dependent manner the duration of cardiac intracellular action potential and contractility. These effects are GABA independent and antagonized selectively by PK 11195 (Mestre et al., 1984). MBR in the aorta and heart seem to be coupled to calcium channels (Mestre et al., 1985, 1986). PK 11195 possesses an antiarrhythmic effect (Mestre et al., 1985). CPZ produces negative chronotropic effects on the heart and is a potent cardiac arrhythmogenic agent (Lathers and Lipka, 1985). It is possible that the cardiac side-effect is associated with the decrease in density of MBR in the heart and may be related to the effect of the drug on cardiac mitochondria (Kitazawa et al., 1981). The unaltered MBR in cardiac tissue of rats following chronic haloperidol treatment, in contrast to the reduction in numbers of MBR induced by CPZ, correlate with the fewer cardiac side-effects of haloperidol observed at therapeutic doses (Brannan et al., 1980).

12.3.2   Stress and anxiety

12.3.2.1   Animal models of stress

The impact of acute and chronic stress on MBR has been investigated in various animal models. Drugan et al. (1986) were the first to demonstrate the involvement of MBR in the physiological response to stress, using inescapable tailshocks in rats as an animal model of stress. Five shocks induced a significant increase of renal MBR, while 80 repeated tailshocks resulted in a significant decrease of MBR density in cerebral cortex, pituitary gland, heart and kidney. The biphasic effects of inescapable shock observed in kidney and cerebral cortex may indicate differences in MBR response to short- versus long-term exposure to stress. Basile *et al.* (1987) showed that acute maximal electroshock increased the density of MBR in mouse cerebral cortex and cardiac ventricles. However, repeated maximal electroshock administration did not alter MBR density in brain and peripheral organs. A significant increase in MBR density of cerebral cortex and kidney as well as platelets and lymphocytes was observed in rats following a single experience of stress caused by forced swimming (Rägo *et al.*, 1989a,b).

Maudsley reactive (MR) rats, which have been bred for a high level of fearfulness, have been compared to Maudsley non-reactive rats (low levels of fearfulness). MBR in the MR rats were decreased in heart and kidney (Drugan *et al.*, 1987). The changes observed in MBR may be related to increased sympathetic discharge or hormonal changes characterizing MR rats.

Another model of stress is food deprivation. Starvation induced in rats a decrease in MBR density in the adrenal, kidney and heart, but not in the hypothalamus and ovary (Weizman *et al.*, 1990). It is possible that metabolic (protein catabolism, metabolic ketoacidosis, natriuresis, etc.) and endocrine (hypercortisolemia) changes associated with starvation are responsible for MBR depletion in these organs. The changes in MBR binding were not accompanied by alterations in cortical and hypothalamic CBR.

Surgery, which is accompanied by transitory alterations in stress hormones (Noel *et al.*, 1972; Cohen *et al.*, 1981), has served as a model for stressful situations. Brain and kidney MBR in rats were investigated on days 1, 3, and 7 following abdominal wall operation (Okun *et al.*, 1988). Significant elevations in the maximal binding capacity of cerebral CBR as well as of brain and renal MBR were observed on days 1 and 3 following the surgery, but not on day 7. The elevation in brain and kidney MBR corresponds to the stages of repair of surgical wounds. (For further details, see Chapter 11.)

12.3.2.2 Stress and anxiety in humans

Platelet MBR have also been found in human studies to be sensitive to stress and anxiety. Increased platelet MBR density was detected in resident physicians exposed to 'examination stress', and a trend toward decrease was observed 10 days later (Karp *et al.*, 1989). A parallel decrease in anxiety scoring was monitored 10 days following the examination and compared to the score obtained immediately after the examination. The alterations in MBR density were not accompanied by alterations in 'stress' hormones (cortisol, prolactin, and growth hormone), which can be explained either by the fact that the study dealt with prolonged stress or by the fact that the hormonal peak release occurred earlier, during the examination. MBR density had not returned to control values 10 days following the examination, which may indicate that stress has a possible long-term effect on these receptors.

Taking the next step, we investigated the impact of repeated stress on platelet MBR in male soldiers exposed to a parachute-training course (Dar et al., 1991). Platelet MBR were significantly decreased after the fourth parachute jump, 8 days after the baseline measurement. The half-life of platelets is approximately 3 days; thus, the binding measured following the fourth parachute jump was due to platelets not present at the time of the first sampling. It is possible that stress-induced changes in MBR occur at the stem cells (megakaryocytes) and are not confined to the peripheral blood platelets. The reduction in MBR density on platelets was associated with a reduction in systolic blood pressure. The neurochemical and neurophysiological changes may be related to a habituation process which involves an adaptive decrease in sympathetic tone.

In conclusion, alterations in the density of MBR seem to be a sensitive indicator of stress. The studies of single and repeated stress indicate a bidirectional effect of stress on MBR, i.e. upregulation after acute stress and downregulation after repeated stress.

MBR seem to be sensitive not only to the natural paradigm of stress, but also to clinical anxiety in psychiatric patients. The involvement of MBR in the pathophysiology and pharmacotherapy of generalized anxiety disorder was evaluated in drug-free outpatients before and after chronic diazepam treatment (Weizman et al., 1987a). Decreased MBR were observed in untreated anxious patients when compared to normal controls. Four weeks of diazepam treatment resulted in an elevation of MBR density, and 1 week of drug withdrawal induced a slight decrease in the MBR density.

Similar results were obtained by Farrarese et al. (1990) in lymphocytes of anxious patients. The latter authors suggest that endogenous anxiogenic neuropeptides such as diazepam-binding inhibitor (Guidotti et al., 1983) may be released in anxiety and be responsible for the MBR alterations in anxious patients. The role of MBR in anxiety is unclear. The decrease in MBR may be attributable to an indirect downregulatory effect of stress hormones that are released in high concentrations during stressful situations.

Ro5-4864 potentiates shock-induced suppression of drinking and reduced activity in the social interaction test (File and Lister, 1983). These proconflict and anxiogenic effects are antagonized by the non-BZ ligand PK 11195 (Mizoule et al., 1985). Furthermore, PK 11195 has been reported in a pilot study to possess anxiolytic properties in man (Papart et al., 1988).

The following step was taken to determine whether chronic diazepam treatment upregulates MBR. Diazepam (0.5 mg/kg) was injected into male Sprague–Dawley rats for 21 days (Weizman and Gavish, 1989), and a significant increase in cardiac and cerebral cortical MBR was obtained. A five day withdrawal resulted in restoration of MBR density to normal range. It appears that the upregulatory effect of diazepam on MBR, as observed in anxious patients (Weizman *et al.*, 1987a), can also be achieved in unstressed normal rats. Thus, stress-induced low MBR levels are not a prerequisite for the observed increase in MBR density. MBR elevation in the heart could be relevant to the reduction of peripheral sympathetic tone (Marthy *et al.*, 1986), due to the general sedative effect of diazepam. This possibility is in accord with the decreased cardiac MBR density after chemical sympathectomy induced by parenteral 6-hydroxydopamine administration (Basile and Skolnick, 1988). However, reserpine injection has no effect on cardiac MBR (Basile and Skolnick, 1988). Thus, the interaction between sympathetic tone and MBR regulation is unclear.

12.3.3   MBR in neurological diseases

12.3.3.1   Parkinson's disease

A reduction in the number of platelet MBR has been observed in patients with Parkinson's disease (Bonuccelli *et al.*, 1991). The reduction in [³H]PK 11195 binding sites was independent of levodopa/carbidopa treatment. The severity or duration of the disease did not correlate with MBR density. The authors suggested that the diminished receptor density values might reflect an impairment in mitochondrial respiratory function and that such a dysfunction could be involved in the pathogenesis of Parkinson's disease.

   Nevertheless, there is no evidence for a concomitant alteration in the CNS, especially in the basal ganglia. Furthermore, platelet MBR density is similar in patients under neuroleptic treatment complicated by the extrapyramidal disorder tardive dyskinesia and in those with uncomplicated neuroleptic treatment (Weizman *et al.*, 1986). Moreover, a preliminary study by McGeer *et al.* (1988) demonstrated increased MBR density in the brain of three patients suffering from Parkinson's disease.

12.3.3.2   Alzheimer's disease

Owen *et al.* (1983) demonstrated increased MBR density in the temporal lobe of patients with Alzheimer's disease. A later study by McGeer *et al.* (1988) showed increased [³H]Ro5-4864 binding sites in Broca's area and in the pre-central and post-central gyri in Alzheimer's disease when compared to normal control subjects. In the latter study, the authors also reported that multi-infarct dementia is not associated with any alteration in brain MBR

(McGeer *et al.*, 1988). Since MBR in the CNS occur mainly on glial cells and brain damage is associated with gliosis and elevated MBR values (Benavides *et al.*, 1987), the increased MBR density in the brains of Alzheimer's patients may be interpreted as a sign of gliosis.

In view of these studies and in order to further evaluate whether the alterations in brain MBR in these patients are also reflected in platelet membranes, we studied [^{3}H]PK 11195 binding to platelet membranes in Alzheimer's patients as compared to multi-infarct demented patients and elderly healthy control subjects (Bidder *et al.*, 1990). The study showed increased platelet MBR density in the Alzheimer's patients, but not in the multi-infarct patients, when compared to the control subjects. It seems that the increase in MBR density in Alzheimer's disease is not confined to brain tissue, but is also reflected in platelet membranes. MBR are known to be sensitive to changes in steroid hormones (Gavish *et al.*, 1987). It is possible that increased hypothalamic–pituitary–adrenal axis activity, as reported in these patients (Coppen *et al.*, 1983; Christie *et al.*, 1987), plays a role in the upregulation of platelet MBR.

12.3.3.3 Carbamazepine in epilepsy

Carbamazepine (CBZ) is a potent anticonvulsant used mainly in the treatment of complex partial seizures of temporal lobe epilepsy. MBR have been reported to be involved in the anticonvulsant activity of CBZ. This suggestion was based on the potency of the drug to displace competitively the *in vitro* binding of Ro5-4864 (Marangos *et al.*, 1983). Ro5-4864 can block the anticonvulsant effect of CBZ on amygdala-kindled seizures, while PK 11195 can reverse this effect (Weiss *et al.*, 1985). However, a subsequent study in rats did not demontrate any alterations in brain MBR following chronic CBZ treatment (Marangos *et al.*, 1985).

We have studied platelet MBR density in patients with temporal lobe epilepsy, before and after 4 weeks of CBZ treatment, in comparison to healthy controls (Weizman *et al.*, 1987b). Epileptic patients before treatment exhibited reduced MBR density, and CBZ treatment induced an upregulation of [^{3}H]PK 11195 binding sites in platelets. The diminished platelet MBR binding in the patients before treatment followed by the increase after CBZ treatment does not necessarily imply parallel changes in brain. Yet the upregulatory effect of the drug may indicate the involvement of MBR in the therapeutic effect of CBZ. Nevertheless, it is possible that the normalization of platelet MBR density is related to the anxiety associated with the stressful convulsive events and the psychotropic properties of CBZ (Uhde *et al.*, 1985), as has also been demonstrated in anxious patients treated with diazepam (Weizman *et al.*, 1987a).

12.3.4 Substance consumption

12.3.4.1 Alcohol

Ethanol, like the benzodiazepines, possesses an anxiolytic and hypnotic effect. There is a cross-tolerance and dependence between the two drugs, which suggests the possibility of common points in the mode of action (Goldstein, 1978). Ethanol stimulates the GABA receptor-coupled transport of chloride ions in synaptoneurosomes from the brain of the rat (Suzdak *et al.*, 1986). A significant increase in brain MBR in rats and mice has been demonstrated following chronic ethanol consumption (Schoemaker *et al.*, 1983b; Tamborska and Marangos, 1986; Syapin and Alkana, 1988).

Suranyi-Cadotte *et al.* (1988) showed decreased platelet MBR density in chronic alcoholic patients, but not in abstinent alcoholics. In a subsequent study we discovered unaltered platelet MBR density in untreated chronic alcoholics, while disulfiram maintenance of detoxified alcoholics resulted in an elevation in [^{3}H]PK 11195 binding capacity (Karp *et al.*, 1991). The cause of the discrepancy between these two human studies is unclear, but may be ascribable to differences in the duration or quantity of alcohol consumption. However, our results are in accord with those of Chesley *et al.* (1990), who demonstrated lack of effect of heavy alcohol intake on platelet MBR in alcoholics. *In vitro* competition studies have shown that disulfiram is a competitive inhibitor at the MBR (Karp *et al.*, 1991). The activity of disulfiram at the MBR may be relevant to the effects of this drug on cell growth, differentiation, and metabolism.

12.3.4.2 Phenobarbital

The pharmacological effects of barbiturates are similar to those of benzodiazepines. Barbiturates allosterically enhance the *in vitro* binding of BZ and GABA receptor agonists in a chloride-dependent and picrotoxin-sensitive manner (Olsen, 1982). Administration of either a sedative/ataxic (20 mg/kg) or hypnotic (60 mg/kg) dose of phenobarbital prior to stress blocks the stress-induced decrease of renal MBR binding capacity in the rat (Drugan *et al.*, 1990). These findings suggest that central nervous mechanisms play a role in the regulation of MBR.

We have studied the effect of chronic pre- and post-natal administration of phenobarbital to mice. Pre-natal administration of the drug did not affect MBR characteristics in the olfactory bulb, heart, and kidney, while in neonates chronic exposure of the drug induced a downregulation of MBR in the heart, but not in the olfactory bulb and kidney. However, the downregulation of cardiac MBR was transitory (Fares *et al.*, 1990).

Chronic phenobarbital administration to adult mice induced a significant increase in the density of MBR in the heart, kidney and cerebellum, but did not affect the density of these receptors in the olfactory bulb (Wiezman *et al.*, 1989). It is possible that the reduced peripheral sympathetic tone secondary to the sedative effect of the drug is responsible for the increased MBR density, as has been demonstrated following chemical sympathectomy (Basile and Skolnick, 1988). The increase of cerebellar MBR may be an indirect index of cerebellar neuronal damage and alterations in glial elements, which contain a high density of these receptors (Benavides *et al.*, 1987.

12.3.4.3 Cocaine abuse

Cocaine acts through specific binding sites in the brain and is linked mainly to the dopaminergic system. The effect of long-term cocaine abuse in humans on platelet MBR has been studied by Chesley *et al.* (1990). They observed increased platelet MBR density in cocaine users when compared to non-users. The mechanism for this effect remains unclear, but might be related to an indirect link between cocaine or dopamine binding sites and MBR. It is noteworthy that chronic neuroleptic treatment induces downregulation of platelet MBR (Gavish *et al.*, 1986b; Weizman *et al.*, 1986). Thus, it seems that long-term administration of dopaminergic agonists (e.g. cocaine) augments MBR binding capacity, while the reverse is true of dopaminergic antagonists (e.g. neuroleptics).

12.4 Endocrinological aspects of MBR

Since MBR were first detected, the following questions have repeatedly been raised: what controls their density? and what is their physiological role? These questions are especially important in view of the fact that these receptors have been detected in almost every organ that has been examined, including the brain. A step toward resolving this enigma has been made in the endocrine field. Thus, MBR have been detected in organs which are subject to hormonal regulation, including both endocrine and non-endocrine organs.

MBR have been localized in the pituitary gland (De Souza *et al.*, 1985), the various organs of the genital axis in the male (De Souza *et al.*, 1985; Katz *et al.*, 1990a) and female (Fares *et al.*, 1987a), the placenta (Fares and Gavish, 1986), the adrenal gland and kidney (Benavides *et al.*, 1983; De Souza *et al.*, 1985; Fares *et al.*, 1989), the liver and pancreas (Braestrup and Squires, 1977; Verma and Snyder, 1989), and the pineal gland (Quirion, 1984; Basile *et al.*, 1986). Also, various organs that are considered to be non-glandular but are under hormonal control, such as the kidney (Gehlert *et al.*, 1983; Beaumont *et al.*, 1984; Fares *et al.*, 1989), the lung (Braestrup and Squires, 1977; Fares *et al.*, 1987b), and the brain (Schoemaker *et al.*, 1981), appear to contain significant levels of MBR.

Studying MBR distribution by autoradiographic techniques has revealed that MBR are located in specific tissues or regions which are relevant to the endocrinological and non-endocrinological functions of these organs. Furthermore, changes in MBR density have been observed mostly in the organs, regions, or tissues that have been found to be directly or indirectly susceptible to the hormones regulating the activity of that particular region or tissue. This observation may suggest hormonal regulation of MBR density. Furthermore, the possible function of MBR has been tested by measuring whether organ-specific activity is altered following exposure of the organ, under *in vivo* or *in vitro* conditions, to various BZ ligands, either alone or in combination with the specific target hormone. Interest in observing and studying the MBR-mediating action of the BZ is also related to the use of diazepam by clinicians. Diazepam reaches micromolar concentration in the plasma, while its affinity to MBR is in the nanomolar range; thus, it has the potential to bind to MBR and to exert a receptor-mediated action in many tissues. Nevertheless, to the best of our knowledge, the presence of MBR is a prerequisite to observing a BZ ligand-dependent action. This phenomenon may suggest that the BZ ligand induces its effect via a receptor-mediated action. A possible deviation from this rule might be seen in the pituitary, as described below.

In the following sections we will present the up-to-date literature, including our own, on the regulation of MBR density, as well as the involvement of BZ-specific ligands in modulating the specific activities of various representative organs.

12.4.1 Pituitary

As indicated in the introduction to this chapter, MBR differ from CBR mainly in their ligand specificity and in the fact that only CBR are coupled to GABA receptors in the brain. Additionally, CBR are located in the CNS, whereas MBR are located in both the peripheral and non-neuronal brain tissues. However, the anterior lobe of the pituitary gland, which is of ectodermal origin, is rich in CBR (Brown and Martin, 1984). Furthermore, clonazepam, but not Ro5-4864, enhances the transient stimulatory effect of muscimol on prolactin release from anterior pituitary slices of male rats under *in vitro* conditions (Anderson and Mitchell, 1984). The inhibitory effect of GABA or dopamine on the release of prolactin and thyroid-stimulating hormone (TSH) from the anterior pituitary cells is enhanced by diazepam (Grandison, 1983).

Higher MBR density has been demonstrated mainly in the posterior lobe of the pituitary (Brown and Martin, 1984; De Souza et al., 1985); however, the reported effect of MBR-specific ligands is in the anterior pituitary lobe. Thus, BZ ligands reduce ACTH release (Bruni et al., 1980) and alter the release of various other hormones, such as prolactin, TSH (Grandison, 1983), growth hormone, and luteinizing hormone (LH) (Verma and Snyder, 1989), that are released from the anterior pituitary lobe. One cannot exclude the possibility that the effect of BZ ligands in the anterior pituitary lobe may be direct and mediated by MBR which exist in lower density in this pituitary lobe. The role of MBR in the posterior pituitary lobe is still unknown. Nevertheless, in intact immature female rats, MBR density in whole pituitary is altered with the animal's age and by hormones which are of pituitary and ovarian origin. Thus, pituitary MBR density is 2-fold higher at 27 days of age as compared to 14 days of age (Fares et al., 1987a). Treatment with pregnant mare serum gonadotropin (PMSG), a hormone which contains follicle-stimulating hormone and LH, increases MBR density in whole female pituitary by 2-fold, whereas treatment with oestradiol-17β (E_2) decreases MBR density by 30% (Fares et al., 1987a). In the male rat, treatment with testosterone does not alter pituitary MBR density (Amiri et al., 1991).

12.4.2 Organs of the genital axis

12.4.2.1 Male

High MBR density has been detected in most of the male genital organs, including the whole vas deferens, prostate, seminal vesicles and Cowper's glands (Katz et al., 1990a), as well as the testis (De Souza et al., 1985). Autoradiographic studies in rat testis have revealed that MBR are located mainly in the interstitial tissues, where the Leydig cells are present, and produce and secrete testosterone under the influence of LH (De Souza et al., 1985; Ritta and Calandra, 1989).

The first indication for the hormonal regulation of MBR density has come from studies in testicular MBR. Hypophysectomy caused a significant depletion of MBR density in the testis (Anholt et al., 1985). Thyroxine treatment for 10 days significantly increased MBR density in the testis and other peripheral organs (Gavish et al., 1986c). On the other hand, 10 days' treatment with E_2 (Gavish et al., 1986a) or 16 days' treatment with testosterone (Amiri et al., 1991) caused a significant reduction in testicular MBR density.

Also, MBR density in the accessory glands of the male genitalia is modulated by hormones. For example, a significant reduction in MBR density in Cowper's gland has been observed following male castration (Weizman et al., 1992). This reduction in Cowper's gland MBR density was not observed in castrated animals following treatment with testosterone (Weizman *et al.*, 1992). In addition, testosterone also increased MBR density in Cowper's glands in the intact rat (Amiri *et al.*, 1991). It may be assumed that a similar phenomenon could be observed in other organs of the male genitalia, such as the prostate, since their development and function are testosterone-dependent (Allen, 1958).

Testicular testosterone secretion is altered by BZ ligands. Earlier studies showed that BZ ligands such as diazepam increase *in vivo* testosterone secretion in the rat (Cook *et al.*, 1979) as well as plasma testosterone in humans (Argüelles and Rosner, 1975). A more direct action of BZ ligands on testosterone secretion has been observed in rat testis. Decapsulated testes were cultured in the absence (basal) and presence of human chorionic gonadotropin. In both cases, the addition of BZ ligands increased the testosterone secretion from these organs (Ritta *et al.*, 1987).

To summarize, studying MBR in the male sex organs has revealed that MBR density is modulated by various hormones that modulate the activity of this genital axis. Moreover, MBR-specific ligands can alter testicular activity in terms of steroid hormone secretion. (The role of MBR ligands in steroidogenesis is reviewed in Chapter 4.)

12.4.2.2 Female

MBR have been localized in the ovary, oviduct and uterus of the female rat (Fares *et al.*, 1987a). In PMSG-stimulated rat ovary, MBR have also been detected in granulosa cells and in theca interstitial cells (S. Bar-Ami, F. Fares, and M. Gavish, unpubl. data). In rabbit ovary, however, although MBR have been demonstrated in theca interstitial cells, MBR were found in granulosa cells only following luteinization (Verma and Snyder, 1989).

In the immature rat ovary, MBR density increases with age (Fares *et al.*, 1987a). Hypophysectomy prevents this age-dependent increase and induces a significant decrease in ovarian MBR density that is associated with follicular atresia (Bar-Ami *et al.*, 1989). The effect of hypophysectomy is totally abolished when rats are treated with PMSG or E_2 (Bar-Ami *et al.*, 1989). Ovarian MBR density in intact rats is significantly increased by PMSG (Fares *et al.*, 1987a) as well as by short-term treatment with progesterone, testosterone, or E_2 (S. Bar-Ami, Z. Amiri, F. Fares and M. Gavish, manuscript submitted). However, while long-term treatment with E_2 increases MBR density in the ovary, long-term treatment with testosterone or progesterone causes a significant decrease in ovarian MBR density (S. Bar-Ami, Z. Amiri, F. Fares and M. Gavish, manuscript submitted).

Combining these various approaches in studying the regulation of MBR in the ovary as a whole, the unavoidable conclusion is that agents which enhance and stimulate ovarian follicular development and differentiation also increase MBR density, whereas inhibition or reduction in these processes is associated with a reduction in MBR density. Furthermore, changes in ovarian PBR density are produced only by ovarian tropic hormones, and not by other hormones such as ACTH (F. Fares, S. Bar-Ami and M. Gavish, unpubl. data). Similarly, in the rat testis, MBR density has been altered by hormones known to affect testicular physiology. This conclusion has received further confirmation by studying MBR density in the ovary of intact adult rats and in human granulosa cells. Thus, in the rat, ovarian MBR density is significantly increased on the day of pro-oestrus, when follicular size is maximal and the follicle is ready to ovulate. Maximum MBR density is acquired 4 h after the onset of LH surge and is minimum in a 4-day cycle on the day of met-oestrus, when the dominant follicles are at their minimum size (Fares et al., 1988).

In human granulosa cells collected after hormonally induced luteinization, MBR density was significantly greater when they were obtained from larger follicles or when proper morphological luteinization of follicular cells was observed (Bar-Ami et al., 1992). Furthermore, when MBR density was studied in individual follicles, a high correlation was found between egg cell performance, in terms of completion of meiotic maturation to the second metaphase stage, fertilization, and embryonic cleavage, and MBR-specific binding in the corresponding granulosa cells (Bar-Ami et al., 1992). MBR density in human ovarian granulosa cells was significantly higher in women with higher plasma E_2 level (>1400 pg/ml) as compared to women with lower plasma E_2 level (<1000 pg/ml). Moreover, granulosa cell MBR density was 1.8-fold higher in women who conceived following treatment by in vitro fertilization and embryo transfer as compared to those who did not conceive following this treatment (Bar-Ami et al., 1992).

The effect of BZ ligands on the steroidogenic activity of human granulosa cells obtained from women subjected to hormonally induced ovulation and luteinization has been studied. PK 11195, as well as Ro5-4864, caused an approximately 2-fold increase in the in vitro secretion of progesterone ($p < 0.02$). Furthermore, the stimulatory effect of these ligands upon granulosa cell progesterone secretion coincided with the increase in [^{3}H]PK 11195 maximal binding capacity that may suggest a receptor-mediated action (Bar-Ami et al., 1988). Thus, in the ovary, MBR density is under endocrine, paracrine and autocrine regulation, and MBR-specific ligands also modulate ovarian follicular cell steroidogenic activity.

MBR density has also been extensively studied in other female sex organs, such as the oviduct and uterus. Autoradiographic studies have revealed that MBR are enriched in the oviduct epithelium and in the uterine epithelium and glands, with somewhat lower levels seen in the smooth muscle of these structures (Verma et al., 1987). MBR density in the oviduct and uterus is under ovarian steroid regulation (Fares *et al.*, 1987a, 1988; Bar-Ami *et al.*, 1989). The increase in MBR density in these organs following PMSG administration or after hypophysectomy is also mediated by ovarian steroid biosynthesis. Recently we have demonstrated a significant increase in MBR-specific binding in human uterine endometrium during the menstrual cycle. Thus, a stepwise increase in MBR density was noted to be maximal on the day of ovulation, when endometrial tissue is highly developed and serum oestrogen is maximal (S. Bar-Ami, Y. Erlik, M. Deutsch, Z. Blumenfeld, Z. Amiri and M. Gavish, in prep.).

Testing the effect of MBR-specific ligands on oxytocin-induced contractions in rat uterus has revealed that diazepam and Ro5-4864, but not clonazepam, inhibit uterine contraction amplitude by 50% (Katz *et al.*, 1990d). It is of interest that the GABA-induced contractions of rat uterine muscle are inhibited by progesterone or progesterone metabolites (Putnam *et al.*, 1991). These various agents that affect uterine contraction should be studied for the possible interaction between GABA and BZ in the peripheral tissues as well.

To sum up, in the ovary, oviduct and uterus, MBR are located in specific regions of the organs which are under specific hormonal control. Secondly, the modulation in MBR density in these organs is correlated with the degree of development and differentiation of the specific tissue containing the MBR.

High MBR density has been detected in human term placenta (Fares and Gavish, 1986). Exposure of placental explants to Ro5-4864 (10^{-8} M) increased progesterone and E_2 secretion 2.4- and 1.4-fold, respectively (Barnea *et al.*, 1989). Treatment of placental explants with diazepam (10^{-7} M) and PK 11195 (10^{-6} M) caused a significant increase in progesterone and E_2 secretion, whereas clonazepam had no effect on steroid secretion (Barnes *et al.*, 1989).

12.4.3 Adrenal

MBR density is highest in the adrenal gland. Autoradiographic studies have revealed that MBR density is localized in the adrenal cortex, mainly in the zona glomerulosa, whereas MBR are scarcely seen in the adrenal medulla (Benavides *et al.*, 1983; De Souza *et al.*, 1985). The zona glomerulosa produces aldosterone, the salt-controlling hormone, and contains the ACTH receptors as well. Thus, both receptors—MBR and ACTH receptors—are in close proximity. Removal of the pituitary gland, thereby eliminating ACTH secretion, causes a significant reduction in adrenal MBR density (Anholt *et al.*, 1985; Fares *et al.*, 1989). Also, indirect interference with ACTH secretion affects MBR density in the adrenal. Administration of cyproterone, an antiandrogenic agent which reduces pituitary ACTH secretion, consequently reduces adrenal MBR density (Amiri *et al.*, 1991).

The effect of hypophysectomy on MBR density in the adrenal cortex is not homogeneous. Thus, following hypophysectomy MBR are conspicuously absent from the zonae fasciculata and reticularis, whereas MBR density in the zona glomerulosa is almost unchanged (Anholt et al., 1985). In fact, both zona fasciculata and zona reticularis function in glucocorticoid biosynthesis, thus explaining the reduction in kidney MBR following hypophysectomy (Anholt et al., 1985) or adrenalectomy (Basile et al., 1985). On the other hand, adrenal MBR density is restored in hypophysectomized animals by the administration of ACTH, which is the specific tropic hormone for this organ (Fares et al., 1989). MBR-specific ligands have tremendous effects on adrenal gland mechanisms that can affect the steroidogenic machinery, such as the cytochrome oxidase system, since porphyrins, putative endogenous MBR ligands, exert a clear effect on these enzymes (Verma and Snyder, 1989).

12.4.4 Other organs

In other organs, MBR have been detected in specific regions or tissues on which specific hormones exert an action. For example, autoradiographic studies in rat kidney indicate that MBR are absent from the cortex and in the inner medulla, but are localized on the outer medulla in the tubular elements, which are identified as the thick ascending limb of the loop of Henle and the collecting ducts (Regan et al., 1981; Gehlert et al., 1983; Beaumont et al., 1984). This region is involved in ionic transport across renal membranes and is highly sensitive to mineralocorticoids such as aldosterone (Regan et al., 1981).

In the rat kidney, MBR density is significantly reduced following hypophysectomy, probably due to the reduction of adrenal hormone levels (Anholt et al., 1985), and by adrenalectomy (Basile et al., 1985). On the other hand, administration of hydrocortisone (Fares et al., 1989) or deoxycorticosterone (Regan et al., 1981) increases MBR density in rat kidney. MBR density in the kidney is also altered after treatment with ligands which affect renal action itself, such as progesterone (Gavish et al., 1987). To the best of our knowledge, a direct effect of MBR-specific ligands on kidney function has not been studied; however, the effect of BZ ligands on electrolyte transport has been demonstrated in other tissues and organs (Verma and Snyder, 1989).

In the above discussion of MBR in the ovary, we have reviewed the literature on a few endocrine organs that have been extensively studied over the last decade. The data collected may shed some light on the multifunctional actions of MBR in various tissues and on possible future directions of the research in this important field.

In summary, our conclusions concerning the endocrinological aspects of MBR are:

(1) Although MBR are ubiquitous, their density is under multifactorial control, including hormonal—autocrine, paracrine and endocrine—regulation.

(2) MBR-specific ligands exert various effects in different organs, including enhancing steroidogenic activity in human ovarian cells and in testicular as well as adrenal cell lines, and inhibiting oxytocin-induced contractions in rat uterine fragments.

(3) Up- or downregulation of MBR density should be considered one of the main mechanisms regulating MBR functionality.

12.5 Other diseases

12.5.1 Hypertension

Platelet MBR were studied in 15 hypertensive patients (12 men and 3 women, mean age 47 $\pm$ 3 years). All the patients suffered from essential hypertension and were free of renal disease, cardiac disease and diabetes mellitus. The blood pressure of all patients was consistently above 160/90 mm Hg, and all had been drug-free for at least 14 days (Benavides *et al.*, 1984). The binding values of the patients were compared to those of 14 normotensive control subjects (6 men and 8 women, mean age 34 $\pm$ 6 years). The maximal binding capacity of [^{3}H] PK 11195 and the affinity of the ligand to MBR were similar in the hypertensive and control subjects. However, in the hypertensive group there were two patients without detectable [^{3}H] PK 11195 binding. The results in human subjects do not accord with the observation of increased platelet MBR density in spontaneously hypertensive rats (Benavides *et al.*, 1984). The discrepancy may be ascribed to the heterogeneity of human hypertension in contrast to the homogeneity of spontaneous hypertension in rats.

12.5.2 Hypercholesterolaemia

MBR are involved in cholesterol transport in the mitochondria (Krueger and Papadopoulos, 1990). Diazepam, which is active at both central and mitochondrial BZ receptors, inhibits *in vitro* cholesterol esterification by arterial acylCoA:cholesterol acyltransferase and plasma lecithin:cholesterol acyltransferase (Beoll, 1984). The ability of diazepam to inhibit these enzyme systems may be relevant to the antiatherosclerotic effect of diazepam in animals (Cornhill *et al.*, 1978), the hypocholesterolaemic effect of the drug, and the elevation of triglycerides and very-low-density lipoproteins associated with the use of BZ in humans (Bell, 1985). The interaction between drugs active at the MBR and cholesterol biosynthesis and metabolism merits further investigation.

Acknowledgement

We thank Miss Ruth Singer for typing and editing the manuscript.

References

Allen, J.M. (1958). The influence of hormones on cell division. II. Time response of several vesicle, coagulating gland, and ventral prostate of castrated male mice to a single injection of testosterone propionate. *Exp. Cell Res.* **14**, 142–148.

Amiri, Z., Weizman, R., Katz, Y., Lochner, A. and Gavish, M. (1991). Testosterone and cyproterone acetate modulate peripheral but not central benzodiazepine receptors in rat. *Brain Res.* **553**, 155–158.

Anderson, R.A. and Mitchell, R. (1984). Benzodiazepines potentiate the effect of muscimol on prolactin secretion in vitro [Abstract]. *Br. J. Pharmacol.* **82**, 343P.

Anholt, R.R.H., Murphy, K.M.M., Mack, G.E. and Snyder, S.H. (1984). Peripheral-type receptors in the central nervous system: localization to olfactory nerves. *J. Neurosci.* **4**, 593–603.

Anholt, R.R.H., De Souza, E.B., Kuhar, M.J. and Snyder, S.H. (1985). Depletion of peripheral-type benzodiazepine receptors after hypophysectomy in rat adrenal gland and testis. *Eur. J. Pharmacol.* **110**, 41–46.

Anholt, R.R.H., Pedersen, P.L., De Souza, E.B. and Snyder S.H. (1986). The peripheral-type benzodiazepine receptor: Localization to the mitochondrial outer membrane. *J. Biol. Chem.* **261**, 576–583.

Argüelles, E.A. and Rosner, J. (1975). Diazepam and plasma testosterone levels. *Lancet* **ii**, 607.

Bar-Ami, S., Fares, F., Bidder, M., Brandes, J.M. and Gavish, M. (1988). Characterization of peripheral-type benzodiazepine binding sites in human granulosa cells: role of follicular growth and oocyte maturity [Abstract]. *Eighth International Congress of Endocrinology*, Kyoto, Japan.

Bar-Ami, S., Fares, F. and Gavish, M. (1989). Effect of hypophysectomy and hormone treatment on the induction of peripheral-type benzodiazepine binding sites in female rat genital axis. *Horm. Metab. Res.* **21**, 106–107.

Bar-Ami, S., Fares, F. and Gavish, M. (1992). Altered peripheral benzodiazepine receptor density in human granulosa-lutein cells in relation to follicular maturity. *Mol. Cell. Endocrinol.* **82**, 285–291.

Barnea, E.R., Fares, F. and Gavish, M. (1989). Modulatory action of benzodiazepines on human term placental steroidogenesis in vitro. *Mol. Cell. Endocrinol.* **64**, 155–159.

Basile, A.S. and Skolnick, P. (1988). Tissue-specific regulation of peripheral-type benzodiazepine receptor density after chemical sympathectomy. *Life Sci.* **42**, 273–276.

Basile, A.S., Paul, S.M. and Skolnick, P. (1985). Adrenalectomy reduces the density of peripheral-type binding sites for benzodiazepines in the rat kidney. *Eur. J. Pharmacol.* **110**, 149–150.

Basile, A.S., Klein, D.C. and Skolnick, P. (1986). Characterization of benzodiazepine receptors in the bovine pineal gland: evidence for the presence of an atypical binding site. *Mol. Brain Res.* **1**, 127–135.

Basile, A.S., Weissman, B.A. and Skolnick, P. (1987). Maximal electroshock increases the density of [³H]Ro5-4864 binding to mouse cerebral cortex. *Brain Res. Bull.* **19**, 1–7.

Beaumont, K., Healy, D.P. and Fanestil, D.D. (1984). Autoradiographic localization of benzodiazepine receptors in the rat kidney. *Am. J. Physiol.* **247**, F718–F724.

Bell, F.P. (1984). Diazepam inhibits cholesterol esterification by arterial ACAT and plasma LCAT, *in vitro*. *Atherosclerosis* **50**, 345–352.

Bell, F.P. (1985). Inhibition of CoA:cholesterol acyltransferase and sterologenesis in rat liver by diazepam, *in vitro*. *Lipids* **20**, 75–79.

Benavides, J., Malgouris, C., Imbault, F., Begassat, F., Uzan, A., Renault, C., Dubroeucq, M.C., Gueremy, C. and Le Fur, G. (1983). Peripheral-type benzodiazepine binding sites in rat adrenal: binding studies with [³H]PK 11195 and autoradiographic localization. *Arch. Int. Pharmacodyn. Thér.* **226**, 38–49.

Benavides, J., Quarteronet, D., Plouin, P.-F., Imbault, F., Phan, T., Uzan, A., Renault, C., Dubroeucq, M.-C., Gueremy, C. and Le Fur, G. (1984). Characterization of peripheral-type benzodiazepine binding sites in human and rat platelets by using [³H]PK 11195: studies in hypertensive patients. *Biochem. Pharmacol.* **33**, 2467–2472.

Benavides, J., Fages, D., Carter, C. and Scatton, B. (1987). Peripheral-type benzodiazepine binding sites are a sensitive indirect index of neuronal damage. *Brain Res.* **421**, 167–172.

Bender, A.S. and Hertz, L. (1985). Pharmacological evidence that the non-neuronal diazepam binding site in primary cultures of glial cells is associated with a calcium channel. *Eur. J. Pharmacol.* **110**, 287–288.

Bidder, M., Ratzoni, G., Weizman, A., Blumensohn, R., Norymberg, M., Tyano, S. and Gavish, M. (1990). Platelet benzodiazepine binding in Alzheimer's disease. *Biol. Psychiat.* **28**, 641–643.

Black, K.L., Ikezaki, K. and Arthur, W.T. (1989). Imaging of brain tumors using peripheral benzodiazepine receptor ligands. *J. Neurosurg.* **71**, 113–118.

Black, K.L., Ikezaki, K., Santori, E.M., Becker, D.P. and Vintens, H.V. (1990). Specific high-affinity binding to peripheral benzodiazepine receptor ligands to brain tumors in rat and man. *Cancer* **65**, 93–97.

Bonuccelli, U., Nuti, A., Del Dotto, P., Piccini, P., Martini, C., Giannacini, G., Lucacchini, A. and Muratorio, A. (1991). Platelet peripheral benzodiazepine receptors are decreased in Parkinson's disease. *Life Sci.* **48**, 1185–1190.

Braestrup, C. and Squires, R.F. (1977). Specific benzodiazepine receptors in rat brain characterized by high-affinity [³H]diazepam binding. *Proc. Natl. Acad. Sci. USA* **74**, 3805–3809.

Brannan, M.D., Riggs, J.J., Hagemann, W.E. and Pruss, T.P. (1980). A comparison of the cardiovascular effects of haloperidol, thioridazine, and chlorpromazine HCl.

Arch. Int. Pharmacodyn. Thér. **244**, 48–57.

Brown, C. and Martin, I.L. (1984). Autoradiographic localisation of benzodiazepine receptors in the rat pituitary gland. *Eur. J. Pharmacol.* **102**, 563–564.

Bruni, G., Dal-Pray, P., Dotti, M.T. and Segri, G. (1980). Plasma ACTH and cortisol levels in benzodiazepine-treated rats. *Pharmacol. Res. Commun.* **12**, 163–165.

Byczkowski, J.Z. and Borysewicz, R. (1979). The action of chlorpromazine and imipramine on rat brain mitochondria. *Gen. Pharmacol.* **10**, 369–372.

Cantor, E.H., Kenessey, A., Semenuk, G. and Spector, S. (1984). Interaction of calcium channel blockers with non-neuronal benzodiazepine binding sites. *Proc. Natl. Acad. Sci. USA* **81**, 1549–1552.

Chesley, S.F., Schatzki, A.D., DeUrrutia, J., Greenblat, D.J., Shader, R.I. and Miller, L.G. (1990). Cocaine augments peripheral benzodiazepine binding in humans. *J. Clin. Psychiat.* **51**, 404–406.

Christie, J.E., Whalley, L.J., Bennie, J., Dick, H., Blackburn, I.M., Blackwood, D.H.R. and Fink, G. (1987). Characteristic plasma hormone changes in Alzheimer's disease. *Br. J. Psychiat.* **150**, 674–681.

Cohen, M.R., Pickar, D. and Dubois, M. (1981). Surgical stress and endorphins. *Lancet* **i**, 213–214.

Cook, P.S., Notelovitz, M., Kalra, P.S. and Kalra, S.P. (1979). Effect of diazepam on serum testosterone and the ventral prostate gland in male rats. *Arch. Androl.* **3**, 31–35.

Coppen, A., Abou-Saleh, M., Milln, P., Metcalf, M., Harwood, J. and Bailey, J. (1983). Dexamethasone suppression test in depression and other psychiatric illness. *Br. J. Psychiat.* **142**, 498–504.

Cornhill, J.F., Levesque, M.J., Nereum, R.M. and Patel, D.J. (1978). The effect of diazepam on the development of experimental atherosclerosis. *Circulation* **58**, II-128.

Dar, D.E., Weizman, A., Karp, L., Grinshpoon, A., Bidder, M., Kotler, M., Tyano, S., Bleich, A. and Gavish, M. (1991). Platelet peripheral benzodiazepine receptors in repeated stress. *Life Sci.* **48**, 341–346.

De Souza, E.B., Anholt, R.R.H., Murphy, K.M.M., Snyder, S.H. and Kuhar, M.J. (1985). Peripheral-type benzodiazepine receptors in endocrine organs: autoradiographic localization in rat pituitary, adrenal, and testis. *Endocrinology* **116**, 567–573.

Drugan, R.C., Basile, A.S., Crawley, J.N., Paul, S.M. and Skolnick, P. (1986). Inescapable shock reduces [^{3}H]Ro5-4864 binding to 'peripheral-type' benzodiazepine receptors in the rat. *Pharmacol. Biochem. Behav.* **24**, 1673–1677.

Drugan, R.C., Basile, A.S., Crawley, J.N., Paul, S.M. and Skolnick, P. (1987). 'Peripheral' benzodiazepine binding sites in Maudsley reactive rats: selective decrease confined to peripheral tissues. *Brain Res. Bull.* **18**, 143–145.

Drugan, R.C., Holmes, P.V. and Stringer, A.P. (1990). Phenobarbital blocks the stress-induced decrease in [^{3}H]Ro5-4864 binding in rat kidney. *Brain Res.* **535**, 151–154.

Fares, F. and Gavish, M. (1986). Characterization of peripheral benzodiazepine binding sites in human term placenta. *Biochem. Pharmacol.* **35**, 227–230.

Fares, F., Bar-Ami, S., Brandes, J.M. and Gavish, M. (1987a). Gonadotropin- and estrogen-induced peripheral-type benzodiazepine binding sites in the hypophyseal-genital axis of rats. *Eur. J. Pharmacol.* **133**, 97–102.

Fares, F., Weizman, A., Zlotogorski, D. and Gavish, M. (1987b). Ontogenesis of peripheral-type benzodiazepine binding sites in rat brain, lung, and heart. *Brain Res.* **408**, 381–384.

Fares, F., Bar-Ami, S., Brandes, J.M. and Gavish, M. (1988). Changes in the density of peripheral benzodiazepine binding sites in genital organs of the female rat during the oestrous cycle. *J. Reprod. Fert.* **83**, 619–625.

Fares, F., Bar-Ami, S., Haj-Yehia, Y. and Gavish, M. (1989). Hormonal regulation of peripheral benzodiazepine binding sites in female rat adrenal gland and kidney. *J. Receptor Res.* **9**, 143–157.

Fares, F., Weizman, A., Pick, C.G., Yanai, J. and Gavish, M. (1990). Effect of prenatal and neonatal chronic exposure to phenobarbital on central and peripheral benzodiazepine receptors. *Brain Res.* **506**, 115–119.

Farrarese, C., Appollonio, I., Frigo, M., Perego, M., Riolti, R., Trabucchi, M. and Frattola, L. (1990). Decreased density of benzodiazepine receptors in lymphocytes of anxious patients: reversal after chronic diazepam treatment. *Acta Psychiat. Scand.* **82**, 169–173.

File, S. and Lister, R.G. (1983). The anxiogenic action of Ro5-4864 is reversed by phenytoin. *Neurosci. Lett.* **35**, 93–96.

Gavish, M. and Wiezman, R. (1989). Effects of chronic chlorpromazine treatment on peripheral benzodiazepine binding sites in the heart, kidney, and cerebral cortex of rats. *J. Neurochem.* **52**, 1553–1558.

Gavish, M., Okun, F., Weizman, A. and Youdim, M.B.H. (1986a). Modulation of peripheral benzodiazepine binding sites following chronic estradiol treatment. *Eur. J. Pharmacol.* **127**, 147–151.

Gavish, M., Weizman, A., Karp, L., Tyano, S. and Tanne, Z. (1986b). Decreased peripheral benzodiazepine binding sites in platelets of neuroleptic-treated schizophrenics. *Eur. J. Pharmacol.* **121**, 275–279.

Gavish, M., Weizman, A., Okun, F. and Youdim, M.B.H. (1986c). Modulatory effects of thyroxine treatment on central and peripheral benzodiazepine receptors in the rat. *J. Neurochem.* **47**, 1106–1110.

Gavish, M., Weizman, A., Youdim, M.B.H. and Okun, F. (1987). Regulation of central and peripheral benzodiazepine receptors in progesterone-treated rats. *Brain Res.* **409**, 386–390.

Gavish, M., Weizman, R. and Tanne, Z. (1988). Effect of chronic haloperidol treatment on peripheral benzodiazepine binding sites in cerebral cortex of rats. *J. Neural Transm.* **74**, 109–116.

Gehlert, D.R., Yamamura, H.I. and Wamsley, J.K. (1983). Autoradiographic localization of 'peripheral' benzodiazepine binding sites in the rat brain and kidney using [^{3}H]Ro5-4864. *Eur. J. Pharmacol.* **95**, 329–330.

Ghadially, F.N. (1975). *Ultrastructural Pathology of Cell*, pp. 122–123. Butterworth, London and Boston.

Gold, J. (1974). Cancer cachexia and gluconeogenesis. *Ann. N.Y. Acad. Sci.* **230**, 103–110.

Goldstein, D.B. (1978). Alcohol-withdrawal reactions in mice: effects of drugs that modify neurotransmission. *J. Pharmacol. Exp. Ther.* **186**, 1–9.

Grandison, L. (1983). Action of benzodiazepines on the neuroendocrine system. *Neuropharmacology* **22**, 1505–1510.

Guidotti, A., Forchetti, C.M., Corda, M.G., Konkel, D., Bennet, C.D. and Costa, E. (1983). Isolation, characterization, and purification to homogeneity of an endogenous polypeptide with agonist action on benzodiazepine receptors. *Proc. Natl. Acad. Sci. USA* **80**, 3531–3533.

Ikezaki, K. and Black, K.L. (1990). Stimulation of cell growth and DNA synthesis by peripheral benzodiazepine. *Cancer Lett.* **49**, 115–120.

Karp, L., Weizman, A., Tyano, S. and Gavish, M. (1989). Examination stress, platelet peripheral benzodiazepine binding sites, and plasma hormone levels. *Life Sci.* **44**,

1077–1082.

Karp, L., Weizman, A., Filman, M., Tyano, S. and Gavish, M. (1991). Peripheral benzodiazepine receptors on platelets in chronic and detoxified alcoholics. *Neuropharmacology* **30**, 665–669.

Katz, Y., Moskovitz, B., Levin, D.R. and Gavish, M. (1989). Absence of peripheral-type benzodiazepine binding sites in renal cell carcinoma: a potential biochemical marker. *Br. J. Urol.* **63**, 124–127.

Katz, Y., Amiri, Z., Weizman, A. and Gavish, M. (1990a). Identification and distribution of peripheral benzodiazepine binding sites in male rat genital tract. *Biochem. Pharmacol.* **40**, 817–820.

Katz, Y., Ben-Baruch, G., Kloog, Y., Menczer, J. and Gavish, M. (1990b). Increased density of peripheral benzodiazepine-binding sites in ovarian carcinoma as compared with benign ovarian tumours and normal ovaries. *Clin. Sci.* **78**, 155–158.

Katz, Y., Eitan, A. and Gavish, M. (1990c). Increase in peripheral benzodiazepine binding sites in colonic adenocarcinoma. *Oncology* **47**, 139–142.

Katz, Y., Fares, F. and Gavish, M. (1990d). Effects of benzodiazepines on oxytocin-induced contractions in rat uterus. *Biochem. Pharmacol. (Life Sci. Adv.)* **9**, 145–148.

Kitazawa, M., Sugiyama, S., Ozawa, T., Miyazaki, Y. and Kotaka, K. (1981). Mechanism of chlorpromazine-induced arrhythmia: arrhythmia and mitochondrial dysfunction. *J. Electrocardiol.* **14**, 219–224.

Krueger, K.E. and Papadopoulos, V. (1990). Peripheral-type benzodiazepine receptors modulate translocation of cholesterol from outer to inner mitochondrial membranes in adrenocortical cells. *J. Biol. Chem.* **265**, 15 015–15 021.

Lathers, C.M. and Lipka, L.J. (1985). Chlorpromazine: cardiac arrhythmogenicity in the cat. *Life Sci.* **38**, 521–538.

Marangos, P.J., Post, R.M., Patel, J., Zander, K., Parma, A. and Weiss, S. (1983). Specific and potent interactions of carbamazepine with brain adenosine receptors. *Eur. J. Pharmacol.* **93**, 175–179.

Marangos, P.J., Weiss, S.R.B., Montgomery, P., Patel, J., Narang, P.K., Cappabianca, A.M. and Post, R.M. (1985). Chronic carbamazepine treatment increases brain adenosine receptors. *Epilepsia* **26**, 493–498.

Marc, V. and Marselli, P.L. (1969). Effect of diazepam on plasma corticosterone levels in the rat. *J. Pharm. Pharmacol.* **21**, 784–786.

Marthy, J., Guazit, R. and Lefevre, P. (1986). Effects of diazepam and midazolam on baroreflex control of heart rate and on sympathetic activity in humans. *Anesth. Analg.* **65**, 113–119.

Matthew, E., Laskin, J.D., Zimmerman, E.A., Weinstein, I.B., Hsu, K.C. and Engelhardt, D.L. (1981). Benzodiazepines have high-affinity binding sites and incude melanogenesis in B16/C3 melanoma cells. *Proc. Natl. Acad. Sci. USA* **78**, 3935–3939.

McGeer, E.G., Singh, E.A. and McGeer, P.L. (1988). Peripheral-type benzodiazepine binding in Alzheimer disease. *Alz. Dis. Assoc. Dis.* **2**, 331–336.

Mestre, M., Carriot, T., Belin, C., Uzan, A., Renault, C., Dubroeucq, M.C., Gueremy, C. and Le Fur, G. (1984). Electrophysiological and pharmacological characterization of peripheral benzodiazepine receptors in a guinea pig heart preparation. *Life Sci.* **35**, 953–962.

Mestre, M., Bouetard, G., Uzan, A., Gueremy, C., Renault, C., Dubroeucq, M.C. and Le Fur, G. (1985). PK 11195, an antagonist of peripheral benzodiazepine receptors, reduces ventricular arrhythmias during myocardial ischemia and reperfusion in the dog. *Eur. J. Pharmacol.* **112**, 257–260.

Mestre, M., Belin, C., Uzan, A., Renault, C., Dubroeucq, M.C., Gueremy, C. and

Le Fur, G. (1986). Modulation of voltage-operated, but not receptor-operated, calcium channels in the rabbit aorta by PK 11195, an antagonist of peripheral-type benzodiazepine receptors. *J. Cardiovasc. Pharmacol.* **8**, 729–734.

Mizoule, J., Gauthier, A., Uzan, A., Renault, C., Dubroeucq, M., Gueremy, C. and Le Fur, G. (1985). Opposite effects of two ligands for peripheral-type benzodiazepine binding sites, PK 11195 and Ro5-4864, in a conflict situation in the rat. *Life Sci.* **36**, 1059–1068.

Mukhin, A.G., Papadopoulos, F., Costa, E. and Krueger, K.E. (1989). Mitochondrial benzodiazepine receptors regulate steroid biosynthesis. *Proc. Natl. Acad. Sci. USA* **86**, 9813–9816.

Noel, G.L., Suh, H.K., Stone, J.C. and Frantz, A.G. (1972). Human prolactin and growth hormone release during surgery and other conditions of stress. *J. Clin. Endocrinol.* **35**, 840–851.

Okun, F., Weizman, R., Katz, Y., Bomzon, A., Youdim, M.B.H. and Gavish, M. (1988). Increase in central and peripheral benzodiazepine receptors following surgery. *Brain Res.* **458**, 31–36.

Olsen, R.W. (1982). Drug interactions at the GABA-receptor–ionophore complex. *Ann. Rev. Pharmacol. Toxicol.* **22**, 245–277.

Owen, F., Poulter, M., Waddington, J.L., Mashal, R.D. and Crow, T.J. (1983). [^{3}H]Ro5-4864 and [^{3}H]flunitrazepam binding in kainate-lesioned rat striatum and in temporal cortex of brains from patients with senile dementia of the Alzheimer type. *Brain Res.* **278**, 373–375.

Papadopoulos, V., Mukhin, A., Costa, E. and Krueger, K.E. (1990). The peripheral-type benzodiazepine receptor is functionally linked to Leydig cell steroidogenesis. *J. Biol. Chem.* **265**, 3772–3779.

Papadopoulos, V., Nowzari, F.B. and Krueger, K.E. (1991). Hormone-stimulated steroidogenesis is coupled to mitochondrial benzodiazepine receptors. *J. Biol. Chem.* **266**, 3682–3687.

Papart, P., Ansseau, M., Certontaine, J.L. and Von Frenckell, R. (1988). Pilot study of 5208 RP (PK 11195), an antagonist of the peripheral-type benzodiazepine binding sites, among inpatients with anxious or depressive symptomatology [Abstract]. *Psychopharmacology* **96** (Suppl. 1), 219.

Putnam, C.P., Brann, D.W., Kolbeck, R.L. and Mahesh, V.D. (1991). Inhibition of uterine contractility by progesterone and progesterone metabolites: mediation by progesterone and gamma-amino butyric acid$_A$ receptor systems. *Biol. Reprod.* **45**, 266–272.

Quirion, R. (1984). High density of [^{3}H]Ro5-4864 peripheral benzodiazepine binding sites in the pineal gland. *Eur. J. Pharmacol.* **102**, 559–560.

Rägo, L., Adojaan, A. and Pokk, P. (1989a). The effect of stress on ω_3 benzodiazepine receptor in rat blood platelets and lymphocytes: the effect of nonbenzodiazepine tranquilizers. In *Molecular Pharmacology of Receptors III* (ed. Allikments, L.H.), pp. 4–16. Tartu University Press, Tartu, Estonia.

Rägo, L., Kiivet, R.-A., Harro, J. and Pold, M. (1989b). Central- and peripheral-type benzodiazepine receptors: similar regulation by stress and GABA receptor agonists. *Pharmacol. Biochem. Behav.* **32**, 879–883.

Regan, J.W., Yamamura, H.I., Yamada, S. and Roeske, W.R. (1981). High-affinity renal [^{3}H]flunitrazepam binding: characterization, localization, and alterations in hypertension. *Life Sci.* **28**, 991–998.

Ritta, M.N. and Calandra, R.S. (1989). Testicular interstitial cells as target for peripheral benzodiazepines. *Neuroendocrinology* **49**, 262–266.

Ritta, M.N., Campos, M.B. and Calandra, R.S. (1987). Effect of GABA and benzodiazepines on testicular androgen production. *Life Sci.* **40**, 791–798.

Schoemaker, H., Bliss, M. and Yamamura, H.I. (1981). Specific high-affinity saturable binding of [^{3}H]Ro5-4864 to benzodiazepine binding sites in the rat cerebral cortex. *Eur. J. Pharmacol.* **71**, 173–175.

Schoemaker, H., Boles, R.G., Horst, W.D. and Yamamura, H.I. (1983a). Specific high-affinity binding sites for [^{3}H]Ro5-4864 in rat brain and kidney. *J. Pharmacol. Exp. Ther.* **225**, 61–69.

Schoemaker, H., Thomas, L.S. and Yamamura, H.I. (1983b). Effect of chronic ethanol consumption on central- and peripheral-type benzodiazepine binding sites in the mouse brain. *Brain Res.* **253**, 347–350.

Shibata, H., Kojima, I. and Ogata, E. (1986). Diazepam inhibits potassium-induced aldosterone secretion in adrenal glomerulosa cells. *Biochem. Biophys. Res. Commun.* **135**, 994–999.

Starosta-Rubinstein, S., Ciliax, B.J., Renny, J.B., McKeever, P. and Young, A.B. (1987). Imaging of a glioma using peripheral benzodiazepine receptor ligands. *Proc. Natl. Acad. Sci. USA* **84**, 891–895.

Suranyi-Cadotte, B., Lafaille, F., Dongier, M., Dumas, M. and Quirion, R. (1988). Decreased density of peripheral benzodiazepine binding sites on platelets of currently drinking but not abstinent alcoholics. *Neuropharmacology* **27**, 443–445.

Suzdak, P.D., Schwartz, R.D., Skolnick, P. and Paul, S.M. (1986). Ethanol stimulates GABA receptor-mediated chloride transport in rat brain synaptoneurosome. *Proc. Natl. Acad. Sci. USA* **83**, 4071–4075.

Syapin, P.J. and Alkana, R.L. (1988). Chronic ethanol exposure increases peripheral-type benzodiazepine receptors in brain. *Eur. J. Pharmacol.* **147**, 101–109.

Tallman, J.F., Paul, S.M., Skolnick, P. and Gallager, D.W. (1980). Receptors for the age of anxiety: pharmacology of benzodiazepine. *Science (Wash. DC)* **207**, 274–281.

Tamborska, E. and Marangos, P.J. (1986). Brain benzodiazepine binding sites in ethanol-dependent and withdrawal states. *Life Sci.* **38**, 465–472.

Tanne, Z., Weizman, R., Karp, L., Katz, Y., Tyano, S. and Gavish, M. (1987). Lack of effect of 28 days of neuroleptic treatment on platelet benzodiazepine binding sites in schizophrenics. *Neuropsychobiology* **17**, 121–123.

Uhde, T.W., Ballanger, J.C. and Post, R.M. (1985). Carbamazepine: treatment of affective illness and anxiety syndromes. In *Psychiatry: The State of the Art*, Vol. 3 (eds Pichot, P., Berner, P., Wolf, R. and Thau, K.), pp. 479–495. Plenum Publishing Company, New York.

Verma, A. and Snyder, S.H. (1989). Peripheral-type benzodiazepine receptors. *Ann. Rev. Pharmacol. Toxicol.* **29**, 307–322.

Verma, A., Trifiletti, R.R., Michael, E.M. and Snyder, S.H. (1987). Peripheral-type benzodiazepine receptor: isolation from outer mitochondrial membrane: porphyrins as endogenous ligands: hormonal associations [Abstract]. *Soc. Neurosci. Abstr.* **13**, 395.

Wang, J.K.T., Morgan, J.I. and Spector, S. (1984a). Benzodiazepines that bind at peripheral sites inhibit cell proliferation. *Proc. Natl. Acad. Sci. USA* **81**, 753–756.

Wang, J.K.T., Morgan, J.I. and Spector, S. (1984b). Differentiation of Friend erythroleukemia cells induced by benzodiazepines. *Proc. Natl. Acad. Sci. USA* **81**, 3770–3772.

Weiss, S.R.B., Post, R.M., Patel, J. and Marangos, P.J. (1985). Differential mediation of the anticonvulsant effects of carbamazepine and diazepam. *Life Sci.* **36**, 2413–2417.

Weizman, R. and Gavish, M. (1989). Chronic diazepam treatment induces an increase in peripheral benzodiazepine binding sites. *Clin. Neuropharmacol.* **12**, 346–351.

Weizman, R., Tanne, Z., Karp, K., Tyano, S. and Gavish, M. (1986). Peripheral-type benzodiazepine-binding sites in platelets of schizophrenics with and without tardive dyskinesia. *Life Sci.* **39**, 549–555.

Weizman, R., Tanne, Z., Granek, M., Karp, L., Golomb, M., Tyano, S. and Gavish, M. (1987a). Peripheral benzodiazepine binding sites on platelet membranes are increased during diazepam treatment of anxious patients. *Eur. J. Pharmacol.* **138**, 289–292.

Weizman, A., Tanne, Z., Karp, L., Martfeld, Y., Tyano, S. and Gavish, M. (1987b). Carbamazepine up-regulates the binding of [^{3}H]PK 11195 to platelets of epileptic patients. *Eur. J. Pharmacol.* **141**, 471–474.

Weizman, A., Fares, F., Pick, C.G., Yanai, J. and Gavish, M. (1989). Chronic phenobarbital administration affects GABA and benzodiazepine receptors in the brain and periphery. *Eur. J. Pharmacol.* **169**, 235–240.

Weizman, A., Bidder, M., Fares, F. and Gavish, M. (1990). Food deprivation modulates γ-aminobutyric acid receptors and peripheral benzodiazepine binding sites in rats. *Brain Res.* **535**, 96–100.

Weizman, A., Amiri, Z., Katz, Y., Snyder, S.H. and Gavish, M. (1992). Testosterone prevents castration-induced reduction in peripheral benzodiazepine receptors in Cowper's gland and adrenal. *Brain Res.* **572**, 72–75.

MITOCHONDRIAL BENZODIAZEPINE RECEPTOR LIGANDS AS INDICATORS OF DAMAGE IN THE CNS: THEIR APPLICATION IN POSITRON EMISSION TOMOGRAPHY

R. Myers

MRC Cyclotron Unit, Hammersmith Hospital, Ducane Road, London W12 0HS, UK

Table of Contents

13.1   Introduction

For more than a century, neurologists and neuropathologists have attempted to define the underlying processes of central nervous system (CNS) dysfunction by using post-mortem or biopsy material. Recent technological advances, however, allow us to follow the time-course of some elements of the lesion process in living humans using *in vivo* imaging. Computerized axial tomography (CT), magnetic resonance imaging (MRI) and positron emission tomography (PET) each contribute to this field as complementary techniques.

PET may be used to visualize and quantify elements of brain dysfunction by producing regionally localized dynamic data following intravenous injection of specific and selective radioligands for neuroreceptors which exhibit a modification of response as a result of that dysfunction. Mitochondrial benzodiazepine receptors (MBR) are expressed at only a very low level in normal brain. However, their numbers are dramatically increased in regions of brain expressing injury. This increase in numbers has frequently been described as indicative of 'neuronal damage' and much indirect evidence has been presented to imply that MBR are associated with 'gliosis' (Benavides *et al.*, 1987). The antagonist, PK 11195 is selective for the MBR, to which it binds with high affinity. Recently, PK 11195 has been labelled with the positron-emitting isotope carbon-11 and its potential evaluated experimentally for use in PET (Junck *et al.*, 1989; Price *et al.*, 1990; Cremer *et al.*, 1992). The distribution of the binding of this radioligand provides an 'image' of a lesion in brain (Benavides *et al.*, 1988). However, if such an image is to be interpreted in terms of pathology, the kinetics of specific binding of the ligand and the identity of the cells on which the receptors are situated, must be established.

This chapter will review the experimental data on the association between MBR expression and CNS injury. The available information on the cellular location of MBR will be considered and the implications for the interpretation of pathology discussed. The use, to date, of MBR ligands for PET studies will be reviewed, the technique will be briefly described and the properties of $[^{11}C]$PK 11195 will be discussed in the context of the specifications for an ideal PET ligand. Finally, the potential contributions of MBR imaging to the understanding of CNS pathology, in particular the inflammatory response, and to the diagnosis of CNS injury will be considered.

13.2 Studies of lesioned brain tissue

13.2.1 An historical perspective

In the late 1970s and early 1980s, many studies were underway to investigate the pharmacology and functions of benzodiazepine receptors using radioligands such as $[^{3}H]$clonazepam and $[^{3}H]$diazepam (e.g. Squires and Braestrup, 1976; Möhler and Okada, 1977). While the former ligand appeared to label only one site, located in the brain, the binding of the latter was more complex, the sites being situated both in brain and in various peripheral organs such as kidney (Braestrup *et al.*, 1977). While clonazepam displaced $[^{3}H]$diazepam binding with high potency from sites in the CNS, it was weaker in its action in the periphery. In contrast, the *p*-chloro-derivative of diazepam, Ro5-4864, had a high affinity for these peripheral sites.

In the CNS, benzodiazepine receptors were shown not to be confined to a single cell type, large numbers being present on cells of both neural and glial origins (Syapin and Skolnick, 1979). McCarthy and Harden (1981), studying the binding of $[^{3}H]$flunitrazepam in cultures of either primarily neuronal cells or primarily glia, concluded that the 'brain specific' benzodiazepine site was similar to that characterized in neuronal cultures, while the benzodiazepine site predominant on the glial cultures was pharmacologically distinct. Gallager and her colleagues (1981) also contrasted the binding of $[^{3}H]$diazepam in 'neuronal' and 'non-neuronal' cell cultures and, in the first published study reporting binding of an MBR ligand in lesioned brain, used membrane preparations from rat brain lesioned by injection of intrastriatal kainic acid. This treatment is known to result in the selective degeneration of neurons at the site of injection accompanied by a proliferation of glial cells (Schwarcz and Coyle, 1977). Gallager *et al.* (1981) demonstrated a four-fold increase in the density of Ro5-4864-sensitive sites in kainate-lesioned striatum compared with control. They concluded that these sites were located principally on non-neuronal elements and were pharmacologically distinct from the clonazepam-sensitive neuronal sites.

The current wave of interest in MBR as indices of CNS damage must, in large part, be attributed to the extensive work of Benavides and his colleagues. In 1983, this group published a series of papers in which they introduced a new ligand, PK 11195 (1-(2-chlorophenyl)-N-methyl-N-(1-methylpropyl)-3-isoquinolinecarboxamide) for this binding site (Le Fur *et al.*, 1983a,b,c; Benavides *et al.*, 1983). These authors showed that PK 11195 was more potent at the MBR than Ro5-4864 in both *in vitro* and *in vivo* studies and determined the *in vitro* binding kinetics of this ligand to rat cortical membranes. The affinity constant of the ligand for the receptor, the K_d, was of the order of 1 nM and temperature insensitive and the B_{max}, or receptor density, was ~120 fmol/mg protein. In addition, Benavides and his colleagues published the first autoradiographic demonstration of the distribution of peripheral benzodiazepine binding sites in normal rat brain, showing high densities of MBR in the glomerular and olfactory nerve layers of the olfactory bulb and the median eminence, in addition to ventricular structures such as the choroid plexus and ependymal cell layer (Benavides *et al.*, 1983). The remainder of the neuropil contained only a low density of MBR. Cat brain (Benavides *et al.*, 1984) showed markedly higher densities of MBR throughout the neuropil, suggesting some inter-species variability, later confirmed in studies by Cymerman *et al.*, in 1986.

The possibility of visualizing and quantifying 'non-neuronal' cells in brain using MBR ligands was subsequently explored by a number of groups using a variety of lesion models and these are summarized in Table 1 and briefly reviewed below.

13.2.2 Studies using animal lesion models

13.2.2.1 Excitotoxic lesions

Lesion models involving the administration of an excitotoxin, either locally in CNS or systemically, have far out-numbered other models in studies of MBR binding. Schoemaker *et al.* (1982), by injection of intrastriatal kainic acid, demonstrated a 10-fold increase in the binding of [^{3}H]Ro5-4864 to homogenates from lesioned striatum compared with the contralateral side, 1 to 6 weeks following the lesion. The authors concluded that '... since kainic acid lesioning results in glial proliferation, it seems likely that a major proportion of the Ro5-4864 sensitive receptor is localized on glial membranes'. Using a similar model, Owen *et al.* (1983) showed a more modest, four-fold increase in binding of [^{3}H]Ro5-4864 compared with control, possibly due to the production of lesions smaller than those in the Schoemaker *et al.* study. However, extension of the work to include post-mortem material from patients with Alzheimer disease gave unimpressive results, a 28% increase in binding,

Table 1 Effect of various excitotoxins on MBR ligand binding in rat brain.

Authors	Excitotoxin	Ligand	Binding to MBR
Gallager et al. (1981)	Intrastriatal kainate	[^{3}H] Diazepam	415% increase
Schoemaker et al. (1982)	Intrastriatal kainate	[^{3}H] Ro5-4864	1000–1200% increase
Owen et al. (1983)	Intrastriatal kainate	[^{3}H] Ro5-4864	382% increase
Gehlert et al. (1985)	Ibotenic acid into piriform cortex	[^{3}H] Ro5-4864	5000% increase
Benavides et al. (1987)	Intrastriatal kainate (4 nmol) AMPA (100 nmol) NMDA (250 nmol) quisqualate (250 nmol)	[^{3}H] Ro5-4864	321% increase 450% increase 525% increase 409% increase
Dubois et al. (1988)	Intrastriatal kainate (12 nmol) NMDA (250 nmol)	[^{3}H] PK 11195	548% increase 500% increase
Price et al. (1990)	Intrastriatal kainate	[^{3}H] PK 11195	>300% increase measured in vivo
Altar and Baudry (1990)	Systemic kainate	[^{3}H] PK 11195	1600% to 5000% increase various regions of ventral limbic forebrain
Baudry and Altar (1991)	Intrahippocampal colchicine	[^{3}H] PK 11195 [^{3}H] Ro5-4864	303% increase

which led the authors to conclude that the 'poor binding characteristics' of radiolabelled Ro5-4864 'may limit its use as an index of gliosis in man'.

Benavides *et al.* (1987) later correlated the binding of [^{3}H]Ro5-4864 with levels of the neuronal marker enzymes choline acetyltransferase (CAT) and glutamate decarboxylase (GAD) in rat striatum lesioned with one of the four excitotoxins, kainic acid (KA), *N*-methyl-D-aspartic acid (NMDA), quisqualic acid or (RS)-α-amino-3-hydroxy-5-methyl-4-isoxazolo-propionic acid (AMPA). Results showed a dose-dependent increase in the levels of MBR sites following injection of each of the excitotoxins. This was matched by a decrease in the activity of both CAT and GAD. The order of potency of the excitotoxins as measured by each of the two responses was the same, i.e. kainate > AMPA > NMDA > quisqualate. Although the relationship between the changes in density of binding sites and loss of enzyme activity differed for the different toxins, alterations in binding site density generally occurred at lower doses of the toxins than those needed to cause changes in the enzyme markers. The authors did not speculate on the various mechanisms of action of the different toxins but simply drew the conclusion, which must form the foundation for this chapter, that peripheral benzodiazepine binding sites '... provide a convenient and sensitive indirect index of cerebral neuronal damage ...'.

13.2.2.2 Models of ischaemia

Dubois *et al.* (1988) made the first excursion from excitotoxic lesion models in a study which included both intrastriatal injection of kainic acid and ischaemia induced by unilateral occlusion of the middle cerebral artery (MCA). These authors compared the binding of [^{3}H]PK 11195 with that of [^{3}H]SCH 23390, a ligand for the D_1 dopamine receptor used to quantify neuronal loss in the striatum. At one week following occlusion of the MCA, the level of MBR as determined by [^{3}H]PK 11195 binding within the core of the infarct was '... barely detectable ...', while a dramatic, 13-fold, increase was seen around the perimeter of the infarct in a region which '... faithfully replicated the histological limits of the consolidated lesion.' Within the nigrostriatal system there was a close correlation between MBR density and loss of density of D_1 receptors. An increase in the density of MBR was observed in ipsilateral substantia nigra as well as striatum, associated with damage to the reciprocal connections between the two sites. Distant lesions, as indicated by increased density of MBR in regions remote from the primary lesion, were also observed in some thalamic structures. The authors noted that the '... detection of unsuspected changes ...' provided an interesting aspect to this marker.

Work from our own laboratory has centred on the use of intrastriatal kainate lesions (Price *et al.*, 1990) and, more recently, a photochemically induced lesion for the production of focal ischaemia (Myers *et al.*, 1991a). In this experimental model in the rat, the photosensitive dye Rose Bengal is injected intravenously and 570 nm laser light is shone via a fibre optic light guide onto the exposed surface of the skull (Watson *et al.*, 1985, 1986; Grome *et al.*, 1988). The light causes the Rose Bengal to release free radicals which in turn cause damage to blood vessel walls resulting in thrombus formation and blockage of surface vessels immediately under the light guide. Blood flow is thus prevented, causing focal ischaemia in a cone-shaped zone of cortex within the perfusion territory of these vessels. The consequent ensuing pathology seen during the first week following lesion induction is typical of that associated with a cortical infarct in human post-mortem brain. The pattern of *ex vivo* binding of [^{3}H]PK 11195 during the first week post-lesioning is reported by Myers *et al.* (1991a). No change in binding was observed prior to 3 days. By day 3, localized regions of increased [^{3}H]PK 11195 binding were visible, particularly in the proximity of the pial membrane. By day 5 a high density of [^{3}H]PK 11195 was apparent around the perimeter of the lesion which, at this time, was cone-shaped with a diameter of about 3.5 mm at the pial surface. At 7 days (Plate 2a) binding was still intense, although the lesion was beginning to shrink in overall size. The maximum ratio of radioactivity (per g of tissue) close to the lesion to that in a similarly sized contralateral region in the same slice was approximately 4:1.

In addition to the binding of [^{3}H]PK 11195 around the primary cortical ischaemic lesion, a discrete region of increased binding was consistently observed at times later than 1 week in the ipsilateral ventrolateral thalamic nucleus (Myers *et al.*, 1991b). There are known reciprocal connections between this nucleus and the lesioned cortical area, and binding of [^{3}H]PK 11195 could be seen along the intervening white matter tracts. As well as highlighting the sensitivity of MBR expression as a means of detecting response to injury, these observations confirm the conclusion of Dubois *et al.* (1988) that MBR ligands may be useful in detecting damage in previously unsuspected regions. The cellular response in these remote lesions (illustrated in Figure 4) is further discussed in Section 13.3.3.3.

13.2.2.3 Tumour studies

Significant number of MBR have been measured in several studies using cultured tumour cell lines (Syapin and Skolnick, 1979; Gallager *et al.*, 1981; Wang *et al.*, 1984). In 1987, Starosta-Rubinstein *et al.* provided the first evidence that MBR ligand binding to glioma cells could provide an effective means of imaging intra-cranial tumours *in vivo*. These authors used either [^{3}H]PK 11195 or [^{3}H]flunitrazepam, blocking the component of central benzodiazepine binding with clonazepam. They showed that binding in rat C_6 glioma and in human glioma, xenografted into rat brain, was almost exclusively to MBR and, in non-necrotic regions of tumour, was many times higher than in normal brain. In areas of tumour necrosis, binding was low. In addition, these authors showed that MBR densities were low in xenografted human pheochromocytoma and absent from human neuroblastoma. Subsequent studies by Olson *et al.* (1988) suggested that PK 11195 was the preferred ligand for use with human tumours (see Section 13.2.3.5).

Studies by Black *et al.* (1989) and Ikezaki *et al.* (1990) using [³H]PK 11195 confirmed the large, though variable, increase in MBR in tumours transplanted into rat brain, compared with normal tissue. The highest densities were recorded in the RG-2 glioma, being 20-fold, compared with 10-fold increases in LK and HK Walker 256 tumours and 3- to 5-fold in the C_6 glioma. These increases were shown to reflect changes in B_{max}, rather than the K_d, i.e. representing an increase in the number of MBR present rather than a change in the affinity of the receptors (Black *et al.*, 1990).

Each of the above studies showed an excellent correlation between tumour topography and MBR ligand binding and the authors proposed radiolabelled ligands for MBR as candidates for studies using positron emission tomography (PET).

13.2.3 Human post-mortem studies

From the earliest studies on MBR, measurements using human post-mortem material have frequently accompanied those using animal lesion models. These studies will be briefly reviewed below, classified by disease.

13.2.3.1 Huntington's disease

Schoemaker *et al.* (1982) likened the pathology following the experimental intrastriatal injection of kainate into rat brain to that seen in the brains of patients with Huntington's disease (HD), each lesion involving selective neuronal death in the striatum accompanied by gliosis. These authors extended their studies on rats (see Section 13.2.2.1) to include material from 14 patients with adult-onset HD and compared their results to those from 18 'controls'. A 'modest' but significant 51% increase in binding of [³H]Ro5-4864 was reported for HD putamen with no significant changes in the other two regions sampled, i.e. caudate nucleus and globus pallidus. There was no correlation with any known ante-mortem factors, such as age, sex or duration of disease, nor with any post-mortem, autopsy or storage variables.

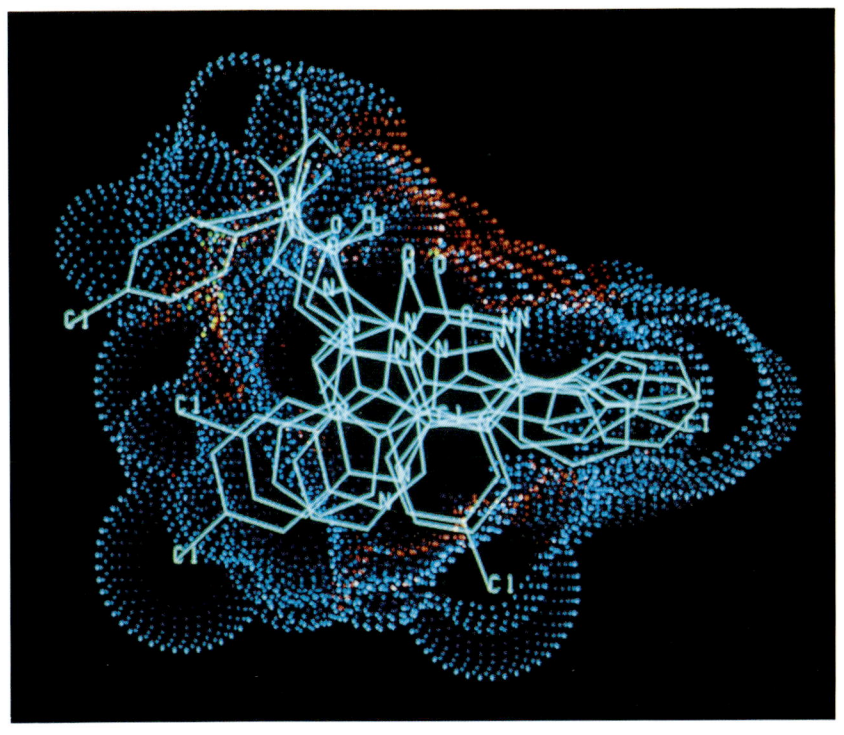

Plate 1 Hydrophobic (in blue) and hydrophilic (in red) regions of the MBR ligand pharmacophore.

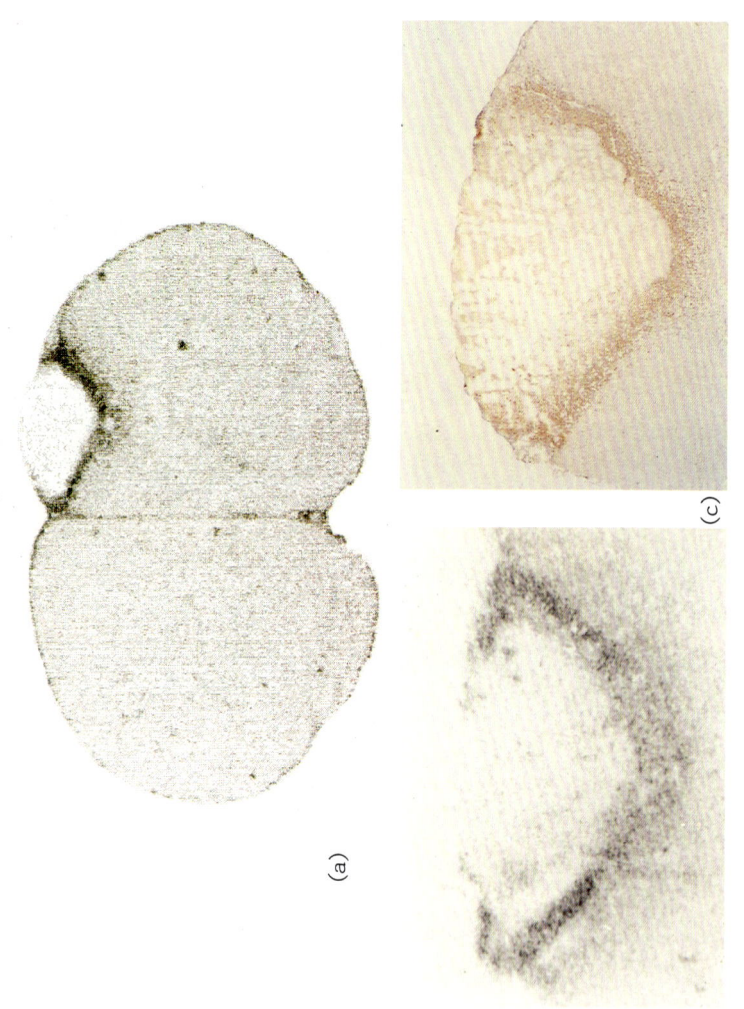

(a)

(b)

(c)

scale bar

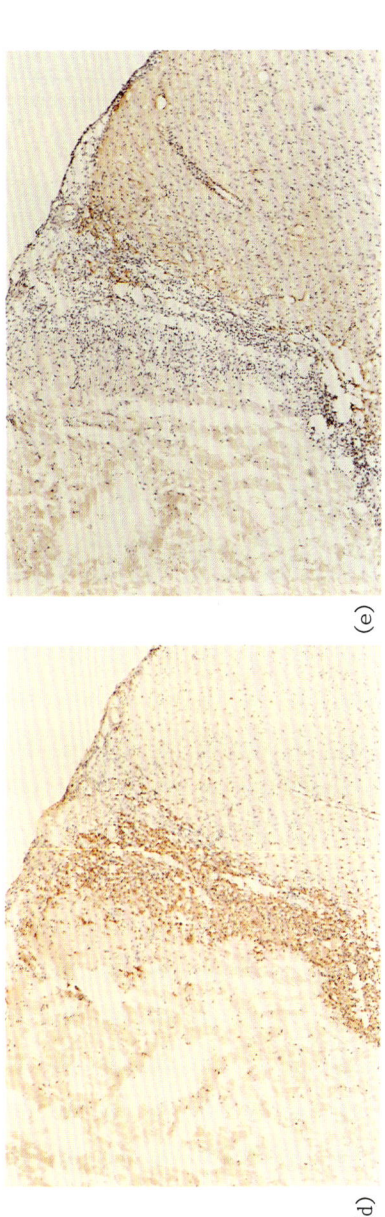

scale bar

(d)

(e)

Plate 2 Focal cortical ischaemia in rat brain 7 days post-lesioning. (a) Digitized image of an autoradiograph of a coronal section of rat brain, assayed 15 min after i.v. injection of 3.7 MBq of [³H]PK 11195 (NEN DuPont UK Ltd). The specific binding around the rim of the cortical lesion is clearly seen. (b) As for (a) but at higher magnification. (c) Photomicrograph of a section serial to that shown in (b). The brown 3,3-diaminobenzidine stain represents ED-1-positive macrophages. (d) Higher magnification view of that shown in (c). (e) A section serial to that shown in (d) in which the brown stain represents GFAP-positive astrocytes. The dense mass of blue counter-stained macrophages corresponds to the ED-1-positive cells in (d). The astrocytes surround the region of macrophage infiltration with little mixing of the two populations and almost no invasion of the macrophage region by astrocytic processes.

The bar represents 760 μm in (b) and (c), and 265 μm in (d) and (e). Reproduced with permission from Myers *et al.* (1991a).

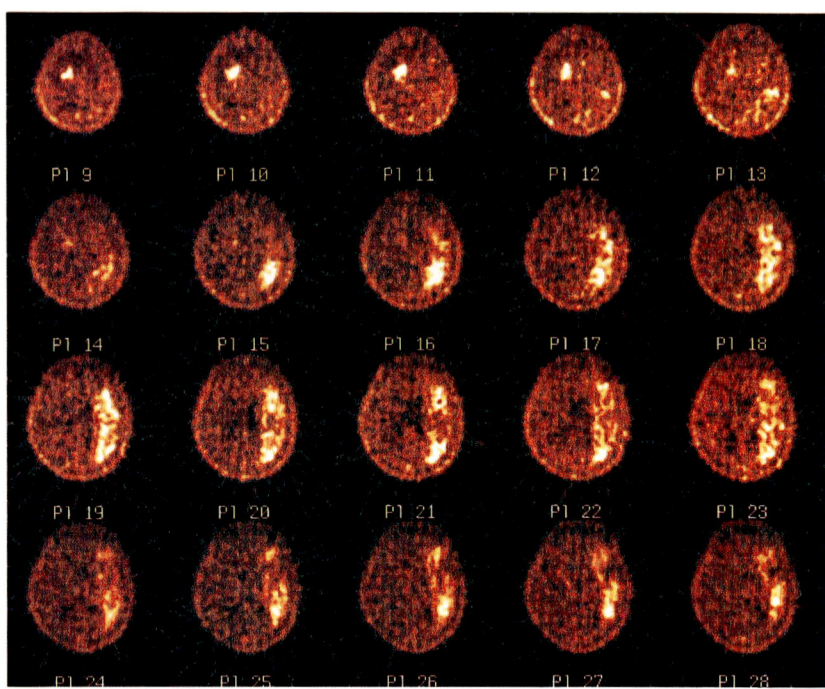

Plate 3 PET images of human brain, 20–60 min after i.v. injection of [^{11}C]PK 11195, 20 days following stroke, as described in the text. Twenty of the 31 transaxial planes collected are shown, representing approximately a 7 cm thickness of brain. High retention of the radioligand, in yellow, is clearly seen surrounding the main lesion in the left cortical territory, on the right of each image. On planes 20 to 22 the 'cold' necrotic core of the infarct can be seen. Planes 9 to 13 show the clinically silent focus of high [^{11}C]PK 11195 retention in the right pre-frontal cortex.

13.2.3.2 Alzheimer's disease

Post-mortem measurements of MBR densities in brains from patients with Alzheimer's disease (AZ) have so far proved inconclusive. The first, by Owen *et al.* (1983), showed only a 28% increase in MBR in the temporal cortex of 10 patients, compared with controls (see Section 13.2.2.1 above). McGeer, E.G. *et al.* (1988), in a more extensive study, measured MBR density in seven regions of brain from 18 AZ patients and although they demonstrated a two- to three-fold increase in Broca's area, post-central and pre-central gyri, changes in all other regions were insignificant. These included the temporal gyri, where the greatest neuronal loss is seen. Furthermore, increases in MBR density did not correlate with activities of the marker enzymes CAT and acetylcholinesterase. The authors concluded that the increases in MBR density were not large enough to be useful in *in vivo* imaging of Alzheimer's disease. It is noteworthy, however, that both these studies were carried out using [^{3}H]Ro5-4864, shown by Olson *et al.* (1988) to have a low affinity for human MBR. A more recent study by Diorio *et al.* (1991) used [^{3}H]PK 11195, rather than [^{3}H]Ro5-4864, to measure the post-mortem MBR density in the brains of 7 patients with AZ. They demonstrated an increase of 120% in MBR density in temporal cortex, reflecting an increase in B_{max} rather than a reduced K_d.

13.2.3.3 Stroke

To date only one published study reports data from post-mortem material from the brains of patients with cerebrovascular disease (Benavides *et al.*, 1988). Using [^{3}H]PK 11195 and *in vitro* autoradiography, these authors reported a seven-fold increase in MBR in the periphery of infarcted areas compared with normal grey matter in brains from five patients, at 7 to 22 days after stroke. There was no correlation of MBR density with the age of the lesion, the areas of increased MBR density corresponding well with the perimeter of the lesion as identified histologically.

13.2.3.4 Multiple sclerosis

In the same study as that reported in Section 13.2.3.3 (Benavides *et al.*, 1988), results from post-mortem material from three patients with multiple sclerosis were reported. In the periphery of active plaques and in the active periphery of chronic plaques, MBR density, as measured using [^{3}H]PK 11195, was increased 10-fold compared to that in normal white matter. There was a close spatial correlation between increased MBR density and the extent of the demyelinative plaques, as demonstrated using Loyez staining.

13.2.3.5 Tumours

Junck *et al.* (1989) compared the binding of [^{3}H]Ro5-4864 with that of [^{3}H]PK 11195 in biopsy specimens from three grade III or IV astrocytomas. Whereas the majority of the small amount of [^{3}H]Ro5-4864 binding was non-specific, the binding of [^{3}H]PK 11195 was shown to be significantly higher than in normal brain and largely specific. In a separate study, Olson *et al.* (1988) measured a K_d value of 14.01 nM for the binding of [^{3}H]PK 11195 in human glioma cell cultures, compared with a K_i value of 1200 nM for the affinity of Ro5-4864 for the PK 11195-sensitive binding site. The pharmacology of the MBR in human glioblastoma was studied in some detail by Broaddus and Bennett (1990). Using [^{3}H]PK 11195, they measured significantly increased binding in six out of six biopsy samples and identified two classes of binding sites, a high affinity (3 nM) site and one with lower affinity (100 nM) which accounted for the majority of the binding under most conditions. MBR in normal cortex fell into a single class with an affinity intermediate in value to those found in glioblastoma. They concluded that MBR imaging in glioblastoma tissue was '... worthy of clinical investigation ...'.

13.2.4 Summary

Taken together, these studies include a wide variety of pathologies but demonstrate that the increase in the density of MBR at the location of a brain lesion can be dramatic and can be clearly and reliably monitored using radiolabelled PK 11195. A conclusion of many of the authors was that the increase in the density of these sites had the potential of providing a means of *in vivo* imaging of brain lesions. However, few of these studies addressed, even indirectly, the question of the identity of the cells in lesioned brain on which MBR are located. The available evidence implicating a specific cell-type will be discussed in the next section.

13.3 The cell-type localization of mitochondrial benzodiazepine receptors

13.3.1 Introduction

As described in the previous section, there has been a general acceptance that MBR sites expressed in CNS damage are located on glial cells (e.g. Syapin and Skolnick, 1979; McCarthy and Harden, 1981; Gallager *et al.*, 1981). The binding of MBR-selective radioligands was seen to be 'non-neuronal' and since the great majority of non-neuronal cells in the CNS are glial cells, the binding was taken to be glial. This concept was reinforced by the lesion studies, neuronal death being accompanied by 'glial' proliferation and an increase in MBR density. However, with *in vivo* imaging of MBR now a reality and with our increasing understanding of the intricacies of the pathological processes in the CNS, it is necessary to clearly define the terms 'glia' and 'gliosis' if any useful interpretation of pathology is to be drawn from the images obtained.

13.3.1.1 Terminology

The word glia is derived from the Greek, meaning 'glue' and is generally applied to all non-neuronal cells in the CNS. These include oligodendrocytes, protoplasmic and fibrous astrocytes and ependymal cells, the so-called macroglia or neuroglia, derived from neuro-ectodermal 'spongioblasts' (Duchen, 1984), and microglia, thought to originate from circulating monocytes which infiltrate the CNS during development (Perry and Gordon, 1991) and now becoming recognized as the resident macrophages of the brain (Perry and Gordon, 1989). The proliferative response of 'glia' to injury is termed 'gliosis', a term frequently applied to the general increase in non-neuronal cellularity seen in response to injury in the CNS. However, as both astrocytes and the microglia contribute to this proliferation, such generalized use of the term may be inappropriate. These two cell types are not only embryologically, but also morphologically and, more importantly, functionally distinct. Their individual responses are characteristic of different stages of pathological development around a lesion and the binding of an MBR ligand to one or other cell type would give different pathological information. In addition, it is necessary to consider the infiltration around a lesion of blood-borne cells such as macrophages, which are known to express large numbers of MBR (Zavala *et al.*, 1984; Ferrarese *et al.*, 1990). It is essential, therefore, to clarify whether MBR are expressed by one, some or all of the cell types involved in the response to injury of the CNS.

13.3.2 Candidate cell types for the location of MBR in lesioned CNS

13.3.2.1 Macrophages

The presence of MBR on mouse and human macrophages in culture is well established (e.g. Zavala *et al.*, 1984; Ferrarese *et al.*, 1990). Dubois *et al.* (1988) suggested that the time-course of the appearance of increased MBR density around the periphery of experimentally-induced ischaemic infarcts corresponded well to the known time course of invasion of the infarcted area by 'haematogenous' cells (Brierley and Brown, 1982), although no attempt was made to positively identify the cells involved. Benavides et al. (1990) used photoaffinity labelling with [³H]PK 14105, a nitrophenyl derivative of PK 11195 which becomes covalently bound to MBR under ultraviolet light (Doble et al., 1987), and high resolution autoradiography to demonstrate the distribution of autoradiographic silver grains over labelled cells. However, the cells themselves were not selectively stained and thus, although the authors reported binding to 'macrophage-like cells', positive identification was not possible.

Work from our own laboratory (Myers et al., 1991a) has directly correlated the binding of [³H]PK 11195 with macrophage infiltration around the focal ischaemic lesion described in Section 13.2.2.2. Macrophages infiltrated the perimeter of the infarct between 2 and 3 days post-lesioning as demonstrated by staining of the antibody ED-1, selective for rat peripheral monocytes (Sminia et al., 1987). At this time, there was a dramatic increase in the binding of [³H]PK 11195 in the region corresponding closely to that occupied by the macrophages. By 7 days, macrophage infiltration had extended around the whole of the infarct and again [³H]PK 11195 binding corresponded closely to ED-1-positive macrophages (Plate 2). Further confirmation of this correlation was obtained using sections from a rat in which experimental allergic encephalo-myelitis (EAE) had been induced. Figure 1 shows the septal region from an EAE rat, illustrating the very close spatial correspondence between [³H]PK 11195 specific binding and ED-1 positive macrophages.

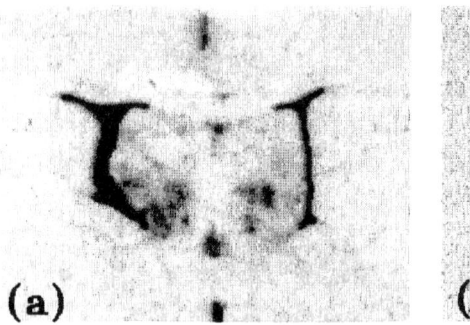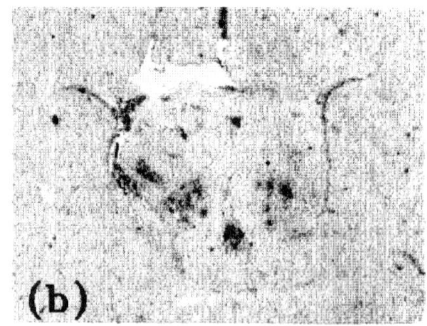

(a)   **(b)**

Figure 1 Coronal sections in the region of the septal nuclei of a rat following induction of experimental allergic encephalomyelitis. (a) Digitized image of an autoradiograph showing [³H]PK 11195 binding. The two large dense regions are the lateral ventricles. Foci of binding can be seen in the intervening septal nuclei. (b) Photomicrograph of a section serial to that in (a). The dark foci represent ED-1-positive macrophages. Note the spatial correlation between the foci in (b) and PK 11195 binding in (a).

13.3.2.2   Astrocytes

In addition to the macrophage response described above, the focal ischaemia produced by the photochemical lesion technique produces an extensive astrocytic gliosis (Manjil *et al.*, 1991). In view of the early reports of the association between astrocytes and MBR, it was therefore expected that a close correspondence might be seen between binding of [^{3}H]PK 11195 and the astrocyte reaction as visualized using immunohistochemical staining for glial fibrillary acidic protein (GFAP), which is specific for astrocytes. No correspondence could, however, be found either temporally or spatially (Myers *et al.*, 1991a).

As discussed in Section 13.2, the early *in vitro* work on MBR identified high densities of MBR in 'non-neuronal', or 'glial' cell cultures. It has been assumed that the majority of the cells in these cultures were astrocytes. Bender and Hertz (1984) claimed an 'unquestionable' astrocyte morphology in 95% and positive staining for glial fibrillary acidic protein (GFAP) of their astrocyte cultures. Cells in these cultures displayed a single class of high affinity, Ro5-4864 sensitive binding site when labelled with [^{3}H]flunitrazepam. However, it is now recognized that brain macrophages, or microglia, are frequent contaminants of primary glial cell cultures (De Groot *et al.*, 1991). Using double-label immunofluorescence techniques, these authors identified up to 46% of the cells in astrocyte-enriched cerebral cultures as brain macrophages. They concluded that the positive identification of astrocytes and brain macrophages in glial cell cultures is 'indispensable'. The purity of the so-called 'glial' cultures of the earlier studies must therefore be in question.

At the present time, in spite of early claims to the contrary, there seems to be little or no firm evidence for the presence of MBR on astrocytes. It is feasible that a combination of high-resolution autoradiography and immunohisto-chemical staining for GFAP could positively identify these sites in astrocytes in tissue sections, but no such studies have yet been reported.

13.3.2.3   Microglia

Recognition of the profound and widespread reaction of microglia to a variety of stimuli has only recently been established. The injection of excitotoxins into the brain induces a vigorous gliosis (Kohler and Schwarcz, 1983; Coffey *et al.*, 1988) and injection of kainic or ibotenic acid is known to induce a dramatic increase in the number of microglia which form a large proportion of the increase in cell density (Murabe *et al.*, 1981; Coffey *et al.*, 1988). Proliferation of hippocampal microglia may be induced by a number of stimuli including intrahippocampal injection of colchicine (Baudry and Altar, 1991), tetanus-toxin induced seizures (Shaw *et al.*, 1990) or transient forebrain ischaemia

(Morioka *et al.*, 1991). In addition, axotomy of the rat facial nerve leads to microglial reaction in its nucleus (Graeber *et al.*, 1988).

In spite of this involvement of microglia in response to chemical or surgical lesioning, little direct evidence has yet been published linking microglial response with MBR density. Such studies are confounded by the fact that microglial and astroglial reactions often occur with temporal and spatial distributions which overlap. This highlights the need for positive, preferably immunohistochemical, identification of the different cell types which, unfortunately, is difficult to combine with ligand autoradiography. Under the conventional conditions used to freeze and section brain for receptor autoradiography, the preservation of microglial morphology is poor, much of the antibody reactivity being lost. It is, however, possible to preserve the integrity of both [^{3}H]PK 11195 binding and immunohistochemical staining of microglia, using an antibody to the CR3 complement receptor, OX-42 (Perry and Gordon, 1987), by the application of fixation protocols designed to preserve antigenicity, such as PLP (paraformaldehyde-lysine-periodate), as proposed by McLean and Nakane (1974).

Following focal ischaemia in the dorsal sensorimotor cortex, induced using the Rose Bengal photochemical lesion model, a widespread microglial response is apparent throughout a large volume of the ipsilateral cortex by 1 day and is still present 3 days after lesioning. The distribution of binding of [^{3}H]PK 11195 corresponds closely to the distribution of reactive microglia identified using OX-42, as seen in Figure 2. By 7 days, this widespread reaction has subsided, reactive microglia being confined to the perimeter of the focal cortical lesion. Conversely, the distribution of GFAP-positive reactive astrocytes does not correlate with MBR density (not shown). It should be noted that these recent, unpublished results conflict with those reported in Section 13.2.2.2 where it was stated that no increase in [^{3}H]PK 11195 binding was observed prior to 3 days (Myers *et al.*, 1991a). The sections from which the images in Figure 2 were taken were prepared using PLP perfusion which, by preserving the binding of [^{3}H]PK 11195, made it possible to demonstrate the association between MBRs and microglial reactivity.

These preliminary results clearly suggest that the expression of MBR by reactive microglia is markedly increased compared with that seen in their resting state. Reactive microglia may be present in sites remote from the primary stimulus and yet their response can be rapid, often within 24 h, being the first observable cellular response following lesioning (Giulian, 1987). By 7 days, when the remote microglial reaction has subsided, MBR expression has returned to baseline levels. Microglia at the periphery of the primary lesion lose their characteristic ramified morphology and round-up to become 'amoeboid' microglia, morphologically indistinct from macrophages. In this state, they continue to express MBR at a high level, although whether this

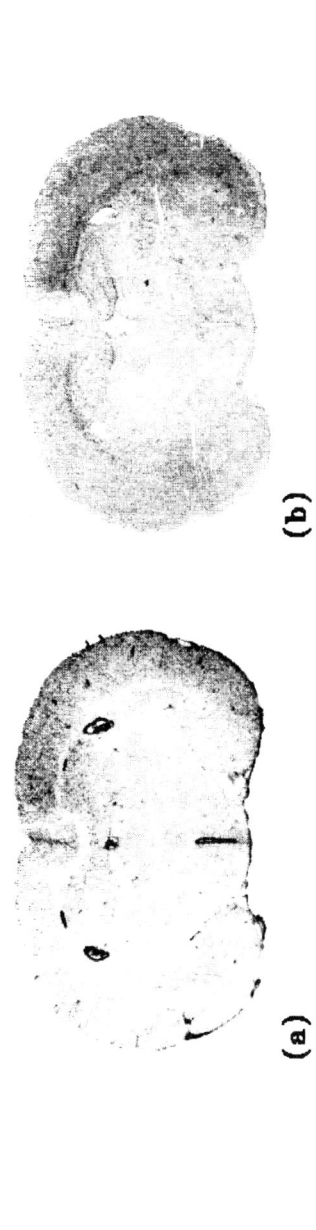

(a)

(b)

Figure 2 Coronal sections of rat brain 3 mm caudal to a cortical lesion similar to that shown in Figure 1, 3 days post-lesioning. (a) Digitized image of an autoradiograph showing widespread binding of [^{3}H]PK 11195 in the cortex ipsilateral to the ischaemic focus. Binding is less dense than around the primary focus, being only ∼ 1.5 times higher than contralateral cortex. (b) Digitized image of a section serial to that in (a) stained for the antibody OX-42, showing extensive microglial activation in the cortex ipsilateral to the main lesion.

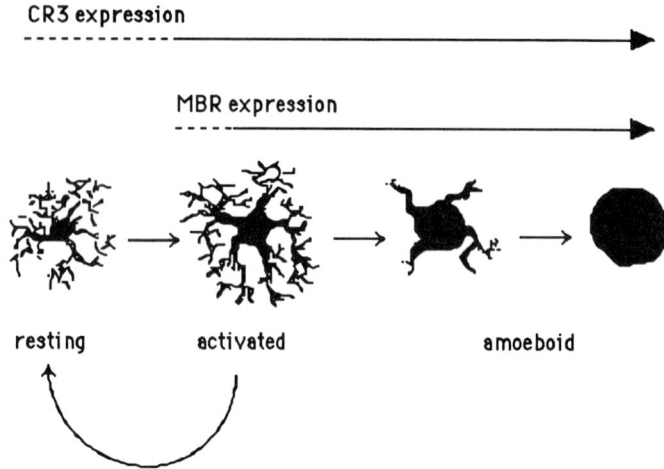

Figure 3 Schematic representation showing the proposed relationship between microglial activation and expression of MBR and CR3 receptors in rat brain following a lesion.

is equivalent to the level of expression by the infiltrating peripheral macrophages is not known. This process is represented schematically in Figure 3.

Further evidence of MBR expression by reactive microglia can be obtained from the remote lesion formed in the ventrolateral thalamic nucleus (VL) following focal ischaemia in the sensorimotor cortex, as mentioned in Section 13.2.2.2. The increased expression of MBR, identified using [^{3}H]PK 11195 (Figure 4a), was not homogeneous within the VL, being more intense in the dorsal part, known to have a majority of fibres directed to the hindlimb motor area (Cicirata *et al.*, 1986). This region of intense MBR expression corresponded to the presence of ED-1 positive macrophages (not shown), while the ventral, less intense, area contained large numbers of reactive, ramified microglia (Figure 4b). Here, expression of MBR was less intense, being about two-thirds of that seen associated with macrophages, but was still significant.

The assumption of macrophage-like characteristics by reacting microglia could be interpreted as reversion to a more primitive form, as it is amoeboid microglia which are active in early brain development (Giulian, 1987). It also provides a rationale for accepting microglia as the non-neuronal population of the CNS which express MBR. However, the trigger for and function of microglial MBR expression is unknown.

13.2.3.4 Summary

It is clear that the widespread reaction of microglia, as visualized by the increased expression of both CR3 antigen and MBR, does not necessarily

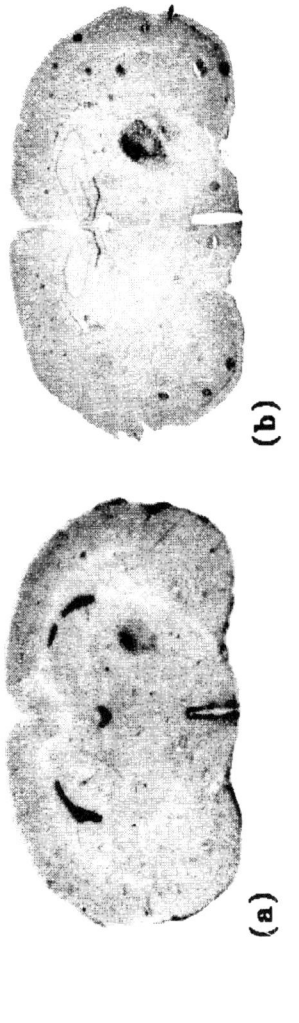

(a)

(b)

Figure 4 Secondary thalamic lesion 21 days following focal cortical ischaemia of the type illustrated in Figure 1. (a) Digitized image of an autoradiograph showing the binding *in vitro* of [³H] PK 11195 in the ventrolateral thalamic nucleus (VL). The binding experiments were carried out using the technique of Dubois *et al.* (1988). (b) Digitized image of a section serial to that in (a) stained for the antibody OX-42 showing extensive microglial activation in the VL. Microscopic examination showed these microglia to be of the ramified form.

correspond to neuronal loss, although it may be representative of a sublethal neuronal reaction which is yet to be defined. While neuronal loss will induce an increase in the binding of PK 11195 locally, mainly associated with infiltrating macrophages, an increase in binding, associated with reactive microglia, may also occur in regions incurring no measurable neuronal damage. Increased MBR density cannot therefore be described solely as a marker of neuronal damage. In view of the association between MBR and both peripheral macrophages and microglia, increased MBR density may be considered to be a marker of inflammation in the CNS.

13.4 Positron emission tomography of the human CNS using MBR ligands

13.4.1 Introduction

In the study of the physiology and pathology of the human CNS using radiotracers, PET represents the most advanced methodology available. Original studies of cerebral blood flow and metabolism have been extended to the quantitation of the specific binding of radiolabelled site-selective ligands, with a view to studying neurotransmitter function and dysfunction in disease.

PET scanning, and the interpretation of the acquired data, is a complex and multidisciplinary process and a detailed account would not be appropriate here. Thus, only a brief description of the basic technique and of the development and use of PET ligands for receptor studies, will be given. Excellent reviews of current PET technology can be found in Jones (1990) or Eriksson et al. (1990) and of the ligands currently used clinically in Mazière and Mazière (1990).

13.4.2 PET

13.4.2.1 The technique of PET

When a positron, a positively-charged particle with a mass equivalent to that of an electron, is emitted within tissue, it travels only a short distance, usually less than 1–2 mm, before it is captured by an electron. Both particles annihilate and the energy released is expressed as two 511 keV γ-rays, emitted at 180° to each other. Two opposed γ-ray sensitive detectors positioned along the line of travel of these photons will detect this coincident release almost simultaneously. In practice, rings of detectors are used; when a coincidence event is registered, the locus of the positron capture can be placed somewhere along the line of sight of the pair of detectors registering the event.

The PET camera used for the studies reported in Section 13.4.4.2, was a CTI/Siemens 953B, comprising a total of 6144 bismuth germanate detector elements (Spinks et al., 1992). The diameter of the field of view is 40 cm with a detector width of 10.8 cm. Coincidence events occurring within a 12 ns time window are recorded between opposing groups of detectors, giving a total of more than 3 million possible lines of response. Coincidence data are recorded in a 'real time sorter' and are reorganized into matrices of angular data or sinograms. These are subsequently reconstructed into an image matrix of $128 \times 128 \times 31$ pixels with a spatial resolution of $8.0 \times 8.0 \times 5.1$ mm full-width half maximum, at the centre of the field of view. During reconstruction of the images, data are corrected for attenuation of signal due to irregularities of tissue density and thickness, by measuring the transmission, through the subject, of radioactivity from a surrounding source. Dynamic scanning is achieved by acquiring data in a sequential series of time 'windows', usually of increasing duration, throughout the scan.

Recorded counts per second per pixel are calibrated by scanning a phantom of known radioactivity and can be directly compared with blood samples taken throughout the scan. Compartmental models can be applied to obtain estimates for kinetic parameters, including the rate constants of transfer between plasma and tissue compartments, giving a means of quantifying pathways of biological interest.

Although the theoretical resolution of scanners such as that described here is better than 6 mm, this is rarely achieved due to poor counting statistics resulting from low levels of radiotracer in the field of view. The short half-life of the radioisotopes used in PET (see Table 2) means that, for most studies, data are acquired over several radioactive half-lives. A recent major advance in PET technology has been removal of the lead collimation between the detector elements, the 'septa', which, although resulting in increased scattered radiation, has also resulted in gains in noise-equivalent counts of three- to five-fold (Spinks et al., 1992).

13.4.2.2 The application of PET

Much of the power of PET lies in its use of organic molecules as tracers. The elements oxygen, carbon, nitrogen and hydrogen make up over 95% of the human body. The first three of these elements have positron-emitting isotopes with half-lives of less than 30 min and this has two important advantages: firstly, organic molecules can be labelled with natural radiotracers as opposed to analogues and, secondly, the short half-lives mean that the radiation dose to the patient is kept to the minimum. A disadvantage is that tracers must be produced at, or close to, the site of use.

The three radioisotopes used for the majority of PET studies are oxygen-15, carbon-11 and fluorine-18, with half lives of approximately 2, 20 and 110 min, respectively. Oxygen-15 is used as either oxygen gas for oxygen metabolism studies, or to radiolabel water or carbon dioxide to measure blood flow, or carbon monoxide to measure blood volume. Fluorine-18 has been widely used to radiolabel fluoro-deoxyglucose for glucose metabolism studies and L-Dopa for measurement of abnormal dopaminergic function in movement disorders such as Parkinson's disease. For the labelling of pharmaceuticals for receptor studies, carbon-11 labelling has proved the most readily achieved. The 20 min half-life of this isotope, however, requires that the syntheses are rapid. In addition, the concentration of cold drug must be sufficiently low that it does not perturb the binding characteristics of the ligand. Thus, the radiolabelled tracers must be produced at very high specific activity. Table 2 summarizes the radioisotopes, tracers and applications currently in use at the MRC Cyclotron Unit at Hammersmith Hospital, London, UK, illustrating the range of studies available in a single research centre. A recent extensive critique of receptor studies in the human brain, focusing largely on PET and on single photon emission computerized tomography (SPECT), has been published by Maziére and Maziére (1990).

13.4.3 PET ligands

13.4.3.1 The properties of PET ligands

Dynamic PET images of a receptor ligand record the 'total' radioactivity in the tissue which includes components of specifically bound, non-specifically bound and free ligand. It is therefore essential to optimize the pharmacological and physiological conditions under which a ligand is used, such that the ratio of specific binding versus non-specific radioactivity is as high as can be achieved. A number of properties have now been identified as desirable or essential if a candidate ligand is to be of use in PET. Accepting that the putative ligand can be labelled with a positron-emitting isotope to give a compound of high specific activity, these properties are summarized in Table 3, subdivided into two broad categories—pharmacological and physiological.

The pharmacological properties include the features of the ligand which can be determined by *in vitro* binding studies. Clearly, the ligand must be highly selective, preferably stereoselective, for a single population of receptors, since the opportunities for using drugs to block sites of unwanted specific binding during PET are limited. The affinity of the ligand for the receptor site should be high, with a K_d preferably in the nanomolar range,

Table 2 Selection of PET tracers and their applications at the MRC Cyclotron Unit, Hammersmith Hospital, London, UK.

Radioisotope	Half-life (min)	Tracer	Application
^{15}O	2.03	$C^{15}O_2$, $H_2^{15}O$	Blood flow
		$^{15}O_2$	Oxygen metabolism
		$C^{15}O$	Blood volume
^{11}C	20.4	^{11}C-Raclopride	Dopamine D_2 receptors
		^{11}C-SCH 23390	Dopamine D_1 receptors
		^{11}C-Deprenyl	Monoamine oxidase B
		^{11}C-Diprenorphine	μ, κ, δ opiate receptors
		^{11}C-s-CGP 12177	Peripheral β-adrenergic receptors
		^{11}C-Flumazenil	Central benzodiazepine receptors
		^{11}C-PK 11195	Peripheral benzodiazepine receptors
^{18}F	109.8	^{18}F-2-Fluoro-2-Deoxyglucose (FDG)	Glucose metabolism
		^{18}F-Dopa	Dopamine storage
		^{18}F-PK 14105	Peripheral benzodiazepine receptors

255

Table 3 Properties of an ideal PET ligand.

	Pharmacological requirements
1.	High affinity for the receptor—K_d in the nM range.
2.	Selective for a single receptor population.
3.	Low non-specific binding.
4.	Antagonist action at the receptor.
	Physiological requirements
5.	Freely crosses the blood–brain barrier.
6.	Non-toxic at low doses.
7.	Minimal metabolism over the time of the scan.
8.	Fast clearance from plasma.
9.	Must be amenable to labelling with a positron-emitting isotope.

as this has been shown to improve the specific versus non-specific signal at low concentrations of ligand. Experience has shown that, with some exceptions, ligands with an antagonist action at the receptor are preferable to those which are agonists. The affinities of agonists are often lower and the higher concentrations of compound required to achieve adequate counts in tissue may in themselves cause a pharmacological effect.

The physiological properties of prospective PET ligands must be determined by *in vivo* or *ex vivo* studies. First and foremost, for CNS studies, the ligand must be able to cross the blood–brain barrier freely. This requires a moderate degree of lipophilicity, although if too high, it will cause an increase in non-specific binding. When metabolism occurs, the fate of the metabolites, particularly radiolabelled metabolites, must be known. If they cross the blood–brain barrier they may contribute to the 'total' counts recorded and add a further degree of complexity to the 'modelling' of the dynamic data. Although only 'tracer' quantities of cold drug are used during a PET study, the compound must, of course, be non-toxic at these doses.

Thus, while suitable *in vitro* ligands now exist for most known subtypes of neuroreceptor, in view of the very stringent requirements for *in vivo* studies, only a small proportion of these ligands can be considered as candidates for PET.

13.4.3.2   PET ligands for the MBR

Each of the three MBR ligands mentioned in this chapter has been labelled with a positron-emitting isotope: Ro5-4864 (Bergstrom *et al.*, 1986) and PK 11195

(Camsonne *et al.*, 1984) with carbon-11 and PK 14105 (Pascali *et al.*, 1990) with fluorine-18. As described in Section 13.2.3.5, the study by Bergstrom *et al.* (1986) using [^{11}C] Ro5-4864 in PET scans of various brain tumours, produced disappointing results, uptake in tumour being lower than in cortical tissue. The subsequent demonstration by Olsen *et al.* (1988) of the relatively low *in vitro* affinity of this ligand, coupled with its agonist nature, makes this result, in retrospect, unsurprising.

Price *et al.* (1990) used the intrastriatal kainate lesion model in rats, as described in Section 13.2 to compare the *ex vivo* binding characteristics of [^{18}F]PK 14105 with those of [^{3}H]PK 11195. The authors concluded that both ligands possessed favourable properties as PET ligands although their binding kinetics differed. [^{18}F]PK 14105 displayed an early specific signal in lesioned regions, enhanced by a rapid clearance of radioactivity from non-lesioned areas. Although the clearance of [^{3}H]PK 11195 was slower, resulting in a more slowly developing specific signal, this signal was longer lasting than that of [^{18}F]PK 14105, the latter being lost by 45 min. The binding characteristics of [^{11}C]PK 11195 in rat brain were later shown to be in excellent agreement with those of [^{3}H]PK 11195, using the focal ischaemia lesion model described in Section 13.2.2.2 (Cremer *et al.*, 1992).

It was suggested by Price *et al.* (1990) that the poorer retention of [^{18}F]PK 14105, compared with [^{11}C]PK 11195, in lesioned rat brain, was due to its lower affinity. Subsequent experiments on human cortical membranes, however, suggested that the difference in the potencies may be much reduced or even reversed. A major advantage of using a fluorinated ligand, as opposed to one labelled with carbon-11 would be its availability for use away from the centre of production and therefore further exploration of [^{18}F]PK 14105 may well prove worthwhile. At present, radiolabelled PK 11195 is the MBR ligand of choice.

[^{11}C]PK 11195 demonstrates several of the characteristics required of a good PET ligand, including high affinity, selectivity and blood–brain barrier permeability (Table 4). A severely limiting factor, however, is the very low extraction of the ligand by the brain from the blood. Junck *et al.* (1989) measured extraction values of <5% and our own unpublished data are in broad agreement. In view of the high lipophilicity of this compound ($\log_p = 3.4$ for octanol:water partition) it is likely that this low extraction is due to binding of the ligand to plasma proteins, thus reducing its availability to the tissues. It is worthy of note that, to date, only racemic mixtures of isomers have been used in binding studies. Where different isomeric forms of a ligand exist, it is often the case that only one isomer is active, the other serving only to 'dilute' the signal. The labelling and testing of the single-isomer forms of various PK 11195 analogues are now underway with a view to further increasing the specific signal in PET.

Table 4 Characteristics of $[^{11}\text{C}]$ PK 11195 as a PET ligand.

1. Affinity—K_d = 1–3 nM
2. Highly selective for MBR
3. 30–40% specific binding in human brain
4. Antagonist action
5. Highly blood–brain barrier permeable; octanol:water coefficient partition $\log_p$ = 3.4
6. Non-toxic
7. 13% metabolized by 5 min, 38% by 60 min[a]
8. 88% bound to plasma proteins[a]
9. Labelled with carbon-11

[a] S. Luthra (unpubl. data).

13.4.3.3 MBR ligand studies in non-human primates

Petit-Taboué *et al.* (1991) characterized the kinetics and specific binding of $[^{11}\text{C}]$PK 11195 in normal baboons using PET. Initial entry of the radioligand into brain was rapid, reaching a plateau by 3 to 5 min. By administering cold PK 11195 either with the radioligand or 8 min later, the authors showed that the brain uptake of the ligand could be significantly altered. The peak radioactive concentration in both cases was more than twice that of control, probably due to blockade of the large numbers of MBR sites present in the peripheral organs and a consequent increase in availability of the tracer to the brain. This profound alteration of the kinetics prevented true comparison of the control with the saturation or displacement time–activity curves and thus accurate calculation of specific binding was not possible. The authors did, however, calculate an estimated value, obtained by scaling the curves to each other, of about 40% specifically bound, a value which was stable from about 15 min.

13.4.4 Clinical PET using MBR ligands

13.4.4.1 Brain tumours

The first successful use of $[^{11}\text{C}]$PK 11195 and PET in human brain was published by Junck *et al.* (1989). These authors compared the uptake of $[^{11}\text{C}]$Ro5-4864 with that of $[^{11}\text{C}]$PK 11195 in various brain tumours. While their results with the former ligand were in agreement with the findings of Bergstrom *et al.* (1986), their results with PK 11195 showed considerably more promise with tumour/grey matter ratios of between 1.5 and 2.0 when the specific activity of the ligand exceeded $\sim 4\,GBq/\mu mol$. Lower specific activities resulted in a reduction or loss of the signal. The extent of binding was unrelated to the grade of tumour and in five or seven tumours scanned using contrast-enhanced CT, the location of the binding corresponded closely to those areas delineated by the contrast enhancement. In the other two tumours, however, high radioactivity was seen outside the areas of high enhancement indicating that the images were not simply illustrating a disrupted blood–brain barrier. The authors concluded that the strong correlation between specific activity and tumour/grey matter ratio suggested that the accumulation of radioactivity represented saturable high affinity binding.

Pappata *et al.* (1991) have recently attempted to test this hypothesis by carrying out a displacement study on a single patient with glioblastoma. Two PET scans were carried out on consecutive weeks. During the first, $[^{11}C]$PK 11195 was given alone while during the second, 56.7 mmol of cold drug was injected 10 min after the $[^{11}C]$PK 11195. The results of the first scan gave a tumour/grey matter ratio in the range of those measured by Junck *et al.* (1989) but the second scan indicated that only 30% of the tumour binding was displaceable. This high degree of non-specifically bound radio-activity may have reflected a large intravascular radioactive pool in tumour. Alternatively, the dose of cold drug used may simply not have been high enough to achieve full displacement, although in view of the results of Junck *et al.* (1989) this seems unlikely.

Whether the binding of PK 11195 in gliomas is associated with reactive or inflammatory cells, or with the tumour cells themselves, remains unclear. While it would appear that MBR binding corresponds closely to tumour topography, a clearer understanding of the cellular location of these sites is required if proper pathological interpretations are to be made.

13.4.4.2 Stroke

A series of PET studies is at present underway at the MRC Cyclotron Unit, to examine the binding of $[^{11}C]$PK 11195 associated with stroke. Five patients have, so far, been scanned, three of whom showed significantly increased binding of the ligand. Scans have been carried out at times between 6 and 61 days post-ictus. Preliminary results indicate that macrophage infiltration in the vicinity of the infarct, as demonstrated by retention of elevated levels of $[^{11}C]$PK 11195, begins during the second week and continues for several weeks.

Plate 3 shows images from part of a serial study carried out on one of the patients, a 55-year-old male, who had suffered an embolic infarction involving the posterior two-thirds of the left MCA territory, sparing the lenticulo-striates, 1 day after aortic valve replacement (Weiller *et al.*, 1991). The patient was scanned at 6, 13 and 20 days post-ictus, with CT scans carried out 7 and 19 days post-ictus, at the latter time, with contrast agent. PET scans to measure regional cerebral blood flow (rCBF) were carried out at the time of each ligand study. At 6 days, a small increase in retention of [^{11}C]PK 11195 was visible around the perimeter of the infarct although it was not quantitatively significant. By 13 days however, a dramatic increase in the retention of [^{11}C]PK 11195 was apparent in a clearly defined region surrounding the ischaemic focus, representing a 2.5-fold increase relative to the contralateral, unaffected side. At 20 days, this quantitative difference between the two sides was retained, as illustrated in Plate 3, although the extent of the region of high activity had begun to consolidate slightly. A central core with low [^{11}C]PK 11195 retention, presumably corresponding to the necrotic focus of the infarct could be clearly seen in three-dimensional reconstructions.

Figure 5 shows time–radioactivity curves taken from a region of high [^{11}C]PK 11195 retention, close to the infarct, and a 'mirror-image' region in the contralateral, unaffected brain. Both regions show an early steep rise in radioactivity, indicating similar delivery of the radioligand (a blood flow and blood volume effect). Following this rise, and subsequent fall over the first 2–3 min of the scan, radioactivity continued to fall in the unaffected brain, while that in the lesioned side rose slowly, reaching a plateau by about 20 min. This difference in the kinetics of the radioligand in the different brain regions is a clear indication that the increased radioactivity associated with the infarct was increased retention of [^{11}C]PK 11195 and not simply increased entry across the blood–brain barrier. Indeed, CT scanning with a contrast agent at 19 days indicated some increased blood–brain barrier permeability but with a poor spatial correlation with the [^{11}C]PK 11195 retention. There was no systematic relationship between rCBF and retention of [^{11}C]PK 11195. These findings are in broad agreement with preliminary results reported by Junck *et al.* (1990) following studies on six stroke patients using [^{11}C]PK 11195, although these authors did not report any values for rCBF in their patients.

Of interest is the observation that on day 13 following the stroke, a previously undetected and clinically silent region in the right pre-frontal cortex now showed a high retention of [^{11}C]PK 11195 and remained clearly visible on day 20 (see Plate 3, planes 9 to 13). This region had a high rCBF at 13 days which returned to normal by day 20. Time–activity curves from this region had the same profile as those from the main lesion suggesting that this represented a second embolic infarct.

Superimposition of rCBF and [^{11}C]PK 11195 PET images demonstrated that the region of high [^{11}C]PK 11195 retention resides at the perimeter of

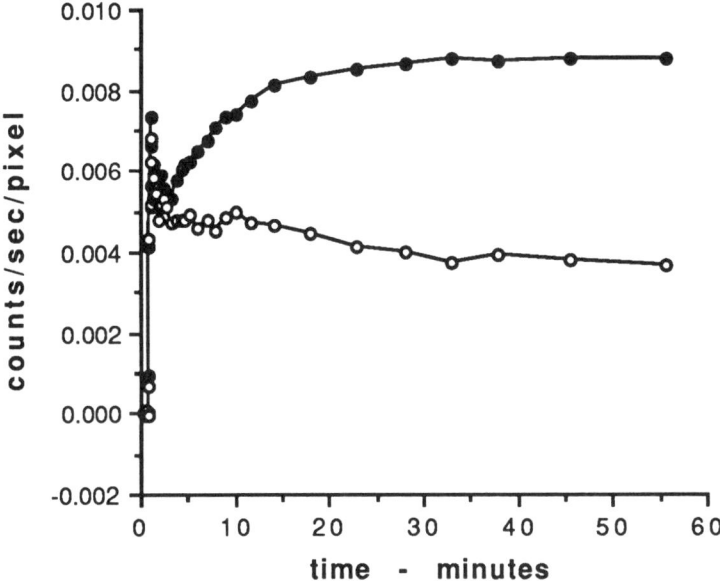

Figure 5 Time–activity curves in humans following an i.v. injection of [¹¹C]PK 11195. The two regions assayed were either close to the infarct (●) or a mirror-image region in the contralateral side of the brain (○). Values shown on the ordinate are counts/s/pixel in the image calibrated against a PET scan image of a phantom containing a known amount of radioactivity. The data have been decay-corrected. Note the rapid rise in radioactivity in both regions, dependent on the blood flow and extraction of the radioligand, followed by the slow accumulation of activity in the region close to the infarct.

the infarcted zone. Viewing the [¹¹C]PK 11195 images alone does not therefore define the region of infarct but rather the location of the brain reaction to the infarct. The patient described above was not scanned after post-ictal day 20, but serial studies on other patients have suggested that at later times post-ictus, towards the end of the second month, the region of [¹¹C]PK 11195 retention is confined to a thin layer closely bounding the focus of the infarct, suggesting a progressive reduction in macrophage infiltration.

13.5 The future potential of MBR imaging

In the foregoing discussion, evidence has been presented to show that radioligands for the MBR, in combination with PET, can be used to image the inflammatory cellular response in CNS. In this section, some of the benefits which may be gained from this unique potential will be explored.

13.5.1 Early detection of neuronal degeneration

In the CNS, the presence of phagocytic cells, either in the form of infiltrating peripheral macrophages or reactive amoeboid microglia, is indicative of the loss of cells or cellular elements following some destructive injury. Clearly, such a process accompanies most, if not all, CNS pathology at some stage in its development. However, observations made post-mortem can only reflect the pathology at the time of death, giving little information on early degenerative changes. McGeer, P.L. *et al.* (1988a,b) have addressed this issue in studies of HLA-DR reactivity in human microglia. This glycoprotein is part of the group II major histocompatibility complex (MHC II) and is expressed on the surface of immunocompetent cells including microglia in the CNS. The authors found large numbers of HLA-DR-positive reactive microglia in the substantia nigra and hippocampus of Parkinson's disease patients and in the hippocampus of Alzheimer's disease patients. They remarked that in the late stages of a disease, when the pathology may reveal only reactive astrocytes, spongiform tissue or simple atrophy, HLA-DR-positive macrophages '... may have come and gone'. However, '... a vigorous HLA-DR reaction may be the strongest premonitory indication of an impending disease'. The authors demonstrated HLA-DR-positive macrophages in a variety of post-mortem tissues and, notably, in some cases '... in areas that would otherwise not have been suspected of being involved in the pathological process...'. The authors conclude that visualization of reactive microglia provides a 'sensitive index of neuropathologic activity'. The potential benefits of serial, non-invasive monitoring of these cells are clear.

13.5.2 Multiple sclerosis

Serial MRI scanning of patients suffering from MS has proved to be a powerful tool in the elucidation of the spatial and temporal relationships between the various pathological features of the syndrome. Scanning protocols are designed either to quantify the T_1 or T_2 relaxation times, dependent on the physical properties of tissue water, or to visualize blood–brain barrier breakdown using a paramagnetic MRI contrast agent, gadolinium diethylene-triamine pentaacetic acid (McDonald and Barnes, 1989). Results have now suggested that a vascular lesion, including blood–brain barrier breakdown and consequent inflammation, is a primary event in MS (Kermode *et al.*,

1990a). Perivascular cellular infiltration is known to be associated with this inflammatory process (Lumsden, 1970) and it is possible that these inflammatory cells contribute to the enhanced MRI signal (Kermode *et al.*, 1990b). Boyle and McGeer (1990) have shown that HLA-DR-positive cells form dense rings around MS plaques and these authors showed that these were analogous to cells seen surrounding EAE lesions in rats when these were visualized using analogous rat antibodies. These findings, and the results of Benavides *et al.* (1988) (see Section 13.2.3.4) and those illustrated in Figure 1 (see Section 13.3.2.1), indicate that the addition of PET to MRI studies could be of considerable value in imaging this cellular component of the MS pathology.

13.5.3 Graft rejection following transplantation

The possibility of treating deficiency in striatal dopamine, such as occurs in Parkinson's disease, by the implantation of foetal donor striatal cells is now being actively explored (see Lindvall, 1991 for a review). As with any technique involving xenogeneic grafting of tissue, rejection by the immune system of the host may be a limiting factor. In spite of previous suggestions to the contrary, the CNS is not an immunologically privileged site and studies of retinal transplants into rat brain have suggested that the implanted tissue is populated with microglia of host origin (Perry and Lund, 1989). This led the authors to suggest that microglia may play an important role in the stability of xenogeneic grafts. Immunohistochemical studies by Poltorak and Freed (1989) have lent further weight to the suggestion that microglia, and not astrocytes as previously suggested, are the antigen-presenting cells of the CNS. The monitoring of the activity of these cells associated with intracerebral transplants could provide valuable information regarding graft stability.

13.5.4 Detection of remote lesions

In Section 13.2.2.2 the potential of MBR ligands to detect lesions remote from the primary site of injury, such as those in the thalamus following cortical injury, was discussed. Although the clinical significance of such lesions is, as yet, unknown, Tamura *et al.* (1991) have reported marked atrophy of the ipsilateral thalamus following cortical infarction from three months to one year in some clinical cases. Shimoyama *et al.* (1988) showed that, in monkey brain, monocyte infiltration into secondary thalamic lesions following cortical ablation was still pronounced at 8 weeks post-lesioning. Such a long-term response to cortical injury may be open to therapy and MBR imaging provides the means of its detection.

13.5.5 The functional significance of MBR expression

13.5.5.1 MBR and the CNS inflammatory response

Inflammatory reactions in the peripheral organs and the CNS are distinct from each other (Andersson *et al.*, 1991). In the former, significant migration of neutrophils and monocytes can be seen within 4 h of injury, whereas in the CNS a significant blood-borne cellular response is not observed for 36–48 h, with an absence of neutrophil involvement. Monocyte recruitment is known to be faster in the peripheral nervous system where they have a significant role in regeneration (Leibovich and Ross, 1975). However, in view of the limited capability of the CNS to regenerate either destroyed neurones or damaged cell processes, Andersson *et al.* (1991) suggest that the significance of the inflammatory reaction in the CNS remains unclear. Several reports have suggested that microglia and macrophages play a significant role in recovery and tissue homeostasis following injury. Amoeboid microglia, for example, have been shown to secrete the cytokine interleukin-1 which stimulates astrogliosis and neo-vascularization (Giulian *et al.*, 1988) and macrophages have been implicated in the regulation of blood flow within a region of selective neuronal death (Iadecola *et al.*, 1990). In contrast, suppression of monocyte activity using chloroquine and colchicine improved recovery of bladder and hindlimb function following a transient ischaemia of the spinal cord which would otherwise have caused a severe impairment of these functions (Giulian and Robertson, 1990). The glucocorticoid dexamethasone, a well-known anti-inflammatory agent, had little effect on recovery possibly due, the authors suggested, to the limited ability of glucocorticoids to inhibit mononuclear phagocyte activity. Such contradictory notions of the implications of macrophage activity in lesioned CNS illustrate the gaps in our knowledge which the imaging of MBR can be used to fill.

13.5.5.2 The function of the MBR

Throughout the foregoing discussion, the MBR has been considered not as a functional entity but merely as a marker, its presence indicating the presence of a particular cell type in a particular state. This is of course an understatement of its importance in the ongoing pathology which it is being used to describe. However, extensive considerations of MBR function may be found elsewhere in this book. (For more information on MBR on immune competent cells and on IL_1 action on MBR, refer to Chapter 11.)

13.5.6 MBR imaging in peripheral organs

This chapter has been concerned exclusively with the use of MBR ligands as markers of damage in the CNS. The normally low level of MBR expression in most of the neuropil provides not only a 'clean' background for the increase in MBR expression associated with lesions, but also an indication that the functioning of this receptor in normal brain is at a very low level. The same is not true in many peripheral organs. In human heart, high densities of MBR are normally expressed, as demonstrated using $[^{11}C]PK$ 11195 and PET (Charbonneau *et al.*, 1986). These receptors have been shown to be coupled to calcium channels, suggesting a role in cardiac function (Mestre *et al.*, 1985). In rat kidney, high densities of binding sites are found in the ascending limb of the loop of Henle and the distal convoluted tubule but not in the proximal tubule (Butlen, 1984; Gehlert *et al.*, 1985), while in human and guinea-pig lung, discrete structures such as submucosal glands showed high levels of binding of $[^{3}H]Ro5-4864$, as did the epithelial lining of the airways (Mak and Barnes, 1990). The induction of chronic inflammatory lesions by local injection of Freund's adjuvant into the legs of rats produced a two-fold increase in MBR binding in the thymus and three-fold increase in the iliac lymph nodes as well as a marked binding associated with various cells in the inflammatory leg infiltrate (Scatton *et al.*, 1990). These observations support the proposed immunomodulatory role of MBR sites (Zavala and Lenfant, 1987).

Studies of the binding of $[^{3}H]PK$ 11195 on biopsy samples from patients have shown that both colonic adenocarcinoma (Katz *et al.*, 1988) and ovarian carcinoma (Katz *et al.*, 1990) showed three- to five-fold increases in MBR densities compared with their corresponding normal tissue, while renal carcinoma showed a marked reduction in MBR binding when compared with normal kidney (Katz *et al.*, 1989).

The imaging of MBR in peripheral organs may well provide valuable information regarding the function and dysfunction of those organs. However, the highly localized distribution of MBR and the possible variety in their functional significance from organ to organ complicates their use as simple macrophage markers. Alterations, increases or reductions, of normal receptor populations would be superimposed on any increase in receptor numbers caused by inflammatory cell response. Whether such information would be of use in imaging would need to be considered in an organ-dependent way.

13.5.7 Diagnosis and research using MBR imaging

13.5.7.1 Diagnosis

$[^{11}C]PK$ 11195 is unlikely ever to find favour as a diagnostic tool as carbon-11 labelled ligands are not well suited to this role. The ~ 20 min half-life of carbon-11 effectively allows no more than $1-2$ h before the level of radioactivity becomes too low for use. The decrease in the specific radioactivity over this time means that lengthy delay in injection after the end of the synthesis of the radiolabelled compound would be deleterious to the signal. Carbon-11 labelled ligands therefore need to be manufactured at the site of the PET scanner and in view of the consequent need for a cyclotron, for the production of the radioisotope, and synthetic chemistry and quality control facilities, this precludes their use other than at large PET centres. In the case of fluorine-18 labelled ligands, however, there is the potential for shipment over short distances providing considerable stimulus for the development of, for example, [^{18}F]PK 14105.

More promising in a diagnostic role is single photon emission computerized tomography (SPECT). The radioisotopes used to label SPECT tracers are generally longer-lived, facilitating production off-site, and lower-cost SPECT scanners are more widely available than PET scanners. However, the attenuation and scattering problems encountered in SPECT prevent full quantitation at this time, limiting its use in receptor research. For qualitative, diagnostic purposes, however, SPECT may prove ideal, drawing on the basic information acquired from PET research. With this in mind iodinated PK 11195 has already been produced and tested as a potential SPECT MBR ligand (Gildersleeve et al., 1989).

13.5.7.2 Research

Confirmation of the relationship between PET images and the pathology associated with them can only come from post-mortem material. It is unlikely that this confirmation will be readily available and suitable model systems are therefore required. Lesion models such as the Rose Bengal focal ischaemia technique, described above, provide suitable pathology and, combined with a dedicated small animal PET scanner, such as that being developed in this laboratory (Rajeswaran et al., 1992) may provide just such a system. A small animal PET scanner of this type would have a spatial resolution of ~ 3 mm and would permit kinetic measurement of ligand binding combined with any of the experimental manipulations now confined to ex vivo studies. At appropriate times, animals could be sacrificed and histology or immunohistochemistry carried out. Analogous studies are already underway using MRI (van Bruggen et al., 1992) and clearly there would be considerable value in combining these techniques.

13.6 Summary and conclusions

Work over a decade, has shown, in a range of CNS pathologies, that the presence of high densites of MBR in the CNS, as visualized using radiolabelled MBR ligands, is indicative of tissue damage. The labelling with positron-emitting isotopes of PK 11195, the present MBR ligand of choice, has enabled the benefits of this finding to be extended to *in vivo* imaging using PET. [^{11}C]PK 11195 works well as a PET radioligand, although further radiochemistry development could improve it still further. Availability of this ligand in an iodine-125 labelled form should extend its use to a diagnostic role in SPECT.

The recognition of MBR as markers of macrophage and microglial reaction to injury in the CNS, rather than of gliosis as previously thought, should better enable detection of these receptors to make constructive contributions in a number of areas of neuropathology, as well as to our understanding of the inflammatory response of the CNS to injury.

Acknowledgements

The author gratefully acknowledges the work of Drs Luisa Manjil, Cornelius Weiller, Stuart Ramsay and Gary Price and the assistance of Dr Adriaan Lammertsma. Sincere thanks are also due to Dr Sajinder Luthra and Miss Safiye Osman for allowing the use of their unpublished data concerning PK 11195 metabolites, and to Dr Susan Hume, Dr Jill Cremer and Professor Richard Frackowiak for their helpful comments on the manuscript.

References

Altar, C.A. and Baudry, M. (1990). Systemic injection of kainic acid: gliosis in olfactory and limbic brain regions quantified with [^{3}H]PK 11195 binding autoradiography. *Exp. Neurol.* **109**, 333–341.

Andersson, P.-B., Perry, V.H. and Gordon, S. (1991). The kinetics and morphological characteristics of the macrophage-microglial response to kainic acid-induced neuronal degeneration. *Neuroscience* **42**, 201–214.

Baudry, M. and Altar, C.A. (1991) Entorhinal cortex lesion or intrahippocampal colchicine injection increases peripheral type benzodiazepine binding sites in rat hippocampus. *Brain Res.* **553**, 215–221.

Benavides, J., Quarteronet, D., Imbault, F., Malgouris, C., Uzan, A., Renault, C., Dubroeucq, M.C., Guérémy, C. and Le Fur, G. (1983). Labelling of 'peripheral-type' benzodiazepine binding sites in the rat brain by using [^{3}H]PK 11195, an isoquinoline carboxamide derivative: Kinetic studies and autoradiographic localization. *J. Neurochem.* **41**, 1744–1750.

Benavides, J., Savaki, H.E., Malgouris, C., Laplace, C., Daniel, M., Begassat, M., Desban, M., Uzan, A., Dubroeucq, M.C., Renault, C., Guérémy, C. and Le Fur, G. (1984). Autoradiographic localization of peripheral benzodiazepine binding sites in the cat brain with [³H]PK 11195. *Brain Res. Bull.* **13**, 69–77.

Benavides, J., Fage, D., Carter, C. and Scatton, B. (1987). Peripheral type benzo-diazepine binding sites are a sensitive indirect index of neuronal damage. *Brain Res.* **421**, 167–172.

Benavides, J., Cornu, P., Dennis, T., Dubois, A., Hauw, J.-J., MacKenzie, E.T., Sazdovitch, V. and Scatton, B. (1988). Imaging of human brain lesions with an ω_3 site radioligand. *Ann. Neurol.* **24**, 708–712.

Benavides, J., Capdeville, C., Dauphin, F., Dubois, A., Duverger, D., Fage, D., Gotti, B., MacKenzie, E.T. and Scatton, B. (1990). The quantification of brain lesions with an ω_3 site ligand: a critical analysis of animal models of cerebral ischaemia and neurodegeneration. *Brain Res.* **522**, 275–289.

Bender, A.S. and Hertz, L. (1984). Flunitrazepam binding to intact and homogenised astrocytes and neurons in primary cultures. *J. Neurochem.* **43**, 1319–1327.

Bergstrom, M., Mosskin, M., Ericson, K., Ehrin, E., Thorell, J.-O., von Holst, H., Norén, G., Persson, A., Halldin, C., Stone-Elander, S. and Collins, V.P. (1986). Peripheral benzodiazepine binding sites in human gliomas evaluated with positron emission tomography. *Acta Radiol. (Scand) (Suppl.)* **369**, 409–411.

Black, K.L., Ikezaki, K. and Toga, A.W. (1989). Imaging of brain tumours using peripheral benzodiazepine receptor ligands. *J. Neurosurg.* **71**, 113–118.

Black, K.L., Ikezaki, K., Santori, E., Becker, D.P. and Vinters, H.V. (1990). Specific high-affinity binding of peripheral benzodiazepine receptor ligands to brain tumours in rat and man. *Cancer* **65**, 93–97.

Boyle, E.A. and McGeer, P.L. (1990). Cellular immune response in multiple sclerosis plaques. *Am. J. Pathol.* **137**, 575–584.

Braestrup, C., Albrechtsen, R. and Squires, R.F. (1977). High densities of benzo-diazepine receptors in human cortical areas. *Nature* **269**, 702–704.

Brierley, J.M. and Brown, A.W. (1982). The origin of lipid phagocytes in the central nervous system. II. The adventitia of blood vessels. *J. Comp. Neurol.* **211**, 407–417.

Broaddus, W.C. and Bennett, J.P. Jr. (1990). Peripheral-type benzodiazepine receptors in human glioblastomas: pharmacological characterization and photoaffinity labelling of ligand recognition site. *Brain Res.* **518**, 199–208.

Butlen, D. (1984). Benzodiazepine receptors along the nephron: [³H]PK 11195 binding in rat tubules. *FEBS Lett.* **169**, 138–142.

Camsonne, R., Crouzel, C., Comar, D., Mazière, M., Prenant, C., Sastre, J., Moulin, M.A. and Syrota, A. (1984). Synthesis of N-[¹¹C]methyl, N-(methyl-1 propyl), (chloro-2 phenyl)-1 isoquinoline carboxamide-3 (PK 11195): a new ligand for peripheral benzodiazepine receptors. *J. Labelled Compd. Radiopharm.* **21**, 985–991.

Charbonneau, P., Syrota, A., Crouzel, C., Valois, J.M., Prenant, C. and Crouzel, M. (1986). Peripheral-type benzodiazepine receptors in the living heart characterised by positron emission tomography. *Circulation* **73**, 476–483.

Cicirata, F., Angaut, P., Cioni, M., Serapide, M.F. and Papale, A. (1986). Functional organization of thalamic projections to the motor cortex. An anatomical and electrophysiological study in the rat. *Neuroscience* **19**, 81–99.

Coffey, P.J., Perry, V.H., Allen, Y., Sinden, J. and Rawlins, J.N.P. (1988). Ibotenic acid induced demyelination in the central nervous system: a consequence of a local inflammatory response. *Neurosci. Lett.* **84**, 178–184.

Cremer, J.E., Hume, S.P., Cullen, B.M., Myers, R., Manjil, L.G., Turton, D.R.,

Luthra, S.K., Bateman, D.M. and Pike, V.W. (1992). The distribution of radioactivity in brains of rats given [N-methyl-^{11}C] PK 11195 *in vivo* after induction of a cortical ischaemic lesion. *Nucl. Med. Biol.* **19**, 159–166.

Cymerman, U., Pazos, A. and Palacios, J.M. (1986). Evidence for species differences in 'peripheral' benzodiazepine receptors: an autoradiographic study. *Neurosci. Lett.* **66**, 153–158.

De Groot, C.J.A., Sminia, T., Dijkstra, C.D., van der Pal, R.H.M. and Lopes-Cardozo, M. (1991). Interferon-γ induced Ia antigen expression on cultured neuroglial cells and brain macrophages from rat spinal cord and cerebrum. *Int. J. Neurosci.* **59**, 53–65.

Diorio, D., Welner, S.A., Butterworth, R.F., Meaney, M.J. and Suranyi-Cadotte, B.E. (1991). Peripheral benzodiazepine binding sites in Alzheimer's disease frontal and temporal cortex. *Neurobiol. Aging* **12**, 255–258.

Doble, A., Burgevin, M.C., Menager, J., Ferris, O., Begassat, F., Renault, C., Dubroeucq, M.C., Guérémy, C., Uzan, A. and Le Fur, G. (1987). Partial purification and pharmacology of peripheral-type benzodiazepine receptors. *J. Recept. Res.* **7**, 55–70.

Dubois, A., Benavides, J., Peny, B., Duverger, D., Fage, D., Gotti, B., MacKenzie, E.T. and Scatton, B. (1988). Imaging of primary and remote ischaemic and excitotoxic brain lesions. An autoradiographic study of peripheral type benzodiazepine binding sites in the rat and cat. *Brain Res.* **445**, 77–90.

Duchen, L.W. (1984). In *Greenfield's Neuropathology* (eds Hume Adams, J., Corsellis, J.A.N. and Duchen, L.W.), pp. 1–52. Edward Arnold, London.

Ferrarese, C., Appollonio, I., Frigo, M., Perego, M., Pierpaoli, C., Trabucchi, M. and Frattola, L. (1990). Characterization of peripheral benzodiazepine receptors in human blood mononuclear cells. *Neuropharmacology* **29**, 375–378.

Eriksson, L., Dahlbom, M., Widén, L. (1990). Positron emission tomography — a new technique for studies of the central nervous system. *J. Microscopy* **157**, 305–333.

Gallager, D.W., Mallorga, P., Oertel, W., Henneberry, R., Tallman, J. (1981). [^{3}H] Diazepam binding in mammalian central nervous system: A pharmacological characterization. *J. Neurosci.* **1**, 218–225.

Gehlert, D.R., Yamamura, H.I. and Wamsley, J.K. (1985). Autoradiographic localization of 'peripheral-type' benzodiazepine binding sites in the rat brain, heart and kidney. *Naunyn-Schmiedeberg's Arch. Pharmacol.* **328**, 454–460.

Gildersleeve, D.L., Lin, T.Y., Wieland, D.M., Ciliax, B.J., Olson, J.M. and Young, A.B. (1989). Synthesis of a high specific activity ^{125}I-labelled analog of PK 11195, potential agent for SPECT imaging of the peripheral benzodiazepine binding site. *Nucl. Med. Biol.* **16**, 423–429.

Giulian, D. (1987). Amoeboid microglia as effectors of inflammation in the central nervous system. *J. Neurosci. Res.* **18**, 155–171.

Giulian, D. and Robertson, C. (1990). Inhibition of mononuclear phagocytes reduces ischaemic injury in the spinal cord. *Ann. Neurol,* **27**, 33–42.

Giulian, D., Woodward, J., Krebs, J.F. and Lachman, L.B. (1988). Intracerebral injections of interleukin-1 stimulate astrogliosis and neovascularisation. *J. Neurosci.* **8**, 2485–2490.

Graeber, M.B., Streit, W.J. and Kreutzberg, G.W. (1988). Axotomy of the rat facial nerve leads to increased CR3 complement receptor expression by activated microglial cells. *J. Neurosci. Res.* **21**, 18–24.

Grome, J.J., Gojowczyk, G., Hofmann, W. and Graham, D.I. (1988). Quantitation of photochemically induced focal cerebral ischaemia in the rat. *J. Cereb. Blood Flow Metab.* **8**, 89–95.

Iadecola, C., Arneric, S.P., Baker, H.D., Callaway, J. and Reis, D.J. (1990). Maintenance of local cerebral blood flow after acute neuronal death: Possible role of non-neuronal cells. *Neuroscience* **35**, 559–575.

Ikezaki, K., Black, K.L., Toga, A.W., Santori, E.M., Becker, D.P. and Smith, M.L. (1990). Imaging peripheral benzodiazepine receptors in brain tumours in rats: *in vitro* binding characteristics. *J. Cereb. Blood Flow Metab.* **10**, 580–587.

Jones, T. (1990). Positron emission tomography. *Clin. Phys. Physiol. Meas.* **11**, Suppl. A, 27–36.

Junck, L., Olson, J.M.M., Ciliax, B.S., Koeppe, R.A., Watkins, G.L., Jewett, D.M., McKeever, P.E., Wieland, D.M., Kilbourn, M.R., Starosta-Rubinstein, S., Mancini, W.R., Kuhl, D.E., Greenberg, H.S. and Young, A.B. (1989). PET imaging of human gliomas with ligands for the peripheral benzodiazepine binding site. *Ann. Neurol.* **26**, 752–758.

Junck, L., Jewett, D.M., Kilbourn, M.R., Young, A.B. and Kuhl, D.E. (1990). PET imaging of cerebral infarcts using a ligand for the peripheral benzodiazepine binding site. *Neurology* **40** (Suppl. 1), 265.

Katz, Y., Eitan, A., Amiri, Z. and Gavish, M. (1988). Dramatic increase in peripheral benzodiazepine binding sites in human colonic adenocarcinoma as compared to normal colon. *Eur. J. Pharmacol.* **148**, 483–484.

Katz, Y., Moskovitz, B., Levin, D.R. and Gavish, M. (1989). Absence of peripheral-type benzodiazepine binding sites in renal carcinoma: a potential biochemical marker. *Br. J. Urol.* **63**, 124–127.

Katz, Y., Ben-Baruch, G., Kloog, Y., Menczer, J. and Gavish, M. (1990). Increased density of peripheral benzodiazepine-binding sites in ovarian carcinomas as compared with benign ovarian tumours and normal ovaries. *Clin. Sci.* **78**, 155–158.

Kermode, A.G., Thompson, A.J., Tofts, P.S., MacManus, D.G., Kendall, B.E., Kingsley, D.P.E., Moseley, I.F., Rudge, P. and McDonald, W.I. (1990a). Breakdown of the blood–brain barrier precedes symptoms and other MRI signs of new lesions in multiple sclerosis. *Brain* **113**, 1477–1489.

Kermode, A.G., Tofts, P.S., Thompson, A.J., MacManus, D.G., Rudge, P., Kendall, B.E., Kingsley, D.P.E., Moseley, I.F., du Boulay, E.P.G.H. and McDonald, W.I. (1990b). Heterogeneity of blood–brain barrier changes in multiple sclerosis: an MRI study with gadolinium-DTPA enhancement. *Neurology* **40**, 229–235.

Kohler, C. and Schwarcz, R. (1983). Comparison of ibotenate and kainate toxicity in the rat brain: a histological study. *Neuroscience* **8**, 819–835.

Le Fur, G., Perrier, M.L., Vaucher, N., Imbault, F., Flamier, A., Benavides, J., Uzan, A., Renault, C., Dubroeucq, M.C. and Gueremy, C. (1983a). Peripheral benzodiazepine binding sites: effect of PK 11195,1-(2-chlorophenyl)-N-methyl-(1-methylpropyl)-3-isoquinolinecarboxamide. I. *In vitro* studies. *Life Sci.* **32**, 1839–1847.

Le Fur, G., Guilloux, F., Rufat, P., Benavides, J., Uzan, A., Renault, C., Dubroeucq, M.C. and Gueremy, C. (1983b). Peripheral benzodiazepine binding sites: effect of PK 11195,1-(2-chlorophenyl)-N-methyl-(1-methylpropyl)-3-isoquinolinecarboxamide. II. *In vivo* studies. *Life Sci* **32**, 1849–1856.

Le Fur, G., Vaucher, N., Perrier, M.L., Flamier, A., Benavides, J., Renault, C., Dubroeucq, M.C., Gueremy, C. and Uzan, A. (1983c). Differentiation between two ligands for peripheral benzodiazepine binding sites [³H]Ro5-4864 and [³H]PK11195 by thermodynamic studies. *Life Sci.* **33**, 449–457.

Leibovich, S.J. and Ross, R. (1975). The role of the macrophage in wound repair. *Am. J. Pathol.* **78**, 71–79.

Lindvall, O. (1991). Prospects of transplantation in human neurodegenerative diseases.

Trends in Neurosciences **14**, 376–384.

Lumsden, C.E. (1970). The neuropathology of multiple sclerosis. In *Handbook of Clinical Neurology* (eds Vinken, P.J. and Bruyn, G.W.), Vol. 9, pp. 217–309. North Holland, Amsterdam.

McCarthy, K.D. and Harden, T.K. (1981). Identification of two benzodiazepine binding sites on cells cultured from rat cerebral cortex. *J. Pharmacol. Exp. Ther.* **261**, 183–191.

McDonald, W.I. and Barnes, D. (1989). Lessons from magnetic resonance imaging in multiple sclerosis. *Trends in Neurosciences* **12**, 376–379.

McGeer, E.G., Singh, E.A., McGeer, P.L. (1988). Peripheral-type benzodiazepine binding in Alzheimer disease. *Alzheimer Dis. Ass. Disorders* **2**, 331–336.

McGeer, P.L., Itagaki, S. and McGeer, E.G. (1988a). Expression of the histocompatibility glycoprotein HLA-DR in neurological disease. *Acta Neuropathol. Berl.* **76**, 550–557.

McGeer, P.L., Itagaki, S., Boyes, B.E. and McGeer, E.G. (1988b). Reactive microglia are positive for HLA-DR in the substantia nigra of Parkinson's and Alzheimer's disease brains. *Neurology* **38**, 1285–1291.

McLean, I.W. and Nakane, P.K. (1974). Periodate-lysine-paraformaldehyde fixative. A new fixative for immunoelectron microscopy. *J. Histochem. Cytochem.* **22**, 1077–1083.

Mak, J.C.W. and Barnes, P.J. (1990). Peripheral type benzodiazepine receptors in human and guinea pig lung: Characterization and autoradiographic mapping. *J. Pharmacol. Exp. Ther.* **252**, 880–885.

Manjil, L.G., Myers, R., Cullen, B.M., Cremer, J.E. and Frackowiak, R.S.J. (1991). Astroglial and microglial reactions following focal cortical ischaemia in the rat brain. *Eur. J. Neurosci. Suppl.* **4**, 102.

Mazière, B. and Mazière, M. (1990). Where have we got to with receptor mapping of the human brain? *Eur. J. Nucl. Med.* **16**, 817–835.

Mestre, M., Carriot, T., Belin, C., Uzan, A., Renault, C., Dubroeucq, M.C., Gueremy, C., Doble, A. and Le Fur, G. (1985). Electrophysiological and pharmacological evidence that peripheral type benzodiazepine receptors are coupled to calcium channels in the heart. *Life Sci.* **36**, 391–400.

Möhler, H. and Okada, T. (1977). Benzodiazepine receptor: demonstration in the central nervous system. *Science* **198**, 849–851.

Morioka, T., Kalehua, A.N. and Streit, W.J. (1991). The microglial reaction in the rat dorsal hippocampus following transient forebrain ischaemia. *J. Cereb. Blood Flow Metab.* **11**, 966–973.

Murabe, Y., Ibata, Y. and Sano, Y. Morphological studies on neuroglia III. Macrophage response and 'microgliocytosis' in kainic-acid induced lesions. *Cell Tissue Res.* **218**, 75–86.

Myers, R., Manjil, L.G., Cullen, B.M., Price, G.W., Frackowiak, R.S.J. and Cremer, J.E. (1991a). Macrophage and astrocyte populations in relation to [^{3}H]PK 11195 binding in rat cerebral cortex following a local ischaemic lesion. *J. Cereb. Blood Flow Metab.* **11**, 314–322.

Myers, R., Manjil, L.G., Frackowiak, R.S.J. and Cremer, J.E. (1991b). [^{3}H]PK 11195 and the localisation of secondary thalamic lesions following focal ischaemia in rat motor cortex. *Neurosci. Lett.* **133**, 20–24.

Olson, J.M., Junck, L., Young, A.B., Penney, J.B. and Mancini, W.R. (1988). Isoquinoline and peripheral-type benzodiazepine binding in gliomas: implications for diagnostic imaging. *Cancer Res.* **48**, 5837–5841.

271

Owen, F., Poulter, M., Waddington, J.L., Mashal, R.D. and Crow, T.J. (1983). [^{3}H]Ro5-4864 and [^{3}H] flunitrazepam binding in kainate-lesioned rat striatum and in temporal cortex of brains from patients with senile dementia of the Alzheimer subtype. *Brain Res.* **278**, 373–375.

Pappata, S., Cornu, P., Samson, Y., Prenant, C., Benavides, J., Scatton, B., Crouzel, C., Hauw, J.J. and Syrota, A. (1991). PET study of carbon-11-PK 11195 binding to peripheral type benzodiazepine sites in glioblastoma: A case report. *J. Nucl. Med.* **32**, 1608–1610.

Pascali, C., Luthra, S.K., Pike, V.W., Price, G.W., Ahier, R.G., Hume, S.P., Myers, R., Manjil, L.G. and Cremer, J.E. (1990). The radiosynthesis of [18F]PK 14105 as an alternative radioligand for peripheral type benzodiazepine binding sites. *Appl. Radiat. Isotopes* **41**, 477–482.

Perry, V.H. and Gordon, S. (1987). Modulation of CD4 antigen on macrophages and microglia in rat brain. *J. Exp. Med.* **166**, 1138–1143.

Perry, V.H. and Gordon, S. (1989). In *Neuroimmune Networks: Physiology and Diseases* (eds Goetzel, E.J. and Spector, N.H.), pp. 119–125. Alan R. Liss Inc., New York.

Perry, V.H. and Gordon, S. (1991). Macrophages and the nervous system. *Int. Rev. Cytol.* **125**, 203–244.

Perry, V.H. and Lund, R.D. (1989). Microglia in retinae transplanted to the central nervous system. *Neuroscience* **31**, 453–462.

Petit-Taboué, M.-C., Baron, J.-C., Barré, L., Travère, J.-M., Speckel, D., Camsonne, R. and MacKenzie, E.T. (1991). Brain kinetics and specific binding of [^{11}C]PK 11195 to ω_3 sites in baboons: positron emission tomography study. *Eur. J. Pharmacol.* **200**, 347–351.

Poltorak, M. and Freed, W.J. (1989). Immunological reactions induced by intracerebral transplantation: evidence that host microglia but not astroglia are the antigen-presenting cells. *Exp. Neurol.* **103**, 222–233.

Price, G.W.P., Ahier, R.G., Hume, S.P., Myers, R., Manjil, L.G., Cremer, J.E., Luthra, S.K., Pascali, C., Pike, V. and Frackowiak, R.S.J. (1990). *In vivo* binding to peripheral benzodiazepine binding sites in lesioned rat brain: Comparison between [^{3}H]PK 11195 and [^{18}F]PK 14105 as markers for neuronal damage. *J. Neurochem.* **55**, 175–185.

Rajeswaran, S., Bailey, D.L., Hume, S.P., Townsend, D.W., Geissbüler, A., Young, J. and Jones, T. (1992). 2D and 3D imaging of small animals and the human radial artery with a high resolution block detector for PET. *IEEE Trans. Med. Imaging* (in press).

Scatton, B., Bénavidès, J., Dubois, A. and Bourdiol, F. (1990). Proliferation of omega-3 binding sites in the immune organs and leg infiltrate of rats with adjuvant induced arthritis. *Int. J. Tissue React.* **12**, 15–20.

Schoemaker, H., Morelli, M., Deshmukh, P. and Yamamura, H.I. (1982). [^{3}H]Ro5-4864 benzodiazepine binding in the kainate lesioned striatum and Huntingdon's diseased basal ganglia. *Brain Res.* **248**, 396–401.

Schwarcz, R. and Coyle, J. T. (1977). Striatal lesions with kainic acid: neurochemical characteristics. *Brain Res.* **127**, 235–249.

Shaw, J.A.G., Perry, V.H. and Mellanby, J. (1990). Tetanus toxin-induced seizures cause microglial activation in rat hippocampus. *Neurosci. Lett.* **120**, 66–69.

Shimoyama, I., Dauth, G.W., Gilman, S., Frey, K.A. and Penney, J.B. (1988). Thalamic, brainstem, and cerebellar glucose metabolism in the hemiplegic monkey. *Ann. Neurol.* **24**, 718–726.

Sminia, T., De Groot, C.J.A., Dijkstra, C.D., Koetsier, J.C. and Polman, C.H. (1987).

Macrophages in the central nervous system of the rat. *Immunobiology* **174**, 43–50.

Spinks, T.J., Jones, T., Bailey, D.L., Townsend, D.W., Grootoonk, S., Bloomfield, P.M., Gilardi, M.-C., Casey, M.E., Sipe, B. and Reed, J. (1992). Physical performance of a positron tomograph for brain imaging with retractable septa. *Phys. Med. Biol.* **37**, 1637–1655.

Squires, R.F. and Braestrup, C. (1976). Benzodiazepine receptors in rat brain. *Nature* **266**, 732–734.

Starosta-Rubinstein, S., Ciliax, B.J., Penney, J.B., McKeever, P. and Young, A.B. (1987). Imaging of a glioma using peripheral benzodiazepine receptor ligands. *Proc. Natl. Acad. Sci. USA* **84**, 891–895.

Syapin, P.J. and Skolnick, P. (1979). Characterization of benzodiazepine binding sites in cultured cells of neural origin. *J. Neurochem.* **32**, 1047–1051.

Tamura, A., Tahira, Y., Nagashima, H., Kirino, T., Gotoh, O., Hojo, S. and Sano, K. (1991). Thalamic atrophy following cerebral infarction in the territory of the middle cerebral artery. *Stroke* **22**, 615–618.

van Bruggen, N., Cullen, B.M., King, M.D., Doran, M., Williams, S.R., Gadian, D.G. and Cremer, J.E. (1992). T2- and diffusion-weighted magnetic resonance imaging of a focal ischaemic lesion in the rat brain. *Stroke* **23**, 576–582.

Wang, J.K.T., Morgan, J.I. and Spector, S. (1984). Benzodiazepines that bind at peripheral sites inhibit cell proliferation. *Proc. Natl. Acad. Sci. USA* **81**, 753–756.

Watson, B.D., Dietrich, W.D., Busto, R., Wachtel, M.S. and Ginsberg, M.D. (1985). Induction of reproducible brain infarction by photochemically initiated thrombosis. *Ann. Neurol.* **17**, 497–504.

Watson, B.D., Prado, R., Ginsberg, M.D. and Green, B.A. (1986). Photochemically induced spinal cord injury in the rat. *Brain Res.* **367**, 296–300.

Weiller, C., Frackowiak, R.S.J., Turton, D.R., Myers, R. and Cremer, J.E. (1991). The evolution of peripheral-type benzodiazepine binding sites in human brain lesions: sequential monitoring with [11C]PK 11195 and positron emission tomography. *Eur. J. Neurosci.* Suppl. **4**, 18.

Zavala, F. and Lenfant, M. (1987). Benzodiazepines and PK 11195 exert immunomodulating activities by binding on a specific receptor on macrophages. *Ann. N.Y. Acad. Sci.* **496**, 240–249.

Zavala, F., Haumont, J. and Lenfant, M. (1984). Interaction of benzodiazepines with mouse macrophages. *Eur. J. Pharmacol* **106**, 561–566.

Index